Surveys of Modern Mathematics
Volume II

Lie Theory and Representation Theory

edited by

Naihong Hu
Bin Shu
Jianpan Wang

International Press
www.intlpress.com

高等教育出版社
HIGHER EDUCATION PRESS

Surveys of Modern Mathematics, Volume II
Lie Theory and Representation Theory

Naihong Hu
Bin Shu
Jianpan Wang

2010 Mathematics Subject Classification. 16T05, 14F10, 16G60, 14L15, 17B10, 17B50, 20J06, 20G05, 20G10, 32C38.

ISBN: 978-1-57146-236-7

Printed in the United States of America.

16 15 14 13 12 1 2 3 4 5 6 7 8 9

SURVEYS OF MODERN MATHEMATICS

On Opening Day of the East China Normal University Summer School on Lie Theory and Representation Theory (II), 13 July 2009

On Closing Day, 31 July 2009

At the Workshop on Lie Theory and Representation Theory (II)

Preface

During the period from July 13 to July 31, 2009, East China Normal University hosted the second workshop and summer school on Lie Theory and Representation Theory. This volume contains the lecture notes of three courses in that summer school, together with the lecture notes of one course given in the first summer school which was held in 2006.

This volume consists of articles by Shun-Jen Cheng and Weiqiang Wang, Rolf Farnsteiner, Daniel K. Nakano, and Toshiyuki Tanisaki. These articles focus on different areas in Lie theory and representation theory. The article jointly by Cheng and Wang introduces some recent developments of representations of Lie superalgebras, explaining how Lie superalgebras of types $\mathfrak{gl}$ and $\mathfrak{osp}$ provide a natural framework for generalized Schur and Howe dualities, and how a super duality gives a conceptual solution to the irreducible character problem for these Lie superalgebras in terms of the classical Kazhdan-Lusztig polynomials.

Farnsteiner's article discusses combinatorial and geometric aspects of representation theory of finite group schemes, and focuses on the "classical" theory of co-commutative Hopf algebras, the defining algebras of affine algebraic group schemes.

Nakano's article gives a survey of recent developments in the representation theory and cohomology theory of reductive algebraic groups, their Frobenius kernels and their associated finite groups of Lie type.

Tanisaki's article presents an overview of the theory of D-modules and its application to representations of Lie algebras.

This volume is well suited for graduate students in the fields of Lie theory and representation theory and related topics, and also for researchers who wish to learn about some current core areas in Lie theory and representation theory and their applications.

At last, we sincerely express our thanks to the Department of Mathematics, the International Exchange Division and the Graduate School of East China Normal University for their financial support to the summer schools and workshops in 2006 and 2009. We are grateful to National Natural Science Foundation of China for financial support (Grant:10926022) in 2009. Our deep appreciation also goes to our colleagues Pei Gu, Youyi Wu and Hongyan Zhang for their assistance in organizing these activities.

Jianpan Wang
Bin Shu
Naihong Hu
In Shanghai
31 October, 2010

Contents

Lie Theory and
Representation Theory
SMM 2, pp. 1–46

Dualities for Lie Superalgebras

Shun-Jen Cheng* and Weiqiang Wang†

Abstract

We explain how Lie superalgebras of types $\mathfrak{gl}$ and $\mathfrak{osp}$ provide a natural framework generalizing the classical Schur and Howe dualities. This exposition includes a discussion of super duality, which connects the parabolic categories O between classical Lie superalgebras and Lie algebras. Super duality provides a conceptual solution to the irreducible character problem for these Lie superalgebras in terms of the classical Kazhdan-Lusztig polynomials.

2000 Mathematics Subject Classification: 17B10.

Keywords and Phrases: Lie superalgebras, Schur duality, Howe duality, super duality, irreducible characters.

0 Introduction

The study of Lie superalgebras, supergroups, and their representations was largely motivated by supersymmetry in mathematical physics, which puts bosons and fermions on the same footing. An earlier achievement is the Cartan-Killing type classification of finite-dimensional simple complex Lie superalgebras by Kac [K1] (also cf. [SNR] for an independent classification of the so-called classical Lie superalgebras). The most important basic classical Lie superalgebras consist of infinite series of types $\mathfrak{sl}, \mathfrak{osp}$. The basic classical Lie superalgebras afford root systems, Dynkin diagrams, Cartan subalgebras, triangular decomposition, Verma modules, category O, and so on. There has been much work on representation theory of Lie superalgebras (in particular, basic classical) in the last three decades, but conceptual approaches have been lacking until recently.

The aim of these lecture notes is to explain three different kinds of dualities for Lie superalgebras:

Schur duality, Howe duality, and Super duality.

*Institute of Mathematics, Academia Sinica, Taipei, E-mail: chengsj@math.sinica.edu.tw
†Department of Mathematics, University of Virginia, Charlottesville, VA 22904. E-mail: ww9c@virginia.edu

In the superalgebra setting, the first (i.e. Schur) duality was formulated by Sergeev, and the latter two dualities have been largely developed by the authors and their collaborators. These lecture notes are also intended to serve as a road map for a forthcoming book by the authors.

The Schur-Sergeev duality is an interplay between Lie superalgebras and the symmetric groups which incorporates the trivial and sign modules in a unified framework. On the algebraic combinatorial level, there is a natural super generalization of the notion of semistandard tableaux which is a hybrid of the traditional version and its conjugate counterpart.

It has been observed that much of the study of the classical invariant theory on polynomial algebras has parallels for exterior algebras, and both admit reformulation and extension in the theory of Howe's reductive dual pairs. Lie superalgebras allow a uniform treatment of Howe duality on the polynomial and exterior algebras.

Super duality has a different flavor. It views representation theories of Lie superalgebras and Lie algebras as two sides of the same coin, and it is an unexpected yet powerful approach developed in the past few years which allow us to overcome various superalgebra difficulties. We give an exposition on the new development on super duality which is an equivalence between parabolic categories $\mathcal{O}$ of Lie superalgebras and Lie algebras. Super duality provides a conceptual solution to the long-standing irreducible character problem for a wide class of modules over (a wide class of) Lie superalgebras in terms of Kazhdan-Lusztig polynomials. This is achieved despite the fact that there are no obvious Weyl groups controlling the linkage for super representation theory.

In Section 1, we give some basic constructions and structures of the general linear and the ortho-symplectic Lie superalgebras. We emphasize the super phenomena that are not observed in the ordinary Lie algebra setting, such as odd roots, non-conjugate Borel subalgebras, and so on. In Section 2, we present Kac's classification of finite-dimensional simple $\mathfrak{g}$-modules [K2]. The classification is very easy for type A, but nontrivial for $\mathfrak{osp}$. In the latter case we explain a new odd reflection approach by Shu and the second author [SW], using a more natural labeling of these modules by hook partitions. We note that odd reflection is also one of the main technical tools in super duality. In addition, we present the *typical* finite-dimensional irreducible character formula, following [K2].

The classical Schur duality relates the representation theory of the general linear Lie algebras and that of the symmetric groups. In Section 3, we explain Sergeev's generalization [Sv1] of Schur duality for the general linear Lie superalgebras $\mathfrak{gl}(m|n)$ (also see Berele and Regev [BeR] for additional insight and detail). More precisely, we establish a double centralizer theorem for the actions of $\mathfrak{gl}(m|n)$ and the symmetric group $\mathfrak{S}_d$ in d letters on the tensor space $(\mathbb{C}^{m|n})^{\otimes d}$. We then provide an explicit multiplicity-free decomposition of the tensor space into a $U(\mathfrak{gl}(m|n)) \otimes \mathbb{C}\mathfrak{S}_d$-modules. We further present a simple formula obtained in our latest work with Lam [CLW] for extremal weights in a simple polynomial $\mathfrak{gl}(m|n)$-module with respect to all Borel subalgebras, which has an explicit diagramatic interpretation from a Young diagram.

Howe's theory of reductive dual pairs [H1, H2] can be viewed as a representation theoretic reformulation and extension of the classical invariant theory (see Weyl [We]). For example, the first fundamental theorem on invariants for classical groups are reformulated in terms of double centralizer properties of two classical Lie groups/algebras. One advantage of Howe duality is that it allows natural generalizations to classical Lie groups/algebras (and superalgebras) other than type A.

We mainly use two examples of dual pairs to illustrate the main ideas of Howe duality and the new phenomena of superalgebra generalizations. For more detailed case study of Howe duality for Lie superalgebras, we refer to the original papers [BPT, CW1, CW2, CW3, CL1, CLZ, CZ2, CKW, LZ, Sv2]. In Section 4, we formulate the $(\mathfrak{gl}(m|n), \mathfrak{gl}(d))$-Howe duality and find the highest weight vectors for each isotypical component in the corresponding multiplicity-free decomposition. In Section 5 we present the $(\mathrm{Sp}(d), \mathfrak{osp}(2m|2n))$-Howe duality and its multiplicity-free decomposition. The application of Howe duality to irreducible characters over Lie superalgebras follows the simpler approach in our work with Kwon [CKW] (which uses Howe duality for infinite-dimensional Lie algebras [Wa]).

We recall some truly super phenomena that have been the main obstacles towards a better understanding of super representation theory:

1. There exist odd roots as well as non-conjugate Borel subalgebras for a Lie superalgebra. A homomorphism between Verma modules may not be injective.
2. The linkage in category $\mathcal{O}$ of modules for a Lie superalgebra is NOT controlled by the Weyl group of $\mathfrak{g}_{\bar{0}}$; see e.g. $\mathfrak{gl}(1|1)$.
3. There is no uniform Weyl-type irreducible finite-dimensional character formula for Lie superalgebras.
4. The super geometry behind super representation theory is still inadequately developed.

In light of these super phenomena, it was a rather unexpected discovery [CWZ, CW4], which was partly inspired by Brundan [Br1], that there exists a (conjectural) equivalence of categories between Lie algebras and Lie superalgebras of type A (at a certain suitable limit at infinity), which was termed *Super Duality*. This conjecture in the full generality of [CW4] has been proved in [CL2], which in particular offers an elementary and conceptual solution to the character problem for all finite-dimensional simple modules and for a large class of infinite-dimensional simple highest weight modules over Lie superalgebras of type A.

Super duality has been subsequently formulated and established between various Lie superalgebras of type $\mathfrak{osp}$ and the corresponding classical Lie algebras in our very recent work with Lam [CLW]. This in particular offers a conceptual solution of the irreducible character problem for a wide class of modules, which include all finite-dimensional irreducibles, of Lie superalgebras of type $\mathfrak{osp}$ in terms of Kazhdan-Lusztig polynomials for classical Lie algebras [KL, BB, BK] (for more on Kazhdan-Lusztig theory see Tanisaki's lectures [Ta]). In addition, it follows easily from the approach of [CL2, CLW] that the $\mathfrak{u}$-homology groups (or Kazhdan-Lusztig polynomials in the sense of Vogan [Vo]) match perfectly between classical Lie superalgebras and the corresponding classical Lie algebras. This generalizes earlier partial results in this direction from Schur or Howe duality approach [CZ1, CK, CKW]. The super duality as outlined above is explained in Section 6.

Let us put the super duality work explained above in perspective. *Finite-dimensional* irreducible characters for $\mathfrak{gl}(m|n)$ have been also obtained earlier in two totally different approaches by [Sva] and [Br1]. The mixed algebraic and geometric approach of Serganova has been extended very recently in [GS] to obtain all irreducible finite-dimensional $\mathfrak{osp}$-characters. Brundan and Stroppel [BrS] also provided another approach to the main results of [Br1] and independently proved a special case of the super duality conjecture in type A as formulated in [CWZ]. All these approaches have brought new and different insights into super representation theory. Our super duality approach has the advantages of explaining the connection with classical Lie algebras and their Kazhdan-Lusztig polynomials, covering infinite-dimensional irreducible characters, and being extendable to general Kac-Moody Lie superalgebras.

A list of symbols is added at the end of the paper to facilitate the reading.

Let us end the Introduction with some remarks on the interrelations among the three dualities.

The $(\mathfrak{gl}(d), \mathfrak{gl}(n))$-Howe duality is equivalent to Schur duality. It follows from the Schur-Sergeev duality that the characters for irreducible polynomial $\mathfrak{gl}(m|n)$-modules are given by the so-called hook Schur functions. On the other hand, the irreducible character formulas for Lie superalgebras of types $\mathfrak{gl}$ or $\mathfrak{osp}$ obtained from Howe duality can be expressed in terms of infinite classical Weyl groups. The appearance of hook Schur functions and infinite Weyl groups in these formulas are conceptually explained from the viewpoint of super duality.

Super duality can be informally interpreted as a categorification of the standard involution on the ring of symmetric functions. It is well known that the ring of symmetric functions in infinitely many variables admits symmetries which are not observed in finitely many variables. Super duality is formulated precisely at the infinite rank limit. On the level of combinatorial parameterizations of highest weights, super duality manifests itself through (variation of) the conjugate of partitions.

Partly due to the time constraint of the lectures, we have left out many interesting topics on super representation theory. We refer to [BL, J] (and more recently [SZ]) for finite-dimensional irreducible characters of atypicality one, to [BKN, DS, Ma, Mu, Pe, PS] for geometric approaches, to [Br2, CWZ2] for further development of the Fock space approach of Brundan for the queer Lie superalgebra $\mathfrak{q}(n)$ and for $\mathfrak{osp}(2|2n)$, to [CK, CKW, CZ1, Ger, San, Sva, Zou] for some cohomological aspects, to [BrK, SW, WZ] for prime characteristic, to [JHKT, Su] for related combinatorial structures; also see [Gor, KW, Naz] for additional work on Lie superalgebras.

Acknowledgment. This paper is a modified and expanded written account of the 8 lectures given by the second author at the summer school of East China Normal University (ECNU), Shanghai, July 2009. We are grateful to Ngau Lam for his collaboration and insight. We thank Bin Shu at ECNU for hospitality and an enjoyable summer school.

1 Lie superalgebra ABC

A vector superspace V is understood as a $\mathbb{Z}_2$-graded vector space $V = V_{\bar{0}} \oplus V_{\bar{1}}$. An element $a \in V_i$ has parity $|a| = i$, and an element in $V_{\bar{0}}$ (respectively, $V_{\bar{1}}$) is called even (respectively, odd).

Definition 1.1. A Lie superalgebra is a vector superspace $\mathfrak{g} = \mathfrak{g}_{\bar{0}} \oplus \mathfrak{g}_{\bar{1}}$ equipped with a bilinear bracket operation $[.,.]$ satisfying $[\mathfrak{g}_i, \mathfrak{g}_j] \subset \mathfrak{g}_{i+j}$, $i, j \in \mathbb{Z}_2$, and the following two axioms: for $\mathbb{Z}_2$-homogeneous $a, b, c \in \mathfrak{g}$,

(1) (Skew-supersymmetry) $[a,b] = -(-1)^{|a|\cdot|b|}[b,a]$.
(2) (Super Jacobi identity) $[a,[b,c]] = [[a,b],c] + (-1)^{|a|\cdot|b|}[b,[a,c]]$.

Remark 1.2. (1) For a Lie superalgebra $\mathfrak{g} = \mathfrak{g}_{\bar{0}} \oplus \mathfrak{g}_{\bar{1}}$, $\mathfrak{g}_{\bar{0}}$ is a Lie algebra and $\mathfrak{g}_{\bar{1}}$ is a $\mathfrak{g}_{\bar{0}}$-module under the adjoint action.

(2) (Sign Rule) As explained in Manin's book [Ma], there is a general heuristic sign rule for superalgebras as follows. *If in some formula for usual algebra there are monomials with interchanged terms, then in the corresponding formula for superalgebra every interchange of neighboring terms, say a and b, is accompanied by the multiplication of the monomials by the factor $(-1)^{|a|\cdot|b|}$.* This is already manifest in the definition of Lie superalgebra and will persist throughout the paper.

Example 1.3. (1) Let $A = A_{\bar{0}} \oplus A_{\bar{1}}$ be an associative superalgebra (i.e. $\mathbb{Z}_2$-graded). Then $(A, [.,.])$ is a Lie superalgebra, where for homogeneous elements $a, b \in A$, we define

$$[a,b] = ab - (-1)^{|a|\cdot|b|}ba.$$

(2) A Lie superalgebra $\mathfrak{g}$ with $\mathfrak{g}_{\bar{1}} = 0$ is just a usual Lie algebra. A Lie superalgebra $\mathfrak{g}$ with purely odd part (i.e. $\mathfrak{g}_{\bar{0}} = 0$) has to be *abelian*, i.e. $[\mathfrak{g}, \mathfrak{g}] = 0$.

1.1 Lie superalgebras of type A and the supertrace

Let $V = V_{\bar{0}} \oplus V_{\bar{1}}$ be a vector superspace. Then $\mathrm{End}(V)$ is naturally an associative superalgebra. The Lie superalgebra $\mathfrak{gl}(V) := (\mathrm{End}(V), [.,.])$ from Example 1.3 (1) is called a *general linear Lie superalgebra*. If $V_{\bar{0}} = \mathbb{C}^m$ and $V_{\bar{1}} = \mathbb{C}^n$, we denote V by $\mathbb{C}^{m|n}$, and $\mathfrak{gl}(V)$ by $\mathfrak{gl}(m|n)$. Note that both $\mathfrak{gl}(m|0) \cong \mathfrak{gl}(0|m)$ are isomorphic to the usual Lie algebra $\mathfrak{gl}(m)$.

The Lie superalgebra $\mathfrak{gl}(m|n)$ consists of block matrices of size $m|n$:

$$g = \begin{pmatrix} a & b \\ c & d \end{pmatrix}. \tag{1.1}$$

Throughout the paper, we choose to parameterize the rows and columns of the matrices by the set

$$I(m|n) = \{\bar{1}, \dots, \bar{m}; 1, \dots, n\}$$

with a total order

$$\bar{1} < \cdots < \bar{m} < 0 < 1 < \cdots < n \tag{1.2}$$

(where 0 is inserted for later convenience). Its even subalgebra consists of matrices of the form

$$\begin{pmatrix} a & 0 \\ 0 & d \end{pmatrix}$$

and is isomorphic to $\mathfrak{gl}(m) \oplus \mathfrak{gl}(n)$.

Example 1.4. For $\mathfrak{g} = \mathfrak{gl}(1|1)$, let

$$e = \begin{pmatrix} 0 & 1 \\ 0 & 0 \end{pmatrix}, \quad f = \begin{pmatrix} 0 & 0 \\ 1 & 0 \end{pmatrix}, \quad h_1 = \begin{pmatrix} 1 & 0 \\ 0 & 0 \end{pmatrix}, \quad h_2 = \begin{pmatrix} 0 & 0 \\ 0 & 1 \end{pmatrix}.$$

Then

$$[e, f] = h_1 + h_2 \text{ (the identity matrix).}$$

The *supertrace*, denoted by str, of (1.1) is defined to be

$$\mathrm{str}(g) = \mathrm{tr}(a) - \mathrm{tr}(d).$$

The *special linear Lie superalgebra* is

$$\mathfrak{sl}(m|n) = \{x \in \mathfrak{gl}(m|n) \mid \mathrm{str}(x) = 0\}.$$

The definitions of supertrace and of the Lie superalgebra $\mathfrak{sl}$ are justified by the following.

Exercise 1.5. Show that $\mathfrak{sl}(m|n) = [\mathfrak{gl}(m|n), \mathfrak{gl}(m|n)]$ and in particular $\mathfrak{sl}(m|n)$ is a Lie subalgebra of $\mathfrak{gl}(m|n)$.

The notion of simple Lie superalgebras is defined in the same way as for Lie algebras. We note that $\mathfrak{sl}(n|n)$ is not a simple Lie superalgebra, as it contains a nontrivial center $\mathbb{C}I_{2n}$.

1.2 The bilinear form

Let $\mathfrak{h}$ denote the Cartan subalgebra of $\mathfrak{gl}(m|n)$ consisting of all diagonal matrices. Note that $\mathfrak{h}$ is an even subalgebra of $\mathfrak{gl}(m|n)$.

Let E_{ij}, for $i, j \in I(m|n)$, denote the standard basis for $\mathfrak{gl}(m|n)$. We define a bilinear form $(\cdot,\cdot)$ on $\mathfrak{g}$ by letting

$$(a, b) = \mathrm{str}(ab), \quad a, b \in \mathfrak{g}.$$

This restricts to a nondegenerate symmetric bilinear form on $\mathfrak{h}$: for $i, j \in I(m|n)$,

$$(E_{ii}, E_{jj}) = \begin{cases} 1 & \text{if } \overline{1} \le i = j \le \overline{m}, \\ -1 & \text{if } 1 \le i = j \le n, \\ 0 & \text{if } i \ne j. \end{cases}$$

Denote by $\{\delta_i, \epsilon_j\}_{i,j}$ the basis of $\mathfrak{h}^*$ dual to $\{E_{\overline{i}\,\overline{i}}, E_{jj}\}_{i,j}$, where $1 \le i \le m$ and $1 \le j \le n$. Under the bilinear form $(\cdot,\cdot)$, we have the identification $\delta_i = (E_{\overline{i},\overline{i}}, \cdot)$ and $\epsilon_j = -(E_{jj}, \cdot)$. Whenever it is convenient we also use the notation

$$\epsilon_{\overline{i}} := \delta_i, \quad \text{for } 1 \le i \le m. \tag{1.3}$$

The form $(\cdot,\cdot)$ on $\mathfrak{h}$ induces a non-degenerate bilinear from on $\mathfrak{h}^*$, which will be denoted by the same notation, as follows: for $i, j \in I(m|n)$,

$$(\epsilon_i, \epsilon_j) = \begin{cases} 1 & \text{if } \overline{1} \le i = j \le \overline{m}, \\ -1 & \text{if } 1 \le i = j \le n, \\ 0 & \text{if } i \neq j. \end{cases} \tag{1.4}$$

1.3 The root system

For the Lie superalgebra $\mathfrak{gl}(m|n)$, we define the root space decomposition, a root system Φ, a set Φ^+ (respectively, Φ^-) of positive (respectively, negative) roots, a set Π of simple roots (in Φ^+), etc. As this can be done in the same way as for semisimple Lie algebras or $\mathfrak{gl}(m)$, we will merely write down the statements for later use.

Now let us make the super phenomenon explicit. A root α is *even* if $\mathfrak{g}_\alpha \subset \mathfrak{g}_{\bar{0}}$, and it is *odd* if $\mathfrak{g}_\alpha \subset \mathfrak{g}_{\bar{1}}$. Denote by $\Phi_{\bar{0}}$ (respectively, $\Phi_{\bar{1}}$) the set of all even (respectively, odd) roots in Φ. Denote

$$\Phi_i^{\pm} = \Phi_i \cap \Phi^{\pm}, \quad \Pi_i = \Phi_i \cap \Pi, \quad i \in \mathbb{Z}_2.$$

With respect to the Cartan subalgebra $\mathfrak{h}$ the Lie superalgebra $\mathfrak{gl}(m|n)$ admits a root space decomposition:

$$\mathfrak{g} = \mathfrak{h} \oplus \bigoplus_{\alpha \in \Phi} \mathfrak{g}_\alpha,$$

with a root system

$$\Phi = \{\epsilon_i - \epsilon_j \mid i, j \in I(m|n), i \neq j\}.$$

The standard set of simple roots is taken to be $\Pi = \Pi_{\bar{0}} \cup \Pi_{\bar{1}}$, where

$$\Pi_{\bar{0}} = \{\epsilon_{\bar{i}} - \epsilon_{\overline{i+1}}\}_{1 \le i \le m-1} \cup \{\epsilon_i - \epsilon_{i+1}\}_{1 \le i \le n-1}, \quad \Pi_{\bar{1}} = \{\epsilon_{\overline{m}} - \epsilon_1\},$$

and the associated standard set of positive roots is

$$\Phi^+ = \{\epsilon_i - \epsilon_j \mid i, j \in I(m|n), i < j\},$$

where the odd roots are $\epsilon_i - \epsilon_j$ with indices $i < 0 < j$. Clearly, $\mathfrak{g}_{\epsilon_i - \epsilon_j} = \mathbb{C}E_{ij}$. It follows by (1.4) that

$$(\delta_i - \epsilon_j, \delta_i - \epsilon_j) = 0,$$

for all the odd roots $\delta_i - \epsilon_j$, where $1 \le i \le m, 1 \le j \le n$. An odd root α with $(\alpha, \alpha) = 0$ is called *isotropic*. The standard Dynkin diagram is:

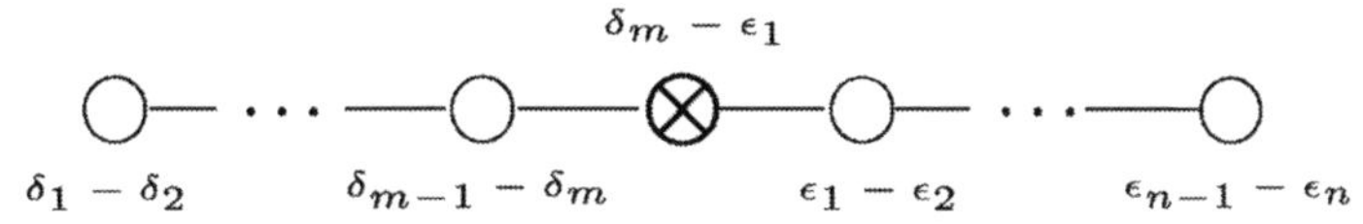

where we have used $\bigotimes$ to denote an isotropic odd simple root.

Remark 1.6. The notion of root systems and Dynkin diagrams makes sense for all the basic classical Lie superalgebras, which consist of $\mathfrak{gl}(m|n), \mathfrak{sl}(m|n), \mathfrak{osp}(m|2n)$ and three exceptional ones (besides the simple Lie algebras).

1.4 Non-conjugate Borel subalgebras and $\epsilon\delta$-sequences

As we have seen above, the bilinear form on the real subspace $\mathfrak{h}^*_{\mathbb{R}}$ spanned by the ϵ_i's is not positive-definite (due to the supertrace), and moreover, there exist isotropic odd roots.

Another distinguished feature of Lie superalgebras is the existence of non-conjugate Borel subalgebras or non-isomorphic Dynkin diagrams (under the Weyl group action).

Lemma 1.7. *Let $\mathfrak{g}$ be a Lie superalgebra with triangular decomposition $\mathfrak{g} = \mathfrak{n}^- \oplus \mathfrak{h} \oplus \mathfrak{n}^+$, which corresponds to the root system $\Phi = \Phi^+ \cup \Phi^-$. Let α be an odd isotropic simple root. Let $\mathfrak{b} = \mathfrak{h} + \mathfrak{n}^+$. Then, $\Phi(\alpha)^+ := (\Phi^+ \backslash \{\alpha\}) \cup \{-\alpha\}$ is a new system of positive roots, whose corresponding set of simple roots is*

$$\Pi(\alpha) = \{\beta \in \Pi \mid (\beta, \alpha) = 0, \beta \neq \alpha\} \cup \{\beta + \alpha \mid \beta \in \Pi, (\beta, \alpha) \neq 0\} \cup \{-\alpha\}.$$

The new Borel subalgebra corresponding to $\Pi(\alpha)$ will be denoted by $\mathfrak{b}(\alpha)$.

Proof. Follows from a straightforward verification. □

The process of obtaining $\Pi(\alpha)$ from Π above will be referred to as an *odd reflection*, and will be denoted by r_α, in accordance with the usual notion of real reflections.

Example 1.8. Associated to $\mathfrak{gl}(1|2)$, we have $\Phi_{\bar{0}} = \{\pm(\epsilon_1 - \epsilon_2)\}$, and $\Phi_{\bar{1}} = \{\pm(\delta_1 - \epsilon_1), \pm(\delta_1 - \epsilon_2)\}$. There are 6 sets of simple roots, that are related by the real and odd reflections as follows. There are three conjugacy classes of Borel subalgebras, and each vertical pair corresponds to such a conjugacy class.

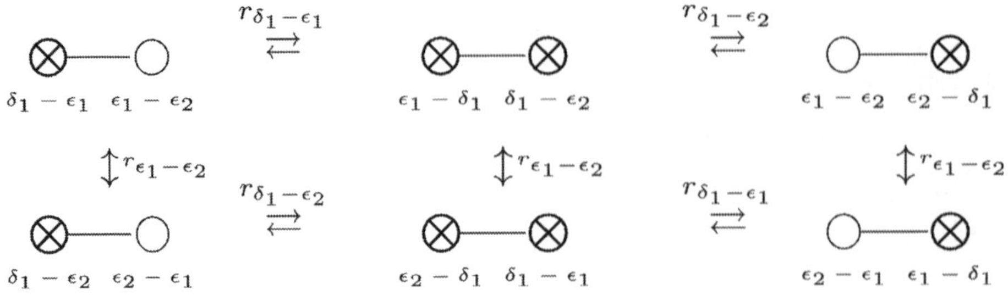

One convenient way to parameterize the conjugacy classes of Borel subalgebras of $\mathfrak{gl}(m|n)$ is via the notion of $\epsilon\delta$-sequences. Keeping (1.3) in mind, we list the simple roots associated to a given Borel subalgebra $\mathfrak{b}$ in order as $\epsilon_{i_1} - \epsilon_{i_2}, \epsilon_{i_2} - \epsilon_{i_3}, \ldots, \epsilon_{i_{m+n-1}} - \epsilon_{i_{m+n}}$, where $\{i_1, i_2, \ldots, i_{m+n}\} = I(m|n)$. Switching the ordered sequence $\epsilon_{i_1}\epsilon_{i_2} \cdots \epsilon_{i_{m+n}}$ to the $\epsilon\delta$-notation by (1.3) and then dropping the indices give us the $\epsilon\delta$-*sequence* associated to $\mathfrak{b}$. Note that the total number of δ's (respectively, ϵ's) is m (respectively, n).

For example, the three conjugacy classes of Borels for $\mathfrak{gl}(1|2)$ above correspond to the three sequences $\delta\epsilon\epsilon$, $\epsilon\delta\epsilon$, $\epsilon\epsilon\delta$, respectively. In more detail, the first sequence $\delta\epsilon\epsilon$ is obtained by removing the indices of $\delta_1\epsilon_1\epsilon_2$ (read off from the upper-left diagram above) or $\delta_1\epsilon_2\epsilon_1$ (from the lower-left diagram above). Also the standard Borel of $\mathfrak{gl}(m|n)$ corresponds to the sequence $\underbrace{\delta\cdots\delta}_{m}\underbrace{\epsilon\cdots\epsilon}_{n}$ while the opposite Borel to the standard one corresponds to $\underbrace{\epsilon\cdots\epsilon}_{n}\underbrace{\delta\cdots\delta}_{m}$.

Exercise 1.9. Let Φ be the roots of $\mathfrak{gl}(m|n)$ with respect to the Cartan subalgebra $\mathfrak{h}$. Prove that the sets of simple roots in Φ are in one-to-one correspondence with the $\epsilon\delta$-sequences with m δ's and n ϵ's in total. In particular, there are $\binom{m+n}{m}$ conjugacy classes of Borel subalgebras.

1.5 The ortho-symplectic Lie superalgebras

Let B be a non-degenerate even supersymmetric bilinear form on a vector superspace $V = V_{\bar{0}} \oplus V_{\bar{1}}$. Here B is *even* if $B(V_i, V_j) = 0$ unless $i = j \in \mathbb{Z}_2$, and B is *supersymmetric* if $B|_{V_{\bar{0}}\times V_{\bar{0}}}$ is symmetric while $B|_{V_{\bar{1}}\times V_{\bar{1}}}$ is skew-symmetric (and hence $\dim V_{\bar{1}}$ is necessarily even).

For $s \in \mathbb{Z}_2$, let

$$\begin{aligned}\mathfrak{osp}(V)_s &= \{g \in \mathfrak{gl}(V)_s \mid B(g(x), y) = -(-1)^{s\cdot|x|}B(x, g(y)), \forall x, y \in V\},\\ \mathfrak{osp}(V) &= \mathfrak{osp}_{\bar{0}}(V) \oplus \mathfrak{osp}_{\bar{1}}(V).\end{aligned}$$

One checks that $\mathfrak{osp}(V)$ is a Lie superalgebra, whose even subalgebra is isomorphic to $\mathfrak{so}(V_{\bar{0}}) \oplus \mathfrak{sp}(V_{\bar{1}})$. When $V = \mathbb{C}^{\ell|2n}$, we write $\mathfrak{osp}(V) = \mathfrak{osp}(\ell|2n)$.

Remark 1.10. One can also define the Lie superalgebra $\mathfrak{spo}(V)$ as the subalgebra of $\mathfrak{gl}(V)$ which preserves a non-degenerate skew-supersymmetric bilinear form on V (here $\dim V_{\bar{0}}$ has to be even). When $V = \mathbb{C}^{2n|\ell}$, we write $\mathfrak{spo}(V) = \mathfrak{spo}(2n|\ell)$.

Exercise 1.11. Show that Lie superalgebras $\mathfrak{osp}(\ell|2n)$ and $\mathfrak{spo}(2n|\ell)$ are isomorphic.

1.6 The Dynkin diagrams

Define the *super transpose* st as follows: for a matrix in the block form (1.1), we let

$$\begin{pmatrix} a & b \\ c & d \end{pmatrix}^{\mathrm{st}} = \begin{pmatrix} a^t & c^t \\ -b^t & d^t \end{pmatrix},$$

where x^t denotes the usual transpose of the matrix x.

Define the $(2n + 2m + 1) \times (2n + 2m + 1)$ matrix in the $(n|n|m|m|1)$-block form

$$\mathfrak{J}_{2n|2m+1} := \begin{pmatrix} 0 & I_n & 0 & 0 & 0 \\ -I_n & 0 & 0 & 0 & 0 \\ 0 & 0 & 0 & I_m & 0 \\ 0 & 0 & I_m & 0 & 0 \\ 0 & 0 & 0 & 0 & 1 \end{pmatrix}, \tag{1.5}$$

where I_n is the $n \times n$ identity matrix. Let $\mathfrak{J}_{2n|2m}$ denote the $(2n+2m) \times (2n+2m)$ matrix obtained from $\mathfrak{J}_{2n|2m+1}$ by deleting the last row and column. Then, for $\ell = 2m$ or $2m+1$, by definition $\mathfrak{spo}(2n|\ell)$ is the subalgebra of $\mathfrak{gl}(2n|\ell)$ which preserves the bilinear form on $\mathbb{C}^{2n|\ell}$ with matrix $\mathfrak{J}_{2n|\ell}$ relative to the standard basis of $\mathbb{C}^{2n|\ell}$, and hence

$$\mathfrak{spo}(2n|\ell) = \{g \in \mathfrak{gl}(2n|\ell) \mid g^{\mathrm{st}} \mathfrak{J}_{2n|\ell} + \mathfrak{J}_{2n|\ell}\, g = 0\}.$$

By a direct computation, $\mathfrak{spo}(2n|2m+1)$ consists of the $(2n+2m+1) \times (2n+2m+1)$ matrices of the following $(n|n|m|m|1)$-block form

$$\begin{pmatrix} d & e & y_1^t & x_1^t & z_1^t \\ f & -d^t & -y^t & -x^t & -z^t \\ x & x_1 & a & b & -v^t \\ y & y_1 & c & -a^t & -u^t \\ z & z_1 & u & v & 0 \end{pmatrix}, \quad b, c \text{ skew-symmetric}, e, f \text{ symmetric}. \tag{1.6}$$

The Lie superalgebra $\mathfrak{spo}(2n|2m)$ consists of matrices (1.6) with the last row and column removed. Here and below, the rows and columns of the matrices $\mathfrak{J}_{2n|\ell}$ and (1.6) (or its modification) are indexed by the finite set $I(2n|\ell)$.

The standard Dynkin diagrams of $\mathfrak{spo}(2n|2m+1)$ and $\mathfrak{spo}(2n|2m)$ are given respectively as follows:

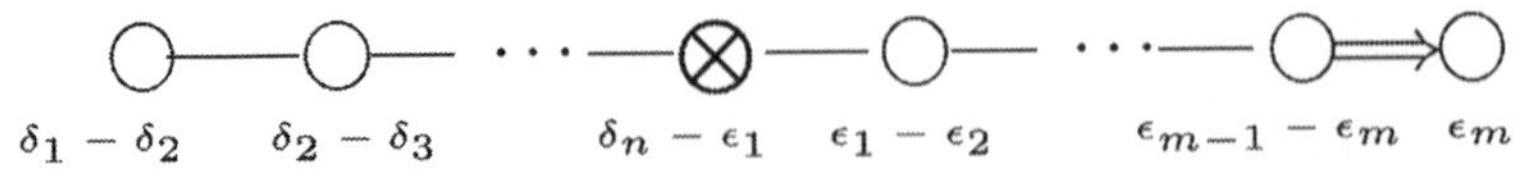

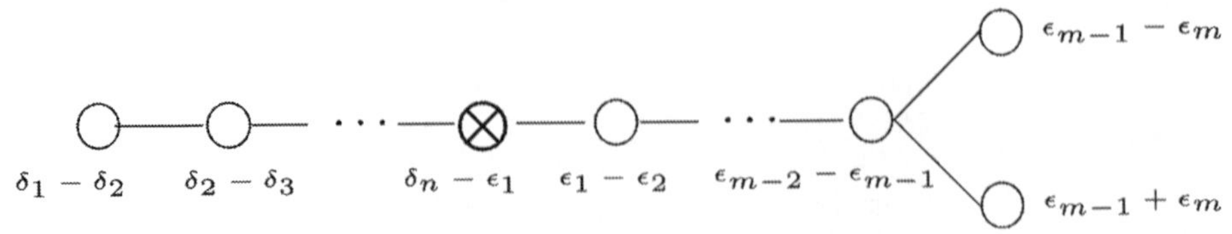

Example 1.12. The Lie superalgebra $\mathfrak{g} = \mathfrak{osp}(1|2)$ has even subalgebra $\mathfrak{g}_{\bar{0}} \cong \mathfrak{sl}(2) = \mathbb{C}\langle e, h, f\rangle$ and $\mathfrak{g}_{\bar{1}}$ isomorphic to the 2-dimensional natural $\mathfrak{sl}(2)$-module $\mathbb{C}\langle E, F\rangle$. Moreoever, $[E, E] = 2e, [F, F] = -2f, [E, F] = h$.

The simple root consists of a (unique) odd non-isotropic root δ, twice of which is an even root. The Dynkin diagram of $\mathfrak{osp}(1|2)$ is denoted by ● (in order to distinguish from an odd simple isotropic root ⊗).

Exercise 1.13. Prove the following identities in $U(\mathfrak{osp}(1|2))$ $(n \in \mathbb{Z}_+)$:

(i) $[E, F^{2n}] = -nF^{2n-1}$,
(ii) $[E, F^{2n+1}] = F^{2n}(h - n)$.

Exercise 1.14. Use Exercise 1.13 to show that the finite-dimensional irreducible representations of $\mathfrak{osp}(1|2)$ are parameterized by the set of highest weights $\{n\delta | n \in$

$\mathbb{Z}_+\}$. Denoting the irreducible module corresponding to $n\delta$ by $L(n\delta)$, show that

$$\mathrm{ch}L(n\delta) = \frac{e^{(n+\frac{1}{2})\delta} - e^{-(n+\frac{1}{2})\delta}}{e^{\frac{1}{2}\delta} - e^{-\frac{1}{2}\delta}}.$$

Exercise 1.15. Consider the element

$$\Omega := h^2 + h + 2FE + 4fe \in U(\mathfrak{osp}(1|2)). \text{ (See Section 2.1)}$$

Use Exercise 1.13 to prove that $[F, \Omega] = [E, \Omega] = 0$, and hence Ω is in the center of $U(\mathfrak{osp}(1|2))$. Conclude from Exercise 1.14 that Ω acts as different scalars on different finite-dimensional irreducible representations and hence that every finite-dimensional $\mathfrak{osp}(1|2)$-module is completely reducible.

2 Finite-dimensional modules of Lie superalgebras

2.1 PBW theorem

The *universal enveloping algebra* $U(\mathfrak{g})$ of a Lie superalgebra $\mathfrak{g} = \mathfrak{g}_{\bar{0}} \oplus \mathfrak{g}_{\bar{1}}$ is an associative superalgebra characterized by a universal property exactly as for Lie algebras.

Assume that $\{x_1, \ldots, x_r\}$ is a basis for $\mathfrak{g}_{\bar{0}}$ and $\{y_1, \ldots, y_s\}$ is a basis for $\mathfrak{g}_{\bar{1}}$. Then the universal enveloping algebra $U(\mathfrak{g})$ admits a PBW basis

$$x_1^{a_1} \cdots x_r^{a_r} y_1^{b_1} \cdots y_s^{b_s}, \qquad \forall a_i \in \mathbb{Z}_+, \forall b_j \in \{0, 1\}. \tag{2.1}$$

Alternatively, if we define the standard PBW filtration $\{F^d U(\mathfrak{g})\}$ on $U(\mathfrak{g})$ by letting $F^d U(\mathfrak{g})$ be the span of elements (2.1) with $\Sigma_i a_i + \Sigma_j b_j \leq d$, then we have the following isomorphism for the associated graded of $U(\mathfrak{g})$ in terms of the symmetric algebra $S(\mathfrak{g}_{\bar{0}})$ and exterior algebra $\wedge(\mathfrak{g}_{\bar{1}})$:

$$\mathrm{gr}^F U(\mathfrak{g}) \cong S(\mathfrak{g}_{\bar{0}}) \otimes \wedge(\mathfrak{g}_{\bar{1}}).$$

Example 2.1. Let $\mathfrak{g} = \mathfrak{g}_{\bar{1}}$ be a purely odd Lie superalgebra. Then its universal enveloping algebra is isomorphic to the exterior algebra $\wedge(\mathfrak{g}_{\bar{1}})$.

2.2 Representations of $\mathfrak{gl}(1|1)$

Recall the basis $\{e, h_1, h_2, f\}$ for $\mathfrak{g} = \mathfrak{gl}(1|1)$ from Example 1.4. Note that the even subalgebra $\mathfrak{gl}(1|1)_{\bar{0}} = \mathfrak{gl}(1) \oplus \mathfrak{gl}(1)$ coincides with the Cartan subalgebra $\mathfrak{h}$, and the Weyl group is trivial in this case.

Given $\lambda = (\lambda_1, \lambda_2) \in \mathbb{C}^2$, we denote by $\mathbb{C}_\lambda := \mathbb{C}v_\lambda$ the one-dimensional $\mathfrak{h}$-module by letting $h_i v_\lambda = \lambda_i v_\lambda$, for $i = 1, 2$. Regarding $\mathbb{C}_\lambda$ as a module over the Borel subalgebra $\mathfrak{h} + \mathbb{C}e$, on which e acts trivially, we define the Verma module (which coincide with Kac module defined below for $\mathfrak{gl}(1|1)$) over $\mathfrak{g}$:

$$K(\lambda) = U(\mathfrak{g}) \otimes_{U(\mathfrak{h}+\mathbb{C}e)} \mathbb{C}_\lambda.$$

By the PBW theorem $K(\lambda)$ can be identified with the following two-dimensional space (where by abuse of notation $1 \otimes v_\lambda$ is denoted by v_λ):

$$K(\lambda) = \mathbb{C}\langle v_\lambda, f v_\lambda \rangle.$$

The $\mathfrak{g}$-module $K(\lambda)$ has a unique simple quotient, denoted by $L(\lambda)$.

The $\mathfrak{g}$-module $K(\lambda)$ is simple if and only if $\lambda_1 \neq -\lambda_2$. This follows from

$$efv_\lambda = (h_1 + h_2)v_\lambda - fev_\lambda = (\lambda_1 + \lambda_2)v_\lambda.$$

If $\lambda_1 \neq -\lambda_2$, then $K(\lambda_1, \lambda_2)$ is the unique simple object in its block (in the category of finite-dimensional $\mathfrak{gl}(1|1)$-modules). This block is semisimple.

For $a \in \mathbb{C}$, we have a non-split short exact sequence of $\mathfrak{g}$-modules:

$$0 \to L(a-1, 1-a) \to K(a, -a) \to L(a, -a) \to 0.$$

The block containing $L(a, -a)$ is not semisimple, and it contains infinitely many simple objects $L(b, -b)$ for $b \in a + \mathbb{Z}$ (the underlying algebra can be described in terms of the A_∞-quiver).

2.3 Finite dimensional simple $\mathfrak{gl}(m|n)$-modules

Let $\mathfrak{g} = \mathfrak{gl}(m|n)$ in this subsection.

Let $\mathfrak{n}^+$ (and respectively, $\mathfrak{n}^-$) be the subalgebra of strictly upper (and respectively, lower) triangular matrices of $\mathfrak{gl}(m|n)$ with respect to the standard basis. Then we have the triangular decomposition

$$\mathfrak{g} = \mathfrak{n}^- \oplus \mathfrak{h} \oplus \mathfrak{n}^+.$$

The even subalgebra admits a compatible triangular decomposition

$$\mathfrak{g}_{\bar{0}} = \mathfrak{n}_{\bar{0}}^- \oplus \mathfrak{h} \oplus \mathfrak{n}_{\bar{0}}^+,$$

where $\mathfrak{n}_{\bar{0}}^\pm = \mathfrak{g}_{\bar{0}} \cap \mathfrak{n}^\pm$.

Moreover, the Lie superalgebra $\mathfrak{g}$ admits a $\mathbb{Z}$-grading

$$\mathfrak{g} = \mathfrak{g}_{-1} \oplus \mathfrak{g}_{\bar{0}} \oplus \mathfrak{g}_1,$$

where $\mathfrak{g}_{-1}$ (respectively, $\mathfrak{g}_1$) is spanned by all E_{ij} with $i > 0 > j$ (respectively, $i < 0 < j$). Note that this $\mathbb{Z}$-grading is compatible with the $\mathbb{Z}_2$-grading, i.e., the degree zero subspace coincides with the $\mathbb{Z}_2$-degree zero subspace. This is equivalent to the fact that $\mathfrak{g}_{\bar{0}}$ is a Levi subalgebra of $\mathfrak{g}$ (corresponding to the removal of the odd simple root from the standard Dynkin diagram of $\mathfrak{g}$).

For $\lambda \in \mathfrak{h}^*$, let $L(\lambda)$ (respectively, $L^0(\lambda)$) be the simple module of $\mathfrak{g}$ (respectively, $\mathfrak{g}_{\bar{0}}$) of highest weight λ. Define the *Kac module* over $\mathfrak{g}$ by

$$K(\lambda) = U(\mathfrak{g}) \otimes_{U(\mathfrak{g}_{\bar{0}} \oplus \mathfrak{g}_1)} L^0(\lambda),$$

which can be identified by the PBW theorem with

$$K(\lambda) = \wedge(\mathfrak{g}_{-1}) \otimes L^0(\lambda). \tag{2.2}$$

Proposition 2.2. *There exists a surjective $\mathfrak{g}$-module homomorphism (unique up to a scalar multiple) $K(\lambda) \twoheadrightarrow L(\lambda)$. Moreover, the following are equivalent:*

(1) *$L(\lambda)$ is finite-dimensional.*
(2) *$L^0(\lambda)$ is finite-dimensional.*
(3) *$K(\lambda)$ is finite-dimensional.*

Proof. The surjective homomorphism follows from the fact that $K(\lambda)$ is a highest weight $\mathfrak{g}$-module of highest weight λ.

(1) $\Rightarrow$ (2). Note that $L^0(\lambda)$ is a quotient of $L(\lambda)$ regarded as $\mathfrak{g}_{\bar{0}}$-module.
(2) $\Rightarrow$ (3). Follows from (2.2).
(3) $\Rightarrow$ (1). Follows from the surjective map $K(\lambda) \twoheadrightarrow L(\lambda)$. □

Arguing as for Lie algebras, every finite-dimensional simple $\mathfrak{g}$-module is a highest weight module $L(\lambda)$ for some λ, and moreover, $L(\lambda) \not\cong L(\mu)$ if $\lambda \neq \mu$. Hence, the above proposition gives a classification of finite-dimensional simple $\mathfrak{g}$-modules.

2.4 Typical irreducible characters

Let $\mathfrak{g} = \mathfrak{gl}(m|n)$ in this subsection.

We will denote by P^+ the set of weights $\lambda \in \mathfrak{h}^*$ such that $L^0(\lambda)$ is finite-dimensional. Recall that the character of $L^0(\lambda)$ for $\lambda \in P^+$ is given by Weyl's formula

$$\mathrm{ch}L^0(\lambda) = \frac{\sum_{\sigma\in W} \mathrm{sgn}(\sigma) e^{\sigma(\lambda+\rho_0)-\rho_0}}{\prod_{\alpha\in\Phi^+_{\bar{0}}}(1-e^{-\alpha})}.$$

Here W ($\cong \mathfrak{S}_m \times \mathfrak{S}_n$) denotes the Weyl group of $\mathfrak{g}_{\bar{0}}$, sgn denotes the sign representation of W, and $\rho_0 = \frac{1}{2}\sum_{\alpha\in\Phi^+_{\bar{0}}} \alpha$.

Also denote by

$$\rho_1 = \frac{1}{2}\sum_{\alpha\in\Phi^+_{\bar{1}}} \alpha, \qquad \rho = \rho_0 - \rho_1.$$

Lemma 2.3. *We have*

(1) $(\rho, \beta) = \frac{1}{2}(\beta, \beta), \forall \beta \in \Pi$.
(2) $\sigma(\rho_1) = \rho_1, \forall \sigma \in W$.

Proof. Part 1 can be verified directly. Part 2 follows from the fact that $\mathfrak{g}_1$ is preserved by the adjoint group $G_{\bar{0}}$ associated to $\mathfrak{g}_{\bar{0}}$ under the adjoint action. □

Corollary 2.4. *We have*

$$\mathrm{ch}K(\lambda) = \frac{\prod_{\alpha\in\Phi^+_{\bar{1}}}(1+e^{-\alpha})}{\prod_{\alpha\in\Phi^+_{\bar{0}}}(1-e^{-\alpha})} \sum_{\sigma\in W} \mathrm{sgn}(\sigma) e^{\sigma(\lambda+\rho)-\rho}. \tag{2.3}$$

Proof. By (2.2), $\mathrm{ch}K(\lambda) = \mathrm{ch}L^0(\lambda)\prod_{\alpha\in\Phi_{\bar 1}^+}(1+e^{-\alpha})$. Now by the Weyl character formula for $L^0(\lambda)$, we have

$$\mathrm{ch}K(\lambda) = \frac{\prod_{\alpha\in\Phi_{\bar 1}^+}(1+e^{-\alpha})}{\prod_{\alpha\in\Phi_{\bar 0}^+}(1-e^{-\alpha})}\sum_{\sigma\in W}\mathrm{sgn}(\sigma)e^{\sigma(\lambda+\rho_0)-\rho_0}.$$

This is equivalent to the formula in (2.3) by the second part of Lemma 2.3. □

Exercise 2.5. Show that

$$\rho_1 = \frac{n}{2}\sum_{i=1}^{m}\delta_i - \frac{m}{2}\sum_{j=1}^{n}\epsilon_j.$$

Again it follows that $\sigma(\rho_1) = \rho_1, \forall\sigma\in W$.

Definition 2.6. A weight $\lambda\in\mathfrak{h}^*$ is called *typical*, if $\prod_{\alpha\in\overline{\Phi}_{\bar 1}^+}(\lambda+\rho,\alpha)\neq 0$. It is called *atypical* if it is not typical.

Theorem 2.7 (Kac). *Let $\lambda\in P^+$. Then λ is typical if and only if* $\mathrm{ch}L(\lambda)$ *is given by the right hand side of* (2.3), *if and only if $K(\lambda)$ is irreducible.*

The example of $\mathfrak{gl}(1|1)$ in Section 2.2 fits well with this theorem.

Sketch of a proof. [K2] Let $v_\lambda\in K(\lambda)$ be a highest weight vector. Denote by e_α, f_α the generators of $\mathfrak{g}_{\pm\alpha}$, where $\alpha\in\Phi^+$.

Assume that S is a nonzero $\mathfrak{g}$-submodule of $K(\lambda)$. One first shows easily that
(i) $\prod_{\alpha\in\Phi_{\bar 1}^+} f_\alpha v_\lambda\in S$.
It follows that
(ii) $\prod_{\beta\in\Phi_{\bar 1}^+} e_\beta\prod_{\alpha\in\Phi_{\bar 1}^+} f_\alpha v_\lambda\in S$.
Then one further shows that (up to a nonzero scalar multiple)
(iii) $\prod_{\beta\in\Phi_{\bar 1}^+} e_\beta\prod_{\alpha\in\Phi_{\bar 1}^+} f_\alpha v_\lambda = \prod_{\alpha\in\Phi_{\bar 1}^+}(\lambda+\rho,\alpha)\, v_\lambda$.

Step (iii) uses the W-invariance of $\Phi_{\bar 1}^+$ and a degree counting argument among others.

From (i) and (iii) it follows that if λ is atypical, then $K(\lambda)$ is irreducible.

Now suppose that $K(\lambda)$ is irreducible. The ad $\mathfrak{g}_0$-invariance of $\mathfrak{g}_{\pm 1}$ implies that if $v_\lambda\in U(\mathfrak{g})\prod_{\alpha\in\Phi_{\bar 1}^+} f_\alpha v_\lambda$, then v_λ is a scalar multiple of $\prod_{\beta\in\Phi_{\bar 1}^+} e_\beta\prod_{\alpha\in\Phi_{\bar 1}^+} f_\alpha v_\lambda$. Thus by (iii) λ is typical. □

2.5 Odd reflections

As usual, we let $\{e_\beta, h_\beta, f_\beta\}$ denote the Chevalley generators for $\beta\in\Phi^+$. The following simple and fundamental lemma for odd reflections has been used by many authors (e.g. [LSS, Appendix], [PS, Lemma 0.3] and [KW, Lemma 1.4]; also compare [DP, (2.12)] for an unconventional definition). Recall the new Borel $\mathfrak{b}(\alpha)$ for a simple isotropic odd root α from Lemma 1.7.

Lemma 2.8. *Let L be a simple $\mathfrak{g}$-module of $\mathfrak{b}$-highest weight λ and let v be a $\mathfrak{b}$-highest weight vector of L. Let α be a simple isotropic odd root.*

(1) *If $\langle \lambda, h_\alpha \rangle = 0$, then L is a $\mathfrak{g}$-module of $\mathfrak{b}(\alpha)$-highest weight λ and v is a $\mathfrak{b}(\alpha)$-highest weight vector.*

(2) *If $\langle \lambda, h_\alpha \rangle \neq 0$, then L is a $\mathfrak{g}$-module of $\mathfrak{b}(\alpha)$-highest weight $\lambda - \alpha$ and $f_\alpha v$ is a $\mathfrak{b}(\alpha)$-highest weight vector.*

Proof. We first observe three simple identities:

(i) $e_\alpha f_\alpha v = [e_\alpha, f_\alpha] v = h_\alpha v = \langle \lambda, h_\alpha \rangle v$.

(ii) $e_\beta f_\alpha v = [e_\beta, f_\alpha] v = 0$ for any $\beta \in \Phi^+ \cap \Phi(\alpha)^+$, since either $\beta - \alpha$ is not a root or it belongs to $\Phi^+ \cap \Phi(\alpha)^+$.

(iii) $f_\alpha^2 v = 0$, since α is an isotropic odd root.

Now, we consider the two cases separately.

(1) Assume that $\langle \lambda, h_\alpha \rangle = 0$. Then we must have $f_\alpha v = 0$, otherwise $f_\alpha v$ would be a $\mathfrak{b}$-singular vector in the simple $\mathfrak{g}$-module V by (i) and (ii). This together with Lemma 1.7 implies that v is a $\mathfrak{b}(\alpha)$-highest weight vector of weight λ in the $\mathfrak{g}$-module V.

(2) Assume that $\langle \lambda, h_\alpha \rangle \neq 0$. Then (i), (ii), (iii) and Lemma 1.7 imply that $f_\alpha v$ is nonzero and it is a $\mathfrak{b}(\alpha)$-highest weight vector of weight $\lambda - \alpha$ in V. □

2.6 Finite-dimensional simple $\mathfrak{osp}$-modules

The case of $\mathfrak{g} = \mathfrak{spo}(2m|2n+1)$ will be treated in detail (while the case of $\mathfrak{g} = \mathfrak{spo}(2m|2n)$ is similar). As usual, we have the triangular decomposition of $\mathfrak{g}$ (with respect to the standard Borel) $\mathfrak{g} = \mathfrak{n}^- \oplus \mathfrak{h} \oplus \mathfrak{n}^+$, which allows us to define the Verma module $\Delta(\lambda)$ associated to $\lambda = \sum_{i=1}^m \lambda_i \delta_i + \sum_{j=1}^n v_j \epsilon_j \in \mathfrak{h}^*$. Then $\Delta(\lambda)$ admits a unique simple quotient $\mathfrak{g}$-module, denoted by $L(\lambda)$.

Clearly a necessary condition for $L(\lambda)$ to be finite-dimensional is that λ is dominant with respect to the even subalgebra $\mathfrak{g}_{\bar{0}} = \mathfrak{sp}(2m) \oplus \mathfrak{so}(2n+1)$. Since $\mathfrak{g}_{\bar{0}}$ is not a Levi subalgebra of $\mathfrak{g}$, the above definition of the Kac module for $\mathfrak{gl}(m|n)$-module is no longer valid for $\mathfrak{g}$. Nevertheless, the notion of "typical weights" in Definition 2.6 still makes sense for $\mathfrak{osp}$ (or any basic classical Lie superalgebra), where we replace $\Phi_{\bar{1}}^+$ in the definition by $\{\alpha \in \Phi_{\bar{1}}^+ | (\alpha, \alpha) = 0\}$. We have the following $\mathfrak{osp}$ counterpart of Theorem 2.7, whose much more involved proof will be skipped.

Theorem 2.9. [K2] *Let $\mathfrak{g} = \mathfrak{osp}(\ell|2n)$, for $\ell = 2m$ or $2m+1$. If the weight $\lambda \in \mathfrak{h}^*$ is typical and $\mathfrak{g}_{\bar{0}}$-dominant integral such that $L(\lambda)$ is finite-dimensional, then* $\mathrm{ch} L(\lambda)$ *is given by the right hand side of* (2.3).

As for Lie algebras, a finite-dimensional simple $\mathfrak{g}$-module has to be a highest weight module (with respect to any Borel), and hence is isomorphic to some $L(\lambda)$, and $L(\lambda) \ncong L(\mu)$ if $\lambda \neq \mu$. However, the classification of finite-dimensional simple $\mathfrak{g}$-modules is non-trivial, partly because the even subalgebra of $\mathfrak{g}$ is not a Levi subalgebra.

Let $\lambda = \sum_{i=1}^m \lambda_i \delta_i + \sum_{j=1}^n v_j \epsilon_j \in \mathfrak{h}^*$. For $\mathfrak{g} = \mathfrak{spo}(2m|2n+1)$, by a $\mathfrak{g}_{\bar{0}}$-*dominant integral* weight λ we mean that $(\lambda_1, \ldots, \lambda_m)$ is a partition, and $v_1 \geq$

$v_2 \geq \cdots \geq v_n \geq 0$, with all v_j in $\mathbb{Z}$ or all v_j in $\frac{1}{2} + \mathbb{Z}$. For $\mathfrak{g} = \mathfrak{spo}(2m|2n)$ with $n \geq 2$, by a $\mathfrak{g}_{\bar{0}}$-*dominant integral* weight λ we mean that $(\lambda_1, \ldots, \lambda_m)$ is a partition, and $v_1 \geq \cdots \geq v_{n-1} \geq |v_n|$, with all v_j in $\mathbb{Z}$ or all v_j in $\frac{1}{2} + \mathbb{Z}$.

Now, if $v_j \in \frac{1}{2} + \mathbb{Z}$, for all $1 \leq j \leq n$ (which will be referred to as a *half-integer weight*), then λ is always typical. We note that a necessary condition for such an $L(\lambda)$ to be finite-dimensional is that $\lambda_m \geq n$ (see Exercise 2.14), and it can be shown that this condition is sufficient as well. It follows that $\mathrm{ch}L(\lambda)$ for such a weight λ is given in Theorem 2.9.

Now we will study in detail the case when $v_j \in \mathbb{Z}$, for all $1 \leq j \leq n$. Let us call such a λ an *integer weight*.

Definition 2.10. A partition $\mu = (\mu_1, \mu_2, \ldots)$ is called an $(m|n)$-*hook partition* if $\mu_{m+1} \leq n$. We will also call μ simply a *hook partition*, when m, n are clear from the context.

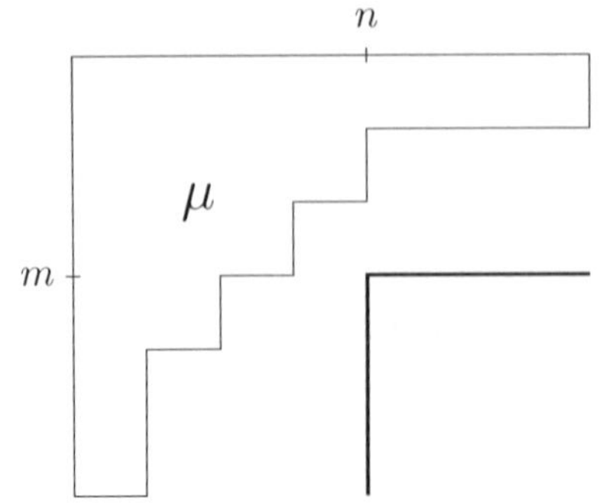

Given an $(m|n)$-hook partition μ, we denote by $\mu^+ = (\mu_{m+1}, \mu_{m+2}, \ldots)$ and write its transpose, which is necessarily of length $\leq n$, as $\nu = (\mu^+)' = (\nu_1, \ldots, \nu_n)$. We define the weights

$$\begin{aligned} \mu^\natural &= \mu_1\delta_1 + \cdots + \mu_m\delta_m + \nu_1\epsilon_1 + \cdots + \nu_{n-1}\epsilon_{n-1} + \nu_n\epsilon_n, \\ \mu^\natural_- &= \mu_1\delta_1 + \cdots + \mu_m\delta_m + \nu_1\epsilon_1 + \cdots + \nu_{n-1}\epsilon_{n-1} - \nu_n\epsilon_n. \end{aligned} \tag{2.4}$$

($\mu^\natural_-$ is only used for $\mathfrak{spo}(2m|2n)$ below.)

Theorem 2.11. *Given any $(m|n)$-hook partition μ, the simple $\mathfrak{spo}(2m|2n+1)$-module $L(\mu^\natural)$ (with respect to the standard Borel) is finite-dimensional. Moreover, these modules form a complete list of non-isomorphic finite-dimensional simple $\mathfrak{spo}(2m|2n+1)$-modules of integer highest weight.*

The above theorem is due to Kac [K2] who formulated the conditions for finite-dimensional highest weight simple $\mathfrak{g}$-modules in terms of Dynkin labels (instead of hook partitions). A different proof using odd reflections is given by Shu and Wang (which also works in characteristic $p > 0$). Let us sketch the idea of a proof following [SW], as the argument therein bears some similarity to the argument used later on for super duality. The same type of argument works for Theorem 2.13 below for $\mathfrak{spo}(2m|2n)$.

Sketch of a proof. Let μ be an $(m|n)$-hook partition. We observe that the simple $\mathfrak{spo}(2M|2n+1)$-module V of highest weight μ (with respect to the standard Borel) is finite-dimensional for $M \geq \ell(\mu)$, since it appears as a subquotient of a suitable tensor product of simple modules of fundamental weights. Then we use odd reflections to change the standard Borel of $\mathfrak{spo}(2M|2n+1)$ to a Borel which is compatible with the standard Borel of $\mathfrak{spo}(2m|2n+1)$ (which is regarded as a subalgebra of $\mathfrak{spo}(2M|2n+1)$). We can show that the highest weight of V with respect to the new Borel is $\mu^\natural$, hence by restriction to $\mathfrak{spo}(2m|2n+1)$ we have proved the first part of the theorem.

A highest weight for a simple finite-dimensional $\mathfrak{g}$-module is necessarily of the form $\mu_1\delta_1+\cdots+\mu_m\delta_m+\nu_1\epsilon_1+\cdots+\nu_n\epsilon_n$, where $(\mu_1,\ldots,\mu_m)$ and $(\nu_1,\ldots,\nu_n)$ are partitions by the dominance condition on the even subalgebra $\mathfrak{g}_{\bar{0}}$. To prove the remaining condition $\nu_1' \leq \mu_m$, it suffices to prove it for $\mathfrak{spo}(2|2n+1)$ (which is a subalgebra of $\mathfrak{spo}(2m|2n+1)$). Let V be a finite-dimensional simple $\mathfrak{spo}(2|2n+1)$-module. Via the sequence of odd reflections $\delta_1-\epsilon_1, \delta_1-\epsilon_2, \ldots, \delta_1-\epsilon_n$ we change the standard Borel to a Borel with an odd non-isotropic simple root δ_1. The dominance condition on the simple root δ_1 (or rather on the even root $2\delta_1$), which is imposed by the finite-dimensionality of V, provides the desired necessary condition. □

Exercise 2.12. Complete the details in the proof of Theorem 2.11.

Similarly, we can classify the finite-dimensional simple $\mathfrak{spo}(2m|2n)$-modules. The case of $\mathfrak{spo}(2m|2)$ is special in that $\mathfrak{g}_{\bar{0}}$ is a Levi subalgebra and its classification of simple modules is very easy as in the case of $\mathfrak{gl}(m|n)$. So we consider only $\mathfrak{spo}(2m|2n)$ for $n \geq 2$ from now on. As before, if λ is a $\mathfrak{g}_{\bar{0}}$-dominant integral half-integer weight, then the irreducible module $L(\lambda)$ is finite-dimensional if and only if the weight satisfies the condition in Exercise 2.14. For the integer highest weights we have the following classification.

Theorem 2.13. [K2] (cf. [SW]) *Let $n \geq 2$. The modules $L(\mu^\natural)$ and $L(\mu^\natural_-)$(with respect to the standard Borel), where μ runs over all $(m|n)$-hook partitions, are all the non-isomorphic finite-dimensional simple $\mathfrak{spo}(2m|2n)$-modules of integer highest weight.*

Exercise 2.14. Let $\mathfrak{g} = \mathfrak{spo}(2m|2n+1)$ or $\mathfrak{g} = \mathfrak{spo}(2m|2n)$ and $\lambda = \sum_{i=1}^m \lambda_i\delta_i + \sum_{j=1}^n v_j\epsilon_j \in \mathfrak{h}^*$ be a $\mathfrak{g}_{\bar{0}}$-dominant integral weight. Assume that $v_j \in \frac{1}{2}+\mathbb{Z}_+$, for all $1 \leq j \leq n$. Prove by a similar argument to Theorem 2.11 that if $L(\lambda)$ is finite-dimensional then $\lambda_m \geq n$.

Exercise 2.15. Recall that $\mathfrak{so}(\ell)$ may be identified with $\wedge^2(\mathbb{C}^\ell)$, and more generally, $\mathfrak{osp}(\ell|2n)$ may be identified with $\wedge^2(\mathbb{C}^{\ell|2n}) = \wedge^2(\mathbb{C}^\ell)\oplus(\mathbb{C}^\ell\otimes\mathbb{C}^{2n})\oplus S^2(\mathbb{C}^{2n})$. Now using the invariant bilinear form to identify $\mathbb{C}^{\ell|2n}$ with $\mathbb{C}^{\ell|0}\oplus\mathbb{C}^{0|n}\oplus\mathbb{C}^{0|n*}$, we define a $\mathbb{Z}$-gradation on $\mathbb{C}^{\ell|2n}$ by setting $\deg v = 0$ for $v \in \mathbb{C}^{\ell|0}$, $\deg v = 1$ for $v \in \mathbb{C}^{0|n}$, and $\deg v = -1$ for $v \in \mathbb{C}^{0|n*}$. This induces a $\mathbb{Z}$-gradation on $\wedge^2(\mathbb{C}^{\ell|2n})$ and hence on $\mathfrak{osp}(\ell|2n)$. Prove that

$$\mathfrak{osp}(\ell|2n) = \bigoplus_{i=-2}^{2} \mathfrak{osp}(\ell|2n)_i,$$

where $\mathfrak{osp}(\ell|2n)_0 \cong \mathfrak{so}(\ell) \oplus \mathfrak{gl}(n)$; as $\mathfrak{osp}(\ell|2n)_0$-modules we have $\mathfrak{osp}(\ell|2n)_1 \cong \mathbb{C}^\ell \otimes \mathbb{C}^n$ and $\mathfrak{osp}(\ell|2n)_2 \cong \mathbb{C} \otimes S^2(\mathbb{C}^n)$.

3 Schur-Sergeev duality

3.1 The formulation

Let $\mathfrak{g} = \mathfrak{gl}(m|n)$. Let $\{e_i | i \in I(m|n)\}$ be the standard basis for the natural $\mathfrak{g}$-module $V = \mathbb{C}^{m|n}$.

Then $V^{\otimes d}$ is naturally a $\mathfrak{g}$-module by letting

$$\begin{aligned}\Phi_d(g).(v_1 \otimes v_2 \otimes \cdots \otimes v_d) =& g.v_1 \otimes \cdots \otimes v_d + (-1)^{|g|\cdot|v_1|} v_1 \otimes g.v_2 \otimes \cdots \otimes v_d \\ &+ \cdots + (-1)^{|g|\cdot(|v_1|+\cdots+|v_{d-1}|)} v_1 \otimes v_2 \otimes \cdots \otimes g.v_d,\end{aligned}$$

where $g \in \mathfrak{g}$ and $v_i \in V$ are assumed to be $\mathbb{Z}_2$-homogeneous.

On the other hand, the action of the symmetric group $\mathfrak{S}_d$ on $V^{\otimes d}$ is determined by

$$\begin{aligned}&\Psi_d((i, i+1)).v_1 \otimes \cdots \otimes v_i \otimes v_{i+1} \otimes \cdots \otimes v_d \\ &\quad = (-1)^{|v_i|\cdot|v_{i+1}|} v_1 \otimes \cdots \otimes v_{i+1} \otimes v_i \otimes \cdots \otimes v_d, \quad 1 \le i \le d-1,\end{aligned}$$

where (i, j) denotes a transposition in $\mathfrak{S}_d$ and v_i, v_{i+1} are $\mathbb{Z}_2$-homogeneous.

Lemma 3.1. *The actions of $(\mathfrak{g}, \Phi_d)$ and $(\mathfrak{S}_d, \Psi_d)$ on $V^{\otimes d}$ commute with each other. Symbolically, we write*

$$\mathfrak{g} \overset{\Phi_d}{\curvearrowright} V^{\otimes d} \overset{\Psi_d}{\curvearrowleft} \mathfrak{S}_d.$$

Exercise 3.2. Verify Lemma 3.1.

We write $\lambda \vdash d$ for a partition λ of d. Denote by $\mathcal{P}(d, m|n)$ the set of all $(m|n)$-hook partitions of size d and let

$$\mathcal{P}(m|n) = \cup_{d \ge 0} \mathcal{P}(d, m|n) = \{\lambda \mid \lambda \text{ are partitions with } \lambda_{m+1} \le n\}.$$

Also set $\mathcal{P}(d, m) = \mathcal{P}(d, m|0)$. We denote by $L(\lambda^\natural)$ for $\lambda \in \mathcal{P}(m|n)$ the simple $\mathfrak{g}$-module of highest weight $\lambda^\natural$ with respect to the standard Borel subalgebra. For a partition $\lambda \vdash d$, we denote by S^λ the Specht module of $\mathfrak{S}_d$. For example, $S^{(d)}$ is the trivial module, and $S^{(1^d)}$ is the sign module sgn. The following superalgebra analog of Schur duality was due to Sergeev [Sv1] (also cf. [BeR]).

Theorem 3.3 (Schur-Sergeev duality). 1. *The images Φ_d and Ψ_d, $\Phi_d(U(\mathfrak{g}))$ and $\Psi_d(\mathbb{C}\mathfrak{S}_d)$, satisfy the double centralizer property, i.e.*

$$\begin{aligned}\Phi_d(U(\mathfrak{g})) =& \mathrm{End}_{\mathfrak{S}_d}(V^{\otimes d}), \\ &\mathrm{End}_{U(\mathfrak{g})}(V^{\otimes d}) = \Psi_d(\mathbb{C}\mathfrak{S}_d).\end{aligned}$$

2. *As a* $U(\mathfrak{gl}(m|n)) \otimes \mathbb{C}\mathfrak{S}_d$*-module, we have*

$$V^{\otimes d} \cong \bigoplus_{\lambda \in \mathcal{P}(d,m|n)} L(\lambda^\natural) \otimes S^\lambda.$$

We refer to a $U(\mathfrak{gl}(m|n)) \otimes \mathbb{C}\mathfrak{S}_d$-module as a $(\mathfrak{gl}(m|n), \mathfrak{S}_d)$-module. Similar conventions also apply to similar setups below.

Remark 3.4. (1) For $n = 0$, the above Sergeev duality reduces to the usual Schur duality. If in addition $d = 2$, $(\mathbb{C}^m)^{\otimes 2} = S^2(\mathbb{C}^m) \oplus \wedge^2(\mathbb{C}^m)$. This fits well with the well-known fact that as $\mathfrak{gl}(m)$-modules $S^2(\mathbb{C}^m)$ and $\wedge^2(\mathbb{C}^m)$ are, respectively, irreducible of highest weights $2\Lambda_1$ and Λ_2, where Λ_i denotes the ith fundamental weight.

(2) For $m = 0$, $\lambda^\natural = \lambda'$ (the conjugate partition), and $S^\lambda = S^{\lambda'} \otimes \mathrm{sgn}$. In this case, the Sergeev duality reduces to the version of Schur duality twisted by the sign representation of $\mathfrak{S}_d$, i.e., as a $(\mathfrak{gl}(m), \mathfrak{S}_d)$-module,

$$(\mathbb{C}^m)^{\otimes d} \cong \bigoplus_{\mu \in \mathcal{P}(d,m)} L(\mu) \otimes S^{\mu'}.$$

(3) If $d \le mn + m + n$, then $\mathcal{P}(d, m|n) = \{\lambda \vdash d\}$, and every simple $\mathfrak{S}_d$-module appears in the Sergeev duality decomposition.

(4) For $d = 2$, the Sergeev duality reduces to the decomposition $(\mathbb{C}^{m|n})^{\otimes 2} = S^2(\mathbb{C}^{m|n}) \oplus \wedge^2(\mathbb{C}^{m|n})$, where S^2 and $\wedge^2$ are understood in the super sense. In particular, as an ordinary vector space,

$$S^2(\mathbb{C}^{m|n}) = S^2(\mathbb{C}^m) \oplus (\mathbb{C}^m \otimes \mathbb{C}^n) \oplus \wedge^2(\mathbb{C}^n).$$

3.2 Proof of Theorem 3.3

Set $\mathcal{A} := \Phi_d(U(\mathfrak{g}))$ and $\mathcal{B} := \Psi(\mathbb{C}\mathfrak{S}_d)$. It is clear that

$$\mathcal{A} \subset \mathrm{End}_{\mathfrak{S}_d}(V^{\otimes d}) = (\mathrm{End}(V^{\otimes d}))^{\mathfrak{S}_d} \cong S^d(\mathrm{End}V),$$

where $S^d(-)$ denotes the d-th symmetric tensor. With extra work, one proves that $S^d(\mathrm{End}V) \subset \mathcal{A}$ (the superalgebra generalization does not cause any extra difficulty). Hence $\Phi_d(U(\mathfrak{g})) = \mathrm{End}_{\mathfrak{S}_d}(V^{\otimes d})$.

Since $\mathcal{B} = \mathbb{C}\mathfrak{S}_d$ is a semisimple algebra, it follows (cf. [GW, Theorem 3.3.7]) that $\mathrm{End}_{U(\mathfrak{g})}(V^{\otimes d}) = \Psi_d(\mathbb{C}\mathfrak{S}_d)$. This proves (1).

Let $W = V^{\otimes d}$. It follows from the double centralizer property and the semisimplicity of $\mathbb{C}\mathfrak{S}_d$ that we have a multiplicity-free decomposition of the $(\mathfrak{gl}(m|n), \mathfrak{S}_d)$-module W:

$$W \equiv V^{\otimes d} \cong \bigoplus_{\lambda \in \mathfrak{P}(d,m|n)} L^{[\lambda]} \otimes S^\lambda,$$

where $L^{[\lambda]}$ is some simple $\mathfrak{gl}(m|n)$-module associated to λ, whose highest weight (with respect to the standard Borel) is to be determined. Also to be determined is the index set $\mathfrak{P}(d, m|n) = \{\lambda \vdash d \mid L^{[\lambda]} \neq 0\}$.

First we need to prepare some notations.

Let $\mathrm{Cp}(m|n)$ be the set of pairs $\nu|\mu$ of compositions $\nu=(\nu_1,\ldots,\nu_m)$ of length $\le m$ and $\mu=(\mu_1,\ldots,\mu_n)$ of length $\le n$, and let

$$\mathrm{Cp}(d,m|n)=\{\nu|\mu\in \mathrm{Cp}(m|n)\mid \sum_i \nu_i+\sum_j \mu_j=d\}.$$

We have the following weight space decomposition (with respect to the Cartan subalgbra of diagonal matrices $\mathfrak{h}\subset \mathfrak{gl}(m|n)$):

$$W=\bigoplus_{\nu|\mu\in \mathrm{Cp}(d,m|n)} W_{\nu|\mu},$$

where $W_{\nu|\mu}$ has a linear basis $e_{i_1}\otimes\cdots\otimes e_{i_d}$, with the indices satisfying the following equality of sets:

$$\{i_1,\ldots,i_d\}=\{\underbrace{\overline{1},\ldots,\overline{1}}_{\nu_1},\ldots,\underbrace{\overline{m},\ldots,\overline{m}}_{\nu_m};\underbrace{1,\ldots,1}_{\mu_1},\ldots,\underbrace{n,\ldots,n}_{\mu_n}\}. \tag{3.1}$$

Let $\mathfrak{S}_{\nu|\mu}:=\mathfrak{S}_{\nu_1}\times\cdots\times\mathfrak{S}_{\nu_m}\times\mathfrak{S}_{\mu_1}\times\cdots\times\mathfrak{S}_{\mu_n}$. The span of the vector $e_{\nu|\mu}:=e_{\overline{1}}^{\otimes\nu_1}\otimes\cdots\otimes e_{\overline{m}}^{\otimes\nu_m}\otimes e_1^{\otimes\mu_1}\otimes\cdots\otimes e_n^{\otimes\mu_n}$ can be identified with the $\mathfrak{S}_{\nu|\mu}$-module $1_\nu\otimes \mathrm{sgn}_\mu$. Since $\mathfrak{S}_d e_{\nu|\mu}$ spans $W_{\nu|\mu}$ we have a surjective $\mathfrak{S}_d$-homomorphism from $\mathrm{Ind}_{\mathfrak{S}_{\nu|\mu}}^{\mathfrak{S}_d}(1_\nu\otimes\mathrm{sgn}_\mu)$ onto $W_{\nu|\mu}$ by Frobenius Reciprocity. By counting the dimensions we have an $\mathfrak{S}_d$-isomorphism:

$$W_{\nu|\mu}\cong \mathrm{Ind}_{\mathfrak{S}_{\nu|\mu}}^{\mathfrak{S}_d}(1_\nu\otimes\mathrm{sgn}_\mu).$$

Let us denote the decomposition of $W_{\nu|\mu}$ into irreducibles by

$$W_{\nu|\mu}=\bigoplus_\lambda K_{\lambda,\nu|\mu}S^\lambda,\qquad \text{for } K_{\lambda,\nu|\mu}\in\mathbb{Z}_+.$$

Let λ be a partition which is identified with its Young diagram. Recall $I(m|n)$ is totally ordered by (1.2). A *hook tableau* T of shape λ, or a *hook λ-tableau* T, is an assignment of an element in $I(m|n)$ to each box of the Young diagram λ satisfying the following conditions:

(1) The numbers are weakly increasing along each row and column.
(2) The numbers from $\{\overline{1},\ldots,\overline{m}\}$ are strictly increasing along each column.
(3) The numbers from $\{1,\ldots,n\}$ are strictly increasing along each row.

Such a T is said to have *content* $\nu|\mu\in\mathrm{Cp}(m|n)$ if $\overline{i}\in I(m|0)$ appears ν_i times and $j\in I(0|n)$ appears μ_j times. Denote by $\mathcal{HT}(\lambda,\nu|\mu)$ the set of hook λ-tableaux of content $\nu|\mu$.

Lemma 3.5. *We have* $K_{\lambda,\nu|\mu}=\#\mathcal{HT}(\lambda,\nu|\mu)$.

Proof. Recall that $\mathrm{Ind}_{\mathfrak{S}_{\nu|\mu}}^{\mathfrak{S}_d}(1_\nu\otimes\mathrm{sgn}_\mu)\cong\bigoplus_\lambda K_{\lambda,\nu|\mu}S^\lambda$.

First assume that $\mu = \emptyset$, and we prove the formula by induction on the length $r = \ell(\nu)$. A hook (=semistandard) tableau T of shape λ and content ν gives rise to a sequence of partitions $\emptyset = \lambda^0 \subset \lambda^1 \subset \cdots \subset \lambda^r = \lambda$ such that λ^i has the shape given by the parts of T with entries $\leq i$, and λ^i/λ^{i-1} has ν_i boxes for each i. This sets up a bijection between $\mathcal{HT}(\lambda, \nu)$ and the set of such sequences of partitions. Denote $d_1 = d - \nu_r$ and $\widetilde{\nu} = (\nu_1, \ldots, \nu_{r-1})$. We have $\mathrm{Ind}_{\mathfrak{S}_{\widetilde{\nu}}}^{\mathfrak{S}_{d_1}} 1_{\mathfrak{S}_{\widetilde{\nu}}} \cong \bigoplus_{\rho \vdash d_1} K_{\rho,\widetilde{\nu}} S^{\rho}$, where $K_{\rho,\widetilde{\nu}} = \#\mathcal{HT}(\rho, \widetilde{\nu})$ by induction hypothesis. Now the induction step is simply the Piere's rule (cf. [Mac, Chapter 1 (5.16)]): for a partition $\rho \vdash d_1$,

$$\mathrm{Ind}_{\mathfrak{S}_{d_1} \times \mathfrak{S}_{\nu_r}}^{\mathfrak{S}_d} (S^{\rho} \otimes 1_{\nu_r}) \cong \bigoplus_{\lambda} S^{\lambda},$$

where λ is such that λ/ρ is a horizontal strip of ν_r boxes.

Then using the above special case as the initial step, we complete the proof in the general case by induction on the length of μ, in which the induction step is exactly the conjugated Piere's rule. □

Lemma 3.6. *Let $\lambda \vdash d$ and $\nu|\mu \in \mathrm{Cp}(d, m|n)$. Then $K_{\lambda,\nu|\mu} = 0$ unless $\lambda \in \mathcal{P}(d, m|n)$.*

Proof. By the identity $K_{\lambda,\nu|\mu} = \#\mathcal{HT}(\lambda, \nu|\mu)$, it suffices to prove that if a hook λ-tableau T of content $\nu|\mu$ exists, then $\lambda_{m+1} \leq n$.

By applying the hook tableau condition (2) to the first column of T, we see that the first entry $k \in I(m|n)$ in row $(m+1)$ satisfies $k > 0$. Applying the hook tableau condition (3) to the $(m+1)$st row, we conclude that $\lambda_{m+1} \leq n$. □

Lemma 3.6 implies that $\mathfrak{P}(d, m|n) \subset \mathcal{P}(d, m|n)$. On the other hand, given $\lambda \in \mathcal{P}(d, m|n)$, clearly a hook λ-tableau exists, e.g., we can fill in the numbers $\overline{1}, \ldots, \overline{m}$ on the first m rows of λ row by row downward, and then for the (possibly) remaining rows of λ, we fill in the numbers $1, \ldots, n$ column by column from left to right. This distinguished λ-tableau will be denoted by T_{λ}. Hence, we have proved that

$$\mathfrak{P}(d, m|n) = \mathcal{P}(d, m|n).$$

For a given $\lambda \in \mathcal{P}(d, m|n)$, we have $L^{[\lambda]} = \bigoplus_{\nu|\mu \in \mathrm{Cp}(d,m|n)} L^{[\lambda]}_{\nu|\mu}$. Among all the contents of hook λ-tableaux, the one for T_{λ} corresponds to a highest weight (by the three hook tableau conditions). Hence, we conclude that $L^{[\lambda]} = L(\lambda^{\natural})$, the simple $\mathfrak{g}$-module of highest weight $\lambda^{\natural}$. This completes the proof of Theorem 3.3.

3.3 The character formula

Clearly, the character of $L(\lambda^{\natural})$

$$\mathrm{ch} L(\lambda^{\natural}) = \sum_{\nu|\mu \in \mathrm{Cp}(m|n)} \dim L(\lambda^{\natural})_{\nu|\mu} \prod_{i,j} x_i^{\nu_i} y_j^{\mu_j}$$

is a polynomial which is symmetric in $\underline{x} = (x_1, \ldots, x_m)$ and $\underline{y} = (y_1, \ldots, y_n)$, respectively. Denote by $m_{\nu}(\underline{x})$ the monomial symmetric polynomial associated to a partition ν.

The following can be read off from the above proof of Theorem 3.3.

Corollary 3.7. *The character of $L(\lambda^\natural)$ is given by:*

$$\mathrm{ch}L(\lambda^\natural) = \sum_{\nu|\mu \in \mathrm{Cp}(d,m|n)} \#\mathcal{HT}(\lambda, \nu|\mu)\, m_\nu(\underline{x}) m_\mu(\underline{y}),$$

where ν and μ above are partitions.

Remark 3.8. The standard basis vectors for $S^2(\mathbb{C}^{m|n})$ are $e_i \otimes e_j + (-1)^{|i||j|} e_j \otimes e_i$, where $i, j \in I(m|n)$ satisfy $i \le j < 0$, $i < 0 < j$, or $0 < i < j$ (cf. Remark 3.4 (4)). They are in bijection with the hook tableaux of shape $\lambda = (2)$:

i	j

This is compatible with the identification $S^2(\mathbb{C}^{m|n}) \cong L(\mathfrak{gl}(m|n), 2\delta_1)$.

Remark 3.9. We sketch below a more standard argument for the standard Schur duality on $(\mathbb{C}^n)^{\otimes d}$, which emphasizes the decomposition of W as a $\mathfrak{g}$-module.

Given a composition (or a partition) μ of d of length $\le n$, we denote by W_μ the μ-weight space of the $\mathfrak{gl}(n)$-module. Clearly W_μ has a basis

$$e_{i_1} \otimes \cdots \otimes e_{i_d}, \quad \text{where } \{i_1, \ldots, i_d\} = \{\underbrace{1, \ldots, 1}_{\mu_1}, \ldots, \underbrace{n, \ldots, n}_{\mu_n}\}. \tag{3.2}$$

As a $(\mathfrak{gl}(n), \mathfrak{S}_d)$-module,

$$W \cong \bigoplus_\lambda L(\lambda) \otimes U^\lambda,$$

where $U^\lambda := \mathrm{Hom}_{\mathfrak{gl}(n)}(L(\lambda), W) \cong W_\lambda^{\mathfrak{n}^+}$ (the space of highest weight vectors in W of weight λ). Only $\lambda \in \mathcal{P}(d, n)$ can be highest weights for $\mathfrak{gl}(n)$ which appear in the decomposition of $(\mathbb{C}^n)^{\otimes d}$, and every such λ indeed appears as it appears as a summand of $\wedge^{\lambda_1'}(\mathbb{C}^n) \otimes \wedge^{\lambda_2'}(\mathbb{C}^n) \otimes \cdots$.

Note that $\mathcal{P}(d, n)$ has two interpretations: one as the polynomial weights for $\mathfrak{gl}(n)$ and the other as compositions of d. A remarkable fact is that the partial order on weights induced by the positive roots of $\mathfrak{gl}(n)$ coincides with the dominance partial order $\trianglerighteq$ on compositions.

Since $W = \oplus_\mu W_\mu = \oplus_{\mu,\lambda:\lambda \trianglerighteq \mu} L(\lambda)_\mu \otimes U^\lambda$, we conclude that, as $\mathfrak{S}_d$-module,

$$W_\mu \cong \bigoplus_{\lambda \trianglerighteq \mu} \dim L(\lambda)_\mu U^\lambda.$$

On the other hand, $\mathfrak{S}_d$ acts on the basis of W_μ (3.2) transitively and the stabilizer of the basis element $e_1^{\otimes \mu_1} \otimes \cdots \otimes e_n^{\otimes \mu_n}$ is the Young subgroup $\mathfrak{S}_\mu = \mathfrak{S}_{\mu_1} \times \cdots \times \mathfrak{S}_{\mu_n}$. Therefore we have

$$W_\mu \cong \mathrm{Ind}_{\mathfrak{S}_\mu}^{\mathfrak{S}_d} 1_\mu = \bigoplus_{\lambda \trianglerighteq \mu} K_{\lambda\mu} S^\lambda,$$

where $K_{\lambda\mu}$ is the Kostka number which satisfies $K_{\lambda\lambda} = 1$.

By the double centralizer property, U^λ has to be an irreducible $\mathfrak{S}_d$-module for each λ. We compare the above two interpretations of W_μ in the special case when μ is dominant (i.e., a partition). One by one downward along the dominance order, this provides the identification $U^\mu = S^\mu$ for every μ, and moreover, we obtain the well-known equality

$$\dim L(\lambda)_\mu = K_{\lambda\mu}.$$

Definition 3.10. Let $\mu \in \mathfrak{h}^*$. The *degree of atypicality* of μ is the maximum cardinality of a set of pairwise orthogonal $\alpha \in \Phi_{\bar{1}}^+$ such that $(\mu + \rho, \alpha) = 0$.

The following observation seems to be new.

Proposition 3.11. *Let $\lambda \in \mathcal{P}(m|n)$. The degree of atypicality of $\lambda^\natural$ equals the minimal number i such that λ contains a partition of rectangular shape $((n-i)^{m-i})$ with $0 \leq i \leq \min\{m, n\}$.*

As a corollary, $\lambda^\natural$ is typical if and only if $\lambda_m \geq n$, as observed earlier in [BeR].

Exercise 3.12. Prove Proposition 3.11.

Remark 3.13. There exists another interesting generalization of Schur duality for the queer Lie superalgebras due to Sergeev [Sv1].

3.4 The extremal weights

Assume that the $\epsilon\delta$-sequence associated to a Borel subalgebra $\mathfrak{b}$ of $\mathfrak{gl}(m|n)$ (cf. Section 1.4) is given by a sequence of d_1 δ's, e_1 ϵ's, d_2 δ's, e_2 ϵ's, $\ldots, d_r$ δ's, e_r ϵ's (possibly $d_1 = 0$ or $e_r = 0$). Associated to an $(m|n)$-hook Young diagram λ, we define a weight $\lambda^{\mathfrak{b}} \in \mathfrak{h}^*$ as follows. Take the first d_1 *row* numbers of λ as the coefficients of the first d_1 δ's. Denote by λ^1 the Young diagram obtained from λ with the first d_1 rows of λ removed. Take the first e_1 *column* numbers of λ^1 as the coefficients of the first e_1 ϵ's. Denote by λ^2 the Young diagram obtained from λ^1 with the first e_1 columns of λ^1 removed. Then take the first d_2 *row* numbers of λ^2 as the coefficients of the following d_2 δ's, and so on, until we reach the empty partition. The resulting weight is denoted by $\lambda^{\mathfrak{b}}$. Below is an example of $\lambda^{\mathfrak{b}}$ for $d_1 = e_1 = 2$ and $d_2 = 1$.

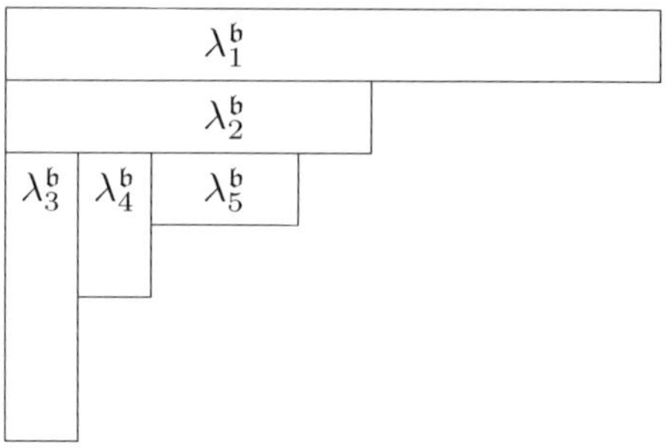

Recall that by convention $L(\lambda^\natural)$ is a highest weight $\mathfrak{g}$-module with respect to the standard Borel subalgebra $\mathfrak{b}^{\mathrm{st}}$.

Theorem 3.14. [CLW] *Let λ be an $(m|n)$-hook partition. Let $\mathfrak{b}$ be an arbitrary Borel subalgebra of $\mathfrak{gl}(m|n)$. Then, the $\mathfrak{b}$-highest weight of the simple $\mathfrak{gl}(m|n)$-module $L(\lambda^\natural)$ is $\lambda^{\mathfrak{b}}$.*

Proof. Let us consider an odd reflection which changes a Borel subalgebra $\mathfrak{b}_1$ to $\mathfrak{b}_2$. Assume the theorem holds for $\mathfrak{b}_1$. We observe by Lemma 2.8 that the statement of the theorem for $\mathfrak{b}_2$ follows from the validity of the theorem for $\mathfrak{b}_1$. The statement of the theorem is apparently consistent with a change of Borel subalgebras induced from a real reflection, and all Borel subalgebras of $\mathfrak{gl}(m|n)$ are linked by a sequence of real and odd reflections. Hence, once we know the theorem holds for one particular Borel subalgebra, it holds for all. We finally note that the theorem holds for the standard Borel subalgebra $\mathfrak{b}^{\rm st}$, which corresponds to the sequence of m δ's followed by n ϵ's. It is clear that $\lambda^{\mathfrak{b}^{\rm st}} = \lambda^\natural$. □

Remark 3.15. A variant of Theorem 3.14 holds for Lie superalgebras of type $\mathfrak{osp}$, see [CLW].

Example 3.16. Let us describe the highest weights in Theorem 3.14 with respect to the three Borel subalgebras of special interest.

(1) As seen above, $\lambda^{\mathfrak{b}^{\rm st}} = \lambda^\natural$.

(2) If we take the opposite Borel subalgebra $\mathfrak{b}^{\rm op}$ corresponding to a sequence of n ϵ's followed by m δ's, then $\lambda^{\mathfrak{b}^{\rm op}} = (\lambda')^\natural$, where $\natural$ applies to an $(n|m)$-hook partition (instead of $(m|n)$-hook partition). This is as expected from Schur-Sergeev duality Theorem 3.3.

(3) In case when $|m-n| \le 1$, we may take a Borel subalgebra $\mathfrak{b}^{\rm o}$ whose simple roots are all odd (or equivalently, the corresponding $\epsilon\delta$- sequence is alternating between ϵ and δ):

$$\bigotimes - \bigotimes - \cdots - \bigotimes - \bigotimes - \bigotimes$$

In this case Theorem 3.14 reduces to [CW3, Theorem 7.1], which states the coefficients of δ and ϵ in $\lambda^{\mathfrak{b}^{\rm o}}$ are given by the *modified Frobenius coordinates* $(p_i|q_i)_{i\ge1}$ of the partition λ (respectively, λ'), when the first simple root is of the form $\delta-\epsilon$ (respectively, $\epsilon-\delta$). Here by modified Frobenius coordinates we mean

$$p_i = \max\{\lambda_i - i + 1, 0\}, \quad q_i = \max\{\lambda'_i - i, 0\}$$

so that $\sum_i (p_i + q_i) = |\lambda|$; "modified" here refers to a shift by 1 from the p_i coordinates defined in [Mac, Chapter 1, Page 3].

For $\lambda = (7,5,4,3,1)$, we have $(p_1,p_2,p_3|q_1,q_2,q_3) = (7,4,2|4,2,1)$.

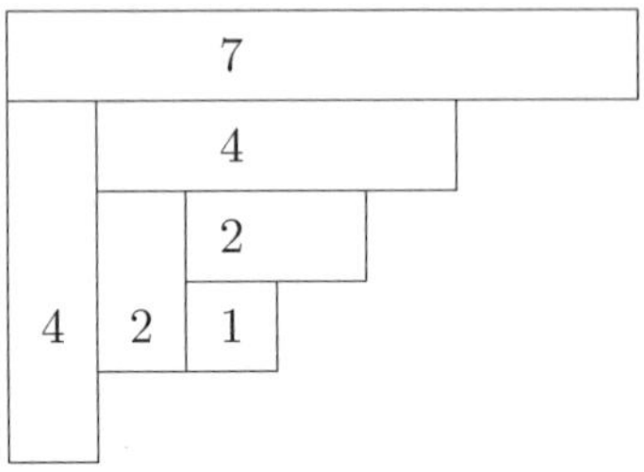

Exercise 3.17. Let $\lambda = (7, 2, 2, 1, 1)$ be a $(1|2)$-hook partition (i.e. $\lambda_2 \le 2$). Write down all the extremal weights with respect to the six Borel subalgebras of $\mathfrak{gl}(1|2)$ from Example 1.8 for $L(\mathfrak{gl}(1|2), \lambda^\natural)$.

4 Howe duality for Lie superalgebras of type $\mathfrak{gl}$

4.1 The $(\mathfrak{gl}(m|n), \mathfrak{gl}(d))$-Howe duality

Let $V = V_{\bar{0}} \oplus V_{\bar{1}}$ be a vector superspace. The symmetric algebra $S(V)$ is understood as $S(V) = S(V_{\bar{0}}) \otimes \wedge(V_{\bar{1}})$. It follows that $S^k(V) = \bigoplus_{\ell=0}^k S^\ell(V_{\bar{0}}) \otimes \wedge^{k-\ell}(V_{\bar{1}})$.

The commuting action of $\mathfrak{gl}(m|n)$ and $\mathfrak{gl}(d)$ on the superspace $\mathbb{C}^d \otimes \mathbb{C}^{m|n}$ induces a corresponding commuting action on its symmetric algebra $S(\mathbb{C}^d \otimes \mathbb{C}^{m|n})$.

Theorem 4.1. [H1, CW1, Sv2]

(1) *The images of $U(\mathfrak{gl}(d))$ and $U(\mathfrak{gl}(m|n))$ in* $\mathrm{End}\big(S(\mathbb{C}^d \otimes \mathbb{C}^{m|n})\big)$ *satisfy the double centralizer property.*

(2) *As a $(\mathfrak{gl}(d), \mathfrak{gl}(m|n))$-module,*

$$S(\mathbb{C}^d \otimes \mathbb{C}^{m|n}) = \sum_{\lambda \in \mathcal{P}(m|n), \ell(\lambda) \le d} L_d(\lambda) \otimes L_{m|n}(\lambda^\natural).$$

Here and below we use subscripts $m|n$ and d to indicate the (super)algebras under consideration.

The part (2) of the theorem is equivalent to

$$S^k(\mathbb{C}^d \otimes \mathbb{C}^{m|n}) = \sum_{\substack{\lambda \vdash k \\ \lambda_{m+1} \le n, \ell(\lambda) \le d}} L_d(\lambda) \otimes L_{m|n}(\lambda^\natural), \quad k \in \mathbb{N}.$$

Proof. Let us prove the equivalent version for $S^k(\mathbb{C}^{m|n} \otimes \mathbb{C}^d)$ for (2). The argument below is well-known (see [H2]).

By the definition of the k-th supersymmetric algebra we have

$$S^k(\mathbb{C}^d \otimes \mathbb{C}^{m|n}) \cong \left((\mathbb{C}^d)^{\otimes^k} \otimes (\mathbb{C}^{m|n})^{\otimes^k}\right)^{\Delta_k},$$

where Δ_k is the diagonal subgroup of $\mathfrak{S}_k \times \mathfrak{S}_k$ and $(-)^{\Delta_k}$ denotes the Δ_k-invariant subspace. By Theorem 3.3 we have therefore

$$\begin{aligned} S^k(\mathbb{C}^d \otimes \mathbb{C}^{m|n}) &\cong \Big(\big(\sum_{|\mu|=k} L_d(\mu) \otimes S^\mu\big) \otimes \big(\sum_{|\lambda|=k} L_{m|n}(\lambda^\natural) \otimes S^\lambda\big)\Big)^{\Delta_k} \\ &\cong \sum_{|\lambda|=|\mu|=k} \big(L_d(\mu) \otimes L_{m|n}(\lambda^\natural)\big) \otimes (S^\mu \otimes S^\lambda)^{\Delta_k} \\ &\cong \sum_{|\lambda|=k} L_d(\lambda) \otimes L_{m|n}(\lambda^\natural). \end{aligned}$$

The last equality follows since the Specht module S^λ is self-contragredient.

Part (1) follows from the decomposition in (2). □

Remark 4.2. In case $n = 0$, we have $\lambda^\natural = \lambda$, and Theorem 4.1 reduces to the classical $(\mathfrak{gl}(d), \mathfrak{gl}(m))$-Howe duality on a symmetric tensor space (see [H2, Section 2.1.2]).

In case $m = 0$, we have $\lambda^\natural = \lambda'$, and Theorem 4.1 reduces to the classical $(\mathfrak{gl}(d), \mathfrak{gl}(n))$-Howe duality on an exterior space $\wedge(\mathbb{C}^d \otimes \mathbb{C}^n)$ (see [H2, Section 4.1.1]).

Remark 4.3. Similarly, using Sergeev's duality for queer Lie superalgebra $\mathfrak{q}(n)$ (see Remark 3.13), we can establish a $(\mathfrak{q}(m), \mathfrak{q}(n))$-Howe duality, see [CW2, Naz, Sv2].

4.2 Hook Schur polynomials as $\mathfrak{gl}(m|n)$-characters

Let $\underline{x} = (x_1, \ldots, x_m)$ and $\underline{y} = (y_1, \ldots, y_n)$ be formal variables. For two partitions λ and μ with $\mu \subseteq \lambda$ denote by $s_{\lambda/\mu}(\underline{x})$ the skew Schur polynomial associated to the skew Young diagram λ/μ such that $\ell(\lambda) \le m$. The skew Schur polynomials specialize to the (usual) Schur polynomials for $\mu = \emptyset$: $s_{\lambda/\emptyset}(\underline{x}) = s_\lambda(\underline{x})$. We refer to Macdonald [Mac] for more on symmetric functions.

The *hook Schur polynomials* $hs_\lambda(\underline{x}, \underline{y})$ in variables $\underline{x}$ and $\underline{y}$ are defined as

$$hs_\lambda(\underline{x}, \underline{y}) = \sum_{\mu \subseteq \lambda} s_\mu(\underline{x}) s_{\lambda'/\mu'}(\underline{y}), \qquad \lambda \in \mathcal{P}(m|n). \tag{4.1}$$

This is one of the several equivalent definitions, see [BeR]. We should compare this definition with the classical identity for (skew) Schur functions

$$s_\lambda(\underline{x}, \underline{y}) = \sum_{\mu \subseteq \lambda} s_\mu(\underline{x}) s_{\lambda/\mu}(\underline{y}). \tag{4.2}$$

The character of the $\mathfrak{gl}(m|n)$-module $L_{m|n}(\lambda^\natural)$ is by definition the trace of the action of the diagonal matrix $\mathrm{diag}(x_1, \ldots, x_m; y_1, \ldots, y_n)$ on $L_{m|n}(\lambda^\natural)$. The following theorem was obtained in [BeR], where the notion of hook Schur polynomials was formulated, and it is actually equivalent to the combinatorial formula given in Corollary 3.7. We offer a different proof below using Howe duality (cf. [CL1]).

Theorem 4.4. *Let $\lambda \in \mathcal{P}(m|n)$. The character of $L_{m|n}(\lambda^\natural)$ is given by*

$$\mathrm{ch} L_{m|n}(\lambda^\natural) = hs_\lambda(\underline{x}, \underline{y}).$$

Proof. Let $\underline{u} = (u_1, \ldots, u_d)$ be another set of formal variables. It is well known that $s_\lambda(\underline{u})$ is the character of the $\mathfrak{gl}(d)$-module $L_d(\lambda)$, i.e. the trace of the diagonal matrix $\mathrm{diag}(u_1, \ldots, u_d)$. Thus, comparing the characters of both sides of Theorem 4.1 (2), we obtain the following combinatorial identity:

$$\sum_{\lambda \in \mathcal{P}(m|n), \ell(\lambda) \le d} s_\lambda(\underline{u})\, \mathrm{ch} L_{m|n}(\lambda^\natural) = \prod_{i \le m, j \le n, k \le d} (1 - x_i u_k)^{-1} (1 + y_j u_k). \tag{4.3}$$

Let $X = (x_1, x_2, \ldots)$ and $Y = (y_1, y_2, \ldots)$ denote two sets of infinitely many variables. Recall the classical Cauchy identity

$$\sum_{\ell(\lambda) \le d} s_\lambda(\underline{u}) s_\lambda(X) = \prod_{k \le d} \prod_{i \ge 1} (1 - x_i u_k)^{-1}, \tag{4.4}$$

which readily follows from Howe duality Theorem 4.1 (by setting $n = 0$ and letting $m \to \infty$).

Consider the involution ω on the ring of symmetric functions in the set of variables $(x_{m+1}, x_{m+2}, \ldots)$. Applying ω to (4.4), replacing $(x_{m+1}, x_{m+2}, \ldots)$ by Y, and using (4.2), we obtain

$$\sum_{\ell(\lambda)\le d} s_\lambda(\underline{u}) hs_\lambda(\underline{x}, Y) = \prod_{i\le m, k\le d, j\ge 1} (1 - x_i u_k)^{-1}(1 + y_j u_k).$$

Now setting $y_{n+1} = y_{n+2} = \cdots = 0$ in the above identity, we obtain that

$$\sum_{\ell(\lambda)\le d} s_\lambda(\underline{u}) hs_\lambda(\underline{x}, \underline{y}) = \prod_{i\le m, j\le n, k\le d} (1 - x_i u_k)^{-1}(1 + y_j u_k).$$

By comparing with (4.3) and noting the linear independence of $\{s_\lambda(\underline{u})\}$, we have proved the theorem. □

Remark 4.5. It follows that

$$hs_\lambda(X, \emptyset) = s_\lambda(X), \quad hs_\lambda(\emptyset, Y) = s_{\lambda'}(Y), \quad hs_\lambda(Y, X) = hs_{\lambda'}(X, Y).$$

Both the usual Cauchy identity and its dual version are obtained from specializations of the corresponding character identity of Theorem 4.1.

4.3 Formulas for highest weight vectors

We let $e^1, \ldots, e^d$ denote the standard basis for the natural $\mathfrak{gl}(d)$-module $\mathbb{C}^d$, and recall that $e_i, i \in I(m|n)$, denote the standard basis for the natural $\mathfrak{gl}(m|n)$-module $\mathbb{C}^{m|n}$. We set

$$x^i_a := e^i \otimes e_{\bar{a}} \ (1 \le a \le m); \quad \eta^i_b := e^i \otimes e_b \ (1 \le b \le n). \tag{4.5}$$

We will denote by $\mathbb{C}[\underline{x}, \underline{\eta}]$ the polynomial superalgebra generated by (4.5). The commuting actions on $\mathbb{C}[\underline{x}, \underline{\eta}]$ of $\mathfrak{gl}(d)$ and $\mathfrak{gl}(m|n)$ may be realized in terms of first order differential operators (4.6) and (4.7) respectively ($1 \le i, i' \le d$ and $1 \le s, s' \le m; 1 \le k, k' \le n$):

$$E^{ii'} := \sum_{j=1}^m x^i_j \frac{\partial}{\partial x^{i'}_j} + \sum_{j=1}^n \eta^i_j \frac{\partial}{\partial \eta^{i'}_j}, \tag{4.6}$$

$$\sum_{j=1}^d x^j_s \frac{\partial}{\partial x^j_{s'}}, \quad \sum_{j=1}^d \eta^j_k \frac{\partial}{\partial \eta^j_{k'}}, \quad \sum_{j=1}^d x^j_s \frac{\partial}{\partial \eta^j_k}, \quad \sum_{j=1}^d \eta^j_k \frac{\partial}{\partial x^j_s}. \tag{4.7}$$

In particular, as $(\mathfrak{gl}(d), \mathfrak{gl}(m|n))$-module, $S(\mathbb{C}^d \otimes \mathbb{C}^{m|n})$ is isomorphic to $\mathbb{C}[\underline{x}, \underline{\eta}]$. The root vectors corresponding to the simple roots of $\mathfrak{gl}(d)$ and $\mathfrak{gl}(m|n)$ are

$$\sum_{j=1}^m x^{i-1}_j \frac{\partial}{\partial x^i_j} + \sum_{j=1}^n \eta^{i-1}_j \frac{\partial}{\partial \eta^i_j}, \tag{4.8}$$

$$\sum_{j=1}^d x^j_{s-1} \frac{\partial}{\partial x^j_s}, \quad \sum_{j=1}^d \eta^j_{k-1} \frac{\partial}{\partial \eta^j_k}, \quad \sum_{j=1}^d x^j_m \frac{\partial}{\partial \eta^j_1}, \tag{4.9}$$

respectively.

Let λ be an $(m|n)$-hook partition of size k and of length at most d. We are looking for the joint highest weight vector (with respect to the standard Borel subalgebras), or equivalently the vector annihilated by (4.8) and (4.9), for the highest weight module $L_d(\lambda) \otimes L_{m|n}(\lambda^\natural)$ of $\mathfrak{gl}(d) \times \mathfrak{gl}(m|n)$ appearing in the decomposition of $\mathbb{C}[\underline{x}, \underline{\eta}]$. Such a vector is unique up to a scalar multiple, thanks to the multiplicity-free decomposition in Part 2 of Theorem 4.1.

For $1 \le r \le m$, define

$$\diamondsuit_r := \det \begin{pmatrix} x_1^1 & x_2^1 & \cdots & x_r^1 \\ x_1^2 & x_2^2 & \cdots & x_r^2 \\ \vdots & \vdots & & \vdots \\ x_1^r & x_2^r & \cdots & x_r^r \end{pmatrix}. \tag{4.10}$$

Let us assume for now that $d > m$ (the case of $d \le m$ is trivially included with much simplification, in which case we will not need the definition of $\diamondsuit_{k,r}$ below). Under this assumption, the condition $\lambda_{m+1} \le n$ is no longer an empty condition. Recall $\lambda' = (\lambda_1', \lambda_2', \ldots)$ denotes the transposed partition of λ. We have $d \ge \lambda_1' \ge \lambda_2' \ge \cdots$ and $m \ge \lambda_{n+1}'$.

Recall that the *row determinant* of an $r \times r$ matrix with possibly non-commuting entries $A = [a_i^j]$ is defined to be

$$\mathrm{rdet} A = \sum_{\sigma \in \mathfrak{S}_r} (-1)^{\ell(\sigma)} a_1^{\sigma(1)} a_2^{\sigma(2)} \cdots a_r^{\sigma(r)}.$$

For $m \le r \le d$, we introduce the following row determinant:

$$\diamondsuit_{k,r} := \mathrm{rdet} \begin{pmatrix} x_1^1 & x_1^2 & \cdots & x_1^r \\ x_2^1 & x_2^2 & \cdots & x_2^r \\ \vdots & \vdots & & \vdots \\ x_m^1 & x_m^2 & \cdots & x_m^r \\ \eta_k^1 & \eta_k^2 & \cdots & \eta_k^r \\ \vdots & \vdots & & \vdots \\ \eta_k^1 & \eta_k^2 & \cdots & \eta_k^r \end{pmatrix}, \quad k = 1, \ldots, n, \tag{4.11}$$

where the last $(r-m)$ rows are filled with the same vector $(\eta_k^1, \eta_k^2, \ldots, \eta_k^r)$.

Remark 4.6. The determinant (4.11) is always nonzero. It reduces to (4.10) when $m = r$, and, up to a scalar multiple, it reduces to $\eta_k^1 \cdots \eta_k^r$ when $m = 0$.

Now let r be specified by the conditions $\lambda_r' > m$ and $\lambda_{r+1}' \le m$. Denote by $\lambda_{\le r}$ the subdiagram of the Young diagram λ which consists of the first r columns of λ, i.e., the columns of length $> m$. A formula for the highest weight vector associated to the Young diagram $\lambda_{\le r}$ is given by the following lemma.

Lemma 4.7. *The vector $\prod_{k=1}^r \diamondsuit_{k,\lambda_k'}$ is a highest weight vector for the highest weight module $L_d(\lambda_{\le r}) \otimes L_{m|n}(\lambda_{\le r}^\natural)$ in the decomposition of $\mathbb{C}[\underline{x}, \underline{\eta}]$.*

Proof. Set $\diamondsuit = \prod_{k=1}^{r} \diamondsuit_{k,\lambda'_k}$, and $\mu = \lambda_{\leq r}$. We verify the following.

(i) $\diamondsuit$ has weight $(\mu, \mu^\natural)$ with respect to the action of $\mathfrak{gl}(d) \times \mathfrak{gl}(m|n)$.

(ii) $\diamondsuit$ is non-zero.

(iii) $\diamondsuit$ is annihilated by the operators in (4.8).

By Exercise 4.8 below, (i)–(iii) imply that $\diamondsuit$ is also a highest weight vector for $\mathfrak{gl}(m|n)$. □

Exercise 4.8. Assume that $v \in \mathbb{C}[\underline{x}, \underline{\eta}]$ has weights λ and $\lambda^\natural$ with respect to $\mathfrak{gl}(d)$ and $\mathfrak{gl}(m|n)$, respectively. Prove that if v is a highest weight vector for $\mathfrak{gl}(d)$, then so is it for $\mathfrak{gl}(m|n)$.

Theorem 4.9. [CW1] *Let λ be a Young diagram of length $\leq d$ such that $\lambda_{m+1} \leq n$. Then, a highest weight vector of weight $(\lambda, \lambda^\natural)$ in the $\mathfrak{gl}(d) \times \mathfrak{gl}(m|n)$-module $\mathbb{C}[\underline{x}, \underline{\eta}]$ is given by*

$$\prod_{k=1}^{r} \diamondsuit_{k,\lambda'_k} \prod_{j=r+1}^{\lambda_1} \diamondsuit_{\lambda'_j},$$

where r is defined by $\lambda'_r > m$ and $\lambda'_{r+1} \leq m$.

Proof. Since the expression is a product of two highest weight vectors, it is a highest weight vector. Also it is easy to verify that it has the correct weight. □

5 Howe duality for Lie superalgebras of type $\mathfrak{osp}$

5.1 General ideas of dual pairs

Let V be a vector (super)space, and $\mathcal{P}(V)$ the polynomial (super)algebra on V. Let $\mathcal{D}(V)$ denote the Weyl (super)algebra of differential operators with polynomial coefficients on V. For example, if $x_1, \ldots, x_m, \xi_1, \ldots, \xi_n$ are the even/odd coordinates on V relative to a homogeneous basis (so that $V \cong \mathbb{C}^{m|n}$), then $\mathcal{D}(V)$ is generated by the even generators $x_i, \partial_i := \partial/\partial x_i$ and the odd generators $\xi_j, \eth_j := \partial/\partial \xi_j$ $(1 \leq i \leq m, 1 \leq j \leq n)$, subject to the only nontrivial (super) commutation relations among the generators:

$$\partial_i x_i - x_i \partial_i = 1, \qquad \eth_j \xi_j + \xi_j \eth_j = 1.$$

Given a reductive group G acting on V, we have a multiplicity-free decomposition over $(G, \mathcal{D}(V)^G)$:

$$\mathcal{P}(V) = \bigoplus_{\lambda \in \mathfrak{P}} L(\lambda) \otimes U^\lambda,$$

where $L(\lambda)$ are pairwise non-isomorphic irreducible G-modules and U^λ are pairwise non-isomorphic irreducible $\mathcal{D}(V)^G$-modules, as λ runs over some set $\mathfrak{P}$ of parameters.

If there exists a generating set $\{T_1, \ldots, T_r\}$ for the (super)algebra $\mathcal{D}(V)^G$, such that $\mathfrak{g}' := \mathbb{C}\text{-span}\langle T_1, \ldots, T_r \rangle$ forms a Lie subalgebra of $\mathcal{D}(V)^G$, then the above U^λ's are pairwise non-isomorphic irreducible $\mathfrak{g}'$-modules. We refer to [GW] for more detailed discussion.

Remark 5.1. For a suitable choice of V and G, there could exist a very canonical candidate for the dual partner $\mathfrak{g}'$, which is suggested by the First Fundamental Theorem of invariant theory [H1, GW]. We will refer to $(G, \mathfrak{g}')$ as a *Howe dual pair* acting on $S(V)$.

Once $G, \mathfrak{g}'$ are fixed, the basic problems in $(G, \mathfrak{g}')$-Howe duality include describing the $\mathfrak{g}'$-module U^λ explicitly (e.g. describing its highest weight if this makes sense) and determining the parameter set $\mathfrak{P}$.

For example, if we let $V = (\mathbb{C}^d \otimes \mathbb{C}^n)^*$ and $G = \mathrm{GL}(d)$, then $\mathcal{P}(V) = S(\mathbb{C}^d \otimes \mathbb{C}^n)$ and we can choose $\mathfrak{g}' = \mathfrak{gl}(n)$. This gives rise to the $(\mathfrak{gl}(d), \mathfrak{gl}(n))$-duality from Remark 4.2.

5.2 The Howe dual pairs $(\mathrm{Sp}(d), \mathrm{osp}(2m|2n))$

In the type A Howe dualities, we can replace the Lie group $\mathrm{GL}(d)$ by its Lie algebra $\mathfrak{gl}(d)$ and vice versa without loss of information. However for more general Howe dualities, it is essential to use Lie groups G as formulated in Section 5.1. Associated to a given Lie algebra $\mathfrak{g}$ (e.g. $\mathfrak{so}(d)$), one associates several different (possibly disconnected) groups G, e.g. $\mathrm{SO}(d), \mathrm{O}(d), \mathrm{Spin}(d), \mathrm{Pin}(d)$. The Howe duality formulation favors $\mathrm{O}(d)$ and $\mathrm{Pin}(d)$ in the sense of Remark 5.1.

To avoid the complication of describing the irreducible representations of different covering groups, we will choose to discuss a Howe duality involving the Lie group $\mathrm{Sp}(d)$, where $d = 2\ell$ must be an even integer. In this case, the finite-dimensional simple modules over the group $\mathrm{Sp}(d)$ and those over the Lie algebra $\mathfrak{sp}(d)$ correspond bijectively, and it makes no difference if we use $\mathfrak{sp}(d)$ to replace $\mathrm{Sp}(d)$. Besides, they admit a nice parametrization in terms of partitions. We denote the $\mathrm{Sp}(d)$-module corresponding to the partition λ by $L(\mathrm{Sp}(d), \lambda)$.

According to [H1], there exists a Howe dual pair $(\mathrm{Sp}(d), \mathfrak{so}(2m))$ on $S(\mathbb{C}^d \otimes \mathbb{C}^m)$, and a Howe dual pair $(\mathrm{Sp}(d), \mathfrak{sp}(2n))$ on $\wedge(\mathbb{C}^d \otimes \mathbb{C}^n)$. By fusing these two dual pairs together, we obtain a Howe dual pair $(\mathrm{Sp}(d), \mathfrak{osp}(2m|2n))$ on $S(\mathbb{C}^d \otimes \mathbb{C}^{m|n})$. Let us explain the commuting actions below.

As before, we identify $S(\mathbb{C}^d \otimes \mathbb{C}^{m|n}) = \mathbb{C}[\underline{x}, \underline{\eta}]$. The action of the Lie algebra $\mathfrak{sp}(d)$ as the subalgebra of $\mathfrak{gl}(d)$ on $\mathbb{C}[\underline{x}, \underline{\eta}]$ lifts to an action of Lie group $\mathrm{Sp}(d)$. On the other hand, the following action of $\mathfrak{gl}(m|n)$ on $\mathbb{C}[\underline{x}, \underline{\eta}]$ is obtained from (4.7) by a shift of scalars on the diagonal matrices:

$$\begin{aligned}
E_{is}^{xx} &= \sum_{j=1}^{d} x_i^j \frac{\partial}{\partial x_s^j} + \frac{d}{2}\delta_{is}, \quad E_{ik}^{x\eta} = \sum_{j=1}^{d} x_i^j \frac{\partial}{\partial \eta_k^j}, \\
E_{ki}^{\eta x} &= \sum_{j=1}^{d} \eta_k^j \frac{\partial}{\partial x_i^j}, \quad E_{tk}^{\eta\eta} = \sum_{j=1}^{d} \eta_t^j \frac{\partial}{\partial \eta_k^j} - \frac{d}{2}\delta_{ik},
\end{aligned} \tag{5.1}$$

where $i, s = 1, \dots, m$ and $k, t = 1, \dots, n$. Introduce the following additional operators

$$I_{is}^{xx}=\sum_{j=1}^{\frac{d}{2}}\left(x_i^j x_s^{d+1-j}-x_i^{d+1-j}x_s^j\right),\quad I_{ik}^{x\eta}=\sum_{j=1}^{\frac{d}{2}}\left(x_i^j\eta_k^j-x_i^{d+1-j}\eta_k^{d+1-j}\right),$$

$$I_{kt}^{\eta\eta}=\sum_{j=1}^{\frac{d}{2}}\left(\eta_k^j\eta_t^{d+1-j}-\eta_k^{d+1-j}\eta_t^j\right),\ \Delta_{is}^{xx}=\sum_{j=1}^{\frac{d}{2}}\left(\frac{\partial}{\partial x_i^j}\frac{\partial}{\partial x_s^{d+1-j}}-\frac{\partial}{\partial x_i^{d+1-j}}\frac{\partial}{\partial x_s^j}\right),\tag{5.2}$$

$$\Delta_{ik}^{x\eta}=\sum_{j=1}^{\frac{d}{2}}\left(\frac{\partial}{\partial x_i^j}\frac{\partial}{\partial\eta_k^{d+1-j}}-\frac{\partial}{\partial x_i^{d+1-j}}\frac{\partial}{\partial\eta_k^j}\right),$$

$$\Delta_{kt}^{\eta\eta}=\sum_{j=1}^{\frac{d}{2}}\left(\frac{\partial}{\partial\eta_k^j}\frac{\partial}{\partial\eta_t^{d+1-j}}-\frac{\partial}{\partial\eta_k^{d+1-j}}\frac{\partial}{\partial\eta_t^j}\right),$$

where $1\le i<s\le m$ and $1\le k\le t\le n$. It is not hard to see that these operators together with (5.1) form a basis for the Lie superalgebra $\mathfrak{osp}(2m|2n)$. Moreover the actions of $\mathfrak{osp}(2m|2n)$ and of $\mathrm{Sp}(d)$ on $\mathbb{C}[\underline{x},\underline{\eta}]$ commute.

Theorem 5.2. [H1] *Let $d=2\ell$. The images of $\mathbb{C}\mathrm{Sp}(d)$ and $U(\mathfrak{osp}(2m|2n))$ in $\mathrm{End}\big(S(\mathbb{C}^d\otimes\mathbb{C}^{m|n})\big)$ satisfy the double centralizer property.*

Proof. The proof is based on the fact that the invariants of the classical group $\mathrm{Sp}(d)$ of the corresponding dual pair in the endomorphism ring of $S(\mathbb{C}^d\otimes\mathbb{C}^{m|n})$ are generated by quadratic invariants. □

Denote by $\mathfrak{u}^+$ (respectively, $\mathfrak{u}^-$) the subalgebra of $\mathfrak{osp}(2m|2n)$ spanned by the Δ (respectively, I) operators in (5.2). The highest weight module $L(\mathfrak{osp}(2m|2n),\mu)$ below is understood to be relative to the Borel subalgebra of $\mathfrak{osp}(2m|2n)$ corresponding to the simple roots listed in the following Dynkin diagram:

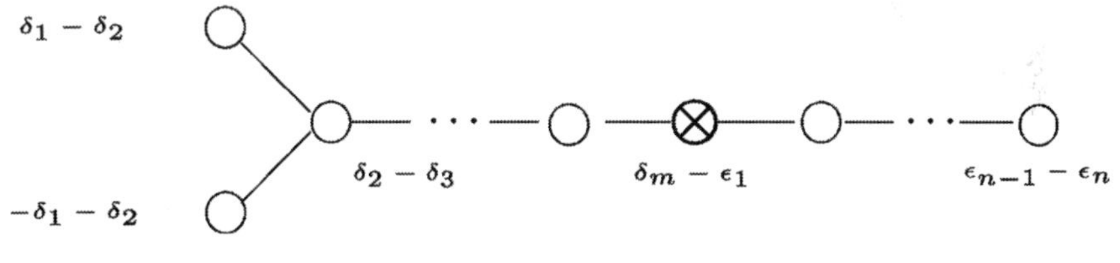

Then $\mathfrak{gl}(m|n)$ is a Levi subalgebra of $\mathfrak{osp}(2m|2n)$ corresponding to the removal of the simple root $-\delta_1-\delta_2$. We have a triangular decomposition of Lie superalgebra $\mathfrak{osp}(2m|2n)=\mathfrak{u}^-\oplus\mathfrak{gl}(m|n)\oplus\mathfrak{u}^+$.

Recall that for $\lambda\in\mathcal{P}(m|n)$, $\lambda^\natural$ is a weight for $\mathfrak{gl}(m|n)$. Then $\lambda^\natural+\ell\mathbf{1}$ can be regarded as a weight for Lie superalgebras $\mathfrak{gl}(m|n)$ and $\mathfrak{osp}(2m|2n)$ (which share the same Cartan subalgebra), where

$$\mathbf{1}=\sum_{i=1}^m\delta_i-\sum_{j=1}^n\epsilon_j.$$

The simplified proof below of the following theorem of [CZ2] is borrowed from [CKW].

Theorem 5.3. *Let $d = 2\ell$. As an $(\mathrm{Sp}(d), \mathfrak{osp}(2m|2n))$-module,*

$$S(\mathbb{C}^d \otimes \mathbb{C}^{m|n}) = \bigoplus_{\lambda \in \mathcal{P}(m|n), \ell(\lambda) \leq \ell} L(\mathrm{Sp}(d), \lambda) \otimes L(\mathfrak{osp}(2m|2n), \lambda^\natural + \ell \mathbf{1}).$$

Proof. An element $f \in \mathbb{C}[\underline{x}, \underline{\eta}]$ is called *harmonic*, if f is annihilated by the subalgebra $\mathfrak{u}^+$. The space of harmonics will be denoted by ${}^{\mathrm{Sp}}H$ and it evidently admits an action of $\mathrm{Sp}(d) \times \mathfrak{gl}(m|n)$. Furthermore, since $S(\mathbb{C}^d \otimes \mathbb{C}^{m|n})$ is a completely reducible $\mathfrak{gl}(m|n)$-module, $L(\mathfrak{osp}(2m|2n), \mu)^{\mathfrak{u}^+}$ is also completely reducible over $\mathfrak{gl}(m|n)$, for any irreducible $\mathfrak{osp}(2m|2n)$-module $L(\mathfrak{osp}(2m|2n), \mu)$ that appears in $S(\mathbb{C}^d \otimes \mathbb{C}^{m|n})$. By irreducibility of $L(\mathfrak{osp}(2m|2n), \mu)$ we must have

$$L(\mathfrak{osp}(2m|2n), \mu)^{\mathfrak{u}^+} \cong L_{m|n}(\mu).$$

So, by Theorem 5.2 $(\mathrm{Sp}(d), \mathfrak{gl}(m|n))$ forms a dual pair on the space of harmonics ${}^{\mathrm{Sp}}H$. Thus, proving the theorem is equivalent to establishing the following decomposition of ${}^{\mathrm{Sp}}H$ as an $\mathrm{Sp}(d) \times \mathfrak{gl}(m|n)$-module:

$${}^{\mathrm{Sp}}H \cong \bigoplus_{\lambda \in \mathcal{P}(m|n), \ell(\lambda) \leq \ell} L(\mathrm{Sp}(d), \lambda) \otimes L_{m|n}(\lambda^\natural + \ell \mathbf{1}). \tag{5.3}$$

We first consider the limit case $n = \infty$ with the space of harmonics denoted by ${}^{\mathrm{Sp}}H^\infty$. Here the only restriction on λ is $\ell(\lambda) \leq \ell$, and we observe that the vector given in Theorem 4.9 associated to such a partition λ is indeed annihilated by $\mathfrak{u}^+$ and hence is a joint $\mathrm{Sp}(d) \times \mathfrak{gl}(m|n)$-highest weight vector of weight $(\lambda, \lambda^\natural + \ell \mathbf{1})$. Hence all the summands on the right hand side of (5.3) occur in the space of harmonics, and in particular, all irreducible representations of $\mathrm{Sp}(d)$ occur. Therefore, we have established (5.3) in this case.

Now consider the finite n case. We may regard $S(\mathbb{C}^d \otimes \mathbb{C}^{m|n}) \subseteq S(\mathbb{C}^d \otimes \mathbb{C}^{m|\infty})$ with compatible actions $\mathfrak{osp}(2m|2n) \subseteq \mathfrak{osp}(2m|2\infty)$. From the formulas of the Δ-operators in (5.2) we see that ${}^{\mathrm{Sp}}H \subseteq {}^{\mathrm{Sp}}H^\infty$. Thus the space ${}^{\mathrm{Sp}}H$ is obtained from ${}^{\mathrm{Sp}}H^\infty$ by setting the variables $\eta_k^i = 0$, for $k > n$. However, it is clear, from the explicit formulas of the joint highest vectors in ${}^{\mathrm{Sp}}H^\infty$, that when setting the variables $\eta_k^i = 0$, for $k > n$, precisely those vectors corresponding to $(m|n)$-hook partitions will survive. This completes the proof. □

5.3 The Howe dual pair $(\mathrm{Sp}(d), \mathfrak{c}_\infty)$

Set $d = 2\ell$. Let $\mathbb{C}^\infty$ be the vector space over $\mathbb{C}$ with basis $\{ e_i \,|\, i \in \mathbb{Z} \}$. Let $\mathfrak{gl}_\infty$ denote the Lie algebra consisting of matrices $(a_{ij})_{i,j \in \mathbb{Z}}$ with finitely many non-zero a_{ij}'s, and $\mathbb{C}^\infty$ is the natural $\mathfrak{gl}_\infty$-module. Then $\mathfrak{gl}_\infty$ has a linear basis given by the matrix units E_{ij} $(i, j \in \mathbb{Z})$. Denote by $\widehat{\mathfrak{gl}}_\infty = \mathfrak{gl}_\infty \oplus \mathbb{C}K$ the central extension of $\mathfrak{gl}_\infty$ by a one-dimensional center $\mathbb{C}K$ given by the 2-cocycle

$$\tau(A, B) := \mathrm{Tr}([J, A]B),$$

where $J = \sum_{i \leq 0} E_{ii}$. The Lie algebra $\widehat{\mathfrak{gl}}_\infty$ admits a $\mathbb{Z}$-grading $\widehat{\mathfrak{gl}}_\infty = \bigoplus_{i \in \mathbb{Z}} \widehat{\mathfrak{gl}}_{\infty, \pm i}$ by letting $\deg E_{ij} = j - i$ and $\deg K = 0$, and this induces a triangular decomposition of $\widehat{\mathfrak{gl}}_\infty$, with $\widehat{\mathfrak{gl}}_{\infty, \pm} = \bigoplus_{i > 0} \widehat{\mathfrak{gl}}_{\infty, \pm i}$.

Exercise 5.4. Show that the cocycle τ on $\mathfrak{gl}_\infty$ is a coboundary and hence $\widehat{\mathfrak{gl}}_\infty$ is isomorphic to $\mathfrak{gl}_\infty \oplus \mathbb{C}K$ as Lie algebras.

Let $\bar{\mathfrak{c}}_\infty$ be the subalgebra of $\mathfrak{gl}_\infty$ preserving the following bilinear form on $\mathbb{C}^\infty$:

$$(e_i|e_j) = (-1)^i \delta_{i,1-j}, \quad i, j \in \mathbb{Z}.$$

Let $\mathfrak{c}_\infty = \bar{\mathfrak{c}}_\infty \oplus \mathbb{C}K$ be the central extension of $\bar{\mathfrak{c}}_\infty$ determined by the restriction of the two-cocycle τ above. Then $\mathfrak{c}_\infty$ is an infinite-rank affine Kac-Moody algebra with Dynkin diagram as follows:

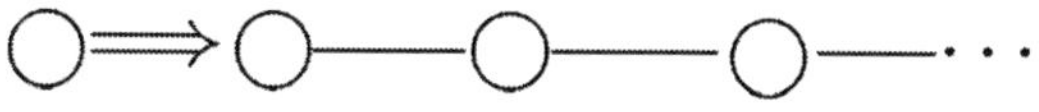

α_0

The Lie algebra $\mathfrak{c}_\infty$ has a natural triangular decomposition induced from $\widehat{\mathfrak{gl}}_\infty$:

$$\mathfrak{c}_\infty = \mathfrak{c}_{\infty,-} \oplus \mathfrak{h} \oplus \mathfrak{c}_{\infty,+},$$

where the Cartan subalgebra of $\mathfrak{c}_\infty$ is spanned by K and

$$\widetilde{E}_i := E_{ii} - E_{1-i,1-i}, \qquad i \in \mathbb{N}.$$

Let $\epsilon_i \in \mathfrak{h}^*$ be so that $\langle \epsilon_i, \widetilde{E}_j \rangle = \delta_{ij}$, for $i, j \in \mathbb{N}$. For $i \in \mathbb{Z}_+$, we denote by $\Lambda_i^{\mathfrak{c}}$ the i-th fundamental weight for $\mathfrak{c}_\infty$. Then $\Lambda_0^{\mathfrak{c}} \in \mathfrak{h}^*$ is determined by $\langle \Lambda_0^{\mathfrak{c}}, \widetilde{E}_i \rangle = 0$ for $i \in \mathbb{N}$ and $\langle \Lambda_0^{\mathfrak{c}}, K \rangle = 1$, and

$$\Lambda_i^{\mathfrak{c}} = \Lambda_0^{\mathfrak{c}} + \epsilon_1 + \cdots + \epsilon_i, \qquad i \geq 1.$$

Let $L(\mathfrak{c}_\infty, \mu)$ denote the irreducible highest weight module with respect to the Borel subalgebra $\mathfrak{h} \oplus \mathfrak{c}_{\infty,+}$ of highest weight μ.

Theorem 5.5. [Wa, Theorem 3.4] *Let $d = 2\ell$. There exists a Howe dual pair $(\mathrm{Sp}(d), \mathfrak{c}_\infty)$ acting on the fermionic Fock space $\mathfrak{F}^\ell$ of ℓ pairs of complex fermions. Moreover, as an $(\mathrm{Sp}(d), \mathfrak{c}_\infty)$-module,*

$$\mathfrak{F}^\ell \cong \bigoplus_{\ell(\lambda) \leq \ell} L(\mathrm{Sp}(d), \lambda) \otimes L(\mathfrak{c}_\infty, \Lambda^{\mathfrak{c}}(\lambda)), \tag{5.4}$$

where $\Lambda^{\mathfrak{c}}(\lambda) := \frac{d}{2}\Lambda_0^{\mathfrak{c}} + \sum_{k \geq 1} \lambda_k' \epsilon_k = \sum_{k=1}^{\frac{d}{2}} \Lambda_{\lambda_k}^{\mathfrak{c}}$.

As all we need later on is the combinatorial identity (5.5) below, we will skip the precise definitions of the fermionic Fock space $\mathfrak{F}^\ell$ and the commuting actions of $\mathrm{Sp}(d)$ and $\mathfrak{c}_\infty$ on $\mathfrak{F}^\ell$ (see [Wa] for detail).

Recalling (4.6) we set $\widetilde{E}^i := E^{ii} - E^{d-i+1,d-i+1}$, for $i = 1, \ldots, \frac{d}{2}$. Computing the trace of $\prod_{n \in \mathbb{N}} x_n^{\widetilde{E}_n} \prod_{i=1}^{\frac{d}{2}} z_i^{\widetilde{E}^i}$ on both sides of (5.4), we have

$$\prod_{i=1}^{\frac{d}{2}} \prod_{n \in \mathbb{N}} (1 + x_n z_i)(1 + x_n z_i^{-1}) = \sum_{\ell(\lambda) \leq \ell} \mathrm{ch} L(\mathrm{Sp}(d), \lambda) \, \mathrm{ch} L(\mathfrak{c}_\infty, \Lambda^{\mathfrak{c}}(\lambda)). \tag{5.5}$$

5.4 Irreducible $\mathfrak{c}_\infty$-characters

Let W (respectively, W_0) be the Weyl group of $\mathfrak{c}_\infty$ (respectively, of the Levi subalgebra of $\mathfrak{c}_\infty$ corresponding to the removal of the simple root α_0). Let W_k^0 be the set of the minimal length representatives of the right coset space $W_0\backslash W$ of length k. Then it is standard to write $W = W_0W^0$ with $W^0 = \bigsqcup_{k\geq 0} W_k^0$. It follows that, for each $w \in W_k^0$, we may find a partition $\lambda_w = ((\lambda_w)_1, (\lambda_w)_2, \ldots)$ such that

$$w(\Lambda^{\mathfrak{c}}(\lambda)+\rho)-\rho = \frac{d}{2}\Lambda_0^{\mathfrak{c}} + \sum_{j>0}(\lambda_w)_j\epsilon_j. \tag{5.6}$$

The following is obtained by applying the Kostant homology formula for integrable modules over Kac-Moody algebras and the Euler-Poincaré principle (cf. [CKW, Proposition 2.5]).

Proposition 5.6. *We have the following character formula:*

$$\mathrm{ch}L(\mathfrak{c}_\infty, \Lambda^{\mathfrak{c}}(\lambda)) = \frac{1}{\prod_{1\leq i\leq j}(1-x_ix_j)}\sum_{k=0}^{\infty}\sum_{w\in W_k^0}(-1)^k s_{\lambda_w}(x_1,x_2,\ldots).$$

Exercise 5.7. Prove the existence of λ_w for each $w \in W^0$ in (5.6).

5.5 The irreducible $\mathfrak{osp}(2m|2n)$-characters

The following character formula for $L(\mathfrak{osp}(2m|2n), \lambda^\natural + \ell\mathbf{1})$ in a different form was first obtained in [CZ2]. The proof here follows [CKW] and is simpler.

Theorem 5.8. *For $\lambda \in \mathcal{P}(m|n)$ such that $\ell(\lambda) \leq \ell$, we have*

$$\mathrm{ch}L(\mathfrak{osp}(2m|2n), \lambda^\natural + \ell\mathbf{1})$$
$$= \left(\frac{y_1\cdots y_m}{x_1\cdots x_n}\right)^{\ell} \frac{\prod_{1\leq i\leq m}\prod_{1\leq s\leq n}(1+y_ix_s)}{\prod_{1\leq i<j\leq m}\prod_{1\leq s\leq t\leq n}(1-y_iy_j)(1-x_sx_t)}\sum_{k=0}^{\infty}\sum_{w\in W_k^0}(-1)^k hs_{\lambda_w}(\underline{x},\underline{y}).$$

Proof. Computing the trace of the operator $\prod_{i,j} y_i^{\widetilde{E}_i} x_j^{\widetilde{E}_{\bar{j}}} \prod_{k=1}^{\frac{d}{2}} z_k^{\widetilde{E}^k}$ on both sides of the isomorphism in Theorem 5.3, where $1 \leq i \leq m$ and $1 \leq j \leq n$, we obtain

$$\prod_{k=1}^{\frac{d}{2}}\prod_{i,j}\frac{(1+x_iz_k^{-1})(1+x_iz_k)}{(1-y_jz_k^{-1})(1-y_jz_k)}$$
$$= \left(\frac{y_1\cdots y_m}{x_1\cdots x_n}\right)^{-\ell}\sum_{\substack{\lambda\in\mathcal{P}(m|n)\\ \ell(\lambda)\leq\ell}} \mathrm{ch}L(\mathrm{Sp}(d),\lambda)\, \mathrm{ch}L(\mathfrak{osp}(2m|2n),\lambda^\natural+\ell\mathbf{1}). \tag{5.7}$$

Replacing $\mathrm{ch}\, L(\mathfrak{c}_\infty, \Lambda^{\mathfrak{c}}(\lambda))$ in (5.5) by the expression in Proposition 5.6, we obtain an identity of symmetric functions in variables $x_1, x_2, \ldots$. Next, we replace x_{n+i}

by y_i for $i \in \mathbb{N}$, and then apply to this identity the involution on the ring of symmetric functions in $y_1, y_2, \ldots$. Finally, we put $y_j = 0$ for $j \geq m+1$. Under the composition of those maps, we obtain a new identity which shares the same left hand side as (5.7). Comparing the right hand sides of this new identity and of (5.7), we obtain the result thanks to the linear independence of $\mathrm{ch}L(\mathrm{Sp}(d), \lambda)$. $\square$

Remark 5.9. The above irreducible character formula can be understood as an alternating sum of the characters of parabolic Vermas over a certain infinite Weyl group. A similar observation was made for type A in [CZ1, Corollary 4.15].

The Kostant $\mathfrak{u}$-cohomology groups for these $\mathfrak{g}$-modules can be further computed and described in terms of the infinite Weyl groups, and in particular, they are multiplicity-free as modules over Levi subalgebras [CKW] (also cf. [CK]). In other words, the corresponding Kazhdan-Lusztig polynomials via $\mathfrak{u}$-cohomology in the sense of Vogan [Vo] are monomials.

6 Super duality

The goal of the remaining part of the lectures is to formulate precisely a direct connection between representation theories of Lie algebras and Lie superalgebras, following [CLW] (cf. [CWZ, CW4, CL2]). This is in part supported by the following.

(1) The $\mathfrak{gl}(1|1)$-principal block can be understood in terms of A_∞-quivers, which is classical in its own way.
(2) There exist intimate connections between finite-dimensional classical Lie superalgebras and infinite Weyl groups (cf. Remark 5.9).
(3) By a purely algebraic approach [Br1], Brundan showed that the category of finite-dimensional $\mathfrak{gl}(m|n)$-modules categorifies the Fock space $\wedge^m \mathbb{V} \otimes \wedge^n \mathbb{V}^*$, where $\mathbb{V}$ denotes the natural $U_q(\mathfrak{gl}(\infty))$-module.
(4) The Brundan-Kazhdan-Lusztig polynomials for some suitable parabolic category $\mathcal{O}$ of super type A are shown in [CWZ, CW4] to coincide with the usual Kazhdan-Lusztig polynomials of type A. For the so-called polynomial representations this was already observed in [CZ1]. Moreover, the super duality conjecture [CW4] (generalizing [CWZ]) on a category equivalence between Lie superalgebras and Lie algebras of type A has been established recently in [CL2].

6.1 Parabolic category $\mathcal{O}$

Assume that $\mathfrak{k}$ is a (possibly Kac-Moody, and possibly infinite rank) Lie (super)algebra, with Cartan subalgebra $\mathfrak{h}$ and root system Φ. Let I be a set of simple roots, and $\mathfrak{b}$ the corresponding Borel subalgebra. Let Y be a subset of I and $\mathfrak{l}_Y$ be the associated Levi subalgebra of $\mathfrak{k}$. Let $\mathfrak{k} = \mathfrak{u}^- \oplus \mathfrak{l}_Y \oplus \mathfrak{u}^+$ be the generalized triangular decomposition. Denote by P the weight lattice for $\mathfrak{k}$, and by $P_Y \subseteq P$ the subset of weights which are Y-dominant and integral.

Given $\lambda \in \mathfrak{h}^*$, we denote by $L(\mathfrak{l}_Y, \lambda)$ the irreducible $\mathfrak{l}_Y$-module of highest weight λ with respect to $\mathfrak{l}_Y \cap \mathfrak{b}$, which extends to an $(\mathfrak{l}_Y + \mathfrak{u}^+)$-module with a trivial action of $\mathfrak{u}^+$. Define the parabolic Verma $\mathfrak{k}$-module

$$\Delta(\lambda) := U(\mathfrak{k}) \otimes_{U(\mathfrak{l}_Y + \mathfrak{u}^+)} L(\mathfrak{l}_Y, \lambda).$$

Let $L(\lambda)$ be the irreducible quotient $\mathfrak{k}$-module of $\Delta(\lambda)$.

Let $\mathcal{O} = \mathcal{O}(\mathfrak{k}, Y, P_Y)$ be the category of $\mathfrak{g}$-modules M such that M is a semisimple $\mathfrak{h}$-module with finite-dimensional weight subspaces M_γ, $\gamma \in \mathfrak{h}^*$, satisfying

(i) M decomposes over $\mathfrak{l}_Y$ into a direct sum of $L(\mathfrak{l}_Y, \mu)$ for $\mu \in P_Y$.
(ii) There exist finitely many weights $\lambda_1, \lambda_2, \ldots, \lambda_k \in P_Y$ (depending on M) such that if γ is a weight in M, then $\gamma \in \lambda_i - \sum_{\alpha \in I} \mathbb{Z}_+ \alpha$, for some i.

The choice of P_Y should be natural and general enough so that $L(\lambda), \Delta(\lambda) \in \mathcal{O}(\mathfrak{k}, Y, P_Y)$ for every $\lambda \in P_Y$. This will be spelled out clearly in the cases of interest later on.

Remark 6.1. In the case when $\mathfrak{k}$ is a finite-dimensional semisimple Lie algebra, or $\mathfrak{gl}(n)$, or of the four classical infinite-rank affine Kac-Moody Lie algebras $\mathfrak{a}_\infty^+$, $\mathfrak{b}_\infty$, $\mathfrak{c}_\infty$, $\mathfrak{d}_\infty$, each block is controlled by the Weyl group W of $\mathfrak{k}$. For the principal block (in the full category $\mathcal{O}$), the transition matrix between $[L(\lambda)]$ and $[\Delta(\mu)]$ is given by the Kazhdan-Lusztig polynomials associated to W (see Tanisaki's lectures [Ta]). Other blocks and parabolic cases are reduced to the principal block.

6.2 The master Lie (super)algebras

Let $(\mathfrak{x})$ denote one of the following four classical Dynkin diagrams of types $\mathfrak{a}, \mathfrak{b}, \mathfrak{c}, \mathfrak{d}$, which will be referred to as a *head diagram*:

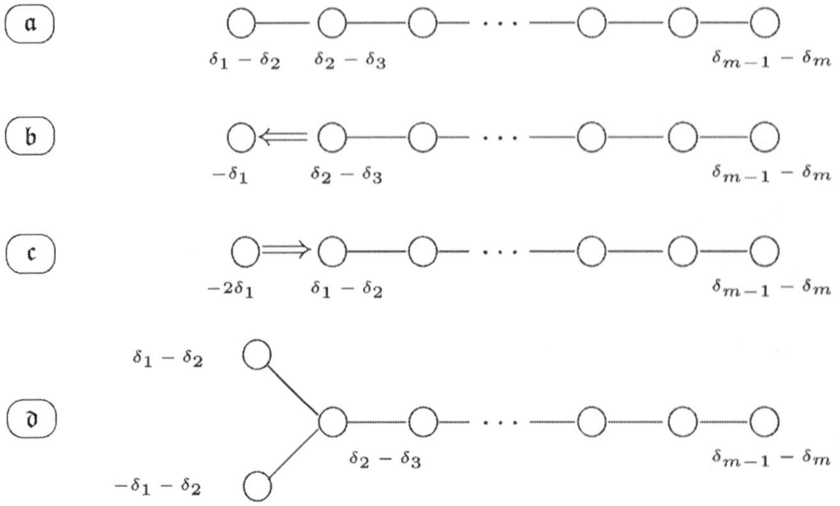

A head diagram $(\mathfrak{x})$ is connected with one of the three tail diagrams to produce three "master" Dynkin diagrams as below, whose corresponding Lie (super)algebras will be denoted by $\mathcal{G} = \mathcal{G}^{\mathfrak{x}}$, $\overline{\mathcal{G}} = \overline{\mathcal{G}}^{\mathfrak{x}}$, and $\widetilde{\mathcal{G}} = \widetilde{\mathcal{G}}^{\mathfrak{x}}$, respectively.

$\mathfrak{g}$: $\overline{\mathfrak{g}}$: $\widetilde{\mathfrak{g}}$:

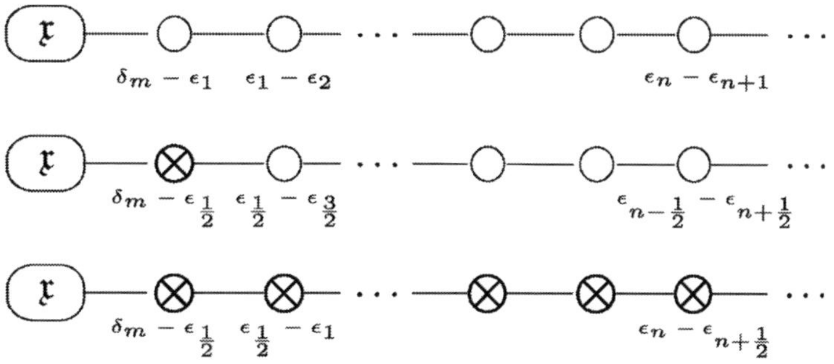

For example, the resulting master Dynkin diagrams for $\mathfrak{g}$ are the four infinite-rank classical Lie algebras $\mathfrak{a}_\infty^+$ (one-sided infinity), $\mathfrak{b}_\infty, \mathfrak{c}_\infty, \mathfrak{d}_\infty$.

6.3 The categories $\mathcal{O}, \overline{\mathcal{O}}, \widetilde{\mathcal{O}}$

Let $\Pi^{\mathfrak{x}}$ denote the standard set of simple roots associated to the diagram $\boxed{\mathfrak{x}}$, and fix $Y^0 \subset \Pi^{\mathfrak{x}}$. Let $\mathfrak{T}^1$, $\overline{\mathfrak{T}}^1$ and $\widetilde{\mathfrak{T}}^1$ be the subsets of the three tail diagrams with the first simple root removed. Let

$$Y = Y^0 \cup \mathfrak{T}^1, \qquad \overline{Y} = Y^0 \cup \overline{\mathfrak{T}}^1, \qquad \widetilde{Y} = Y^0 \cup \widetilde{\mathfrak{T}}^1.$$

Let P (respectively, $\overline{P}$, $\widetilde{P}$) be the lattice which consists of the integral span of ϵ_i, where i runs over $I(m|\mathbb{N})$ (respectively, $I(m|\frac{1}{2}+\mathbb{N})$, $I(m|\frac{1}{2}\mathbb{N})$). Here $I(m|\mathbb{N}) = \{\bar{1}, \dots, \bar{m}\} \cup \mathbb{N}$, and the other two sets are similarly defined.

Define $P_Y = \cup_{d\in\mathbb{Z}} P_Y^d$, where

$$P_Y^d = \Big\{\lambda = \sum_{i\in I(m|\mathbb{N})} \lambda_i \epsilon_i \in P \mid \lambda \text{ is } Y\text{-dominant and integral; } \lambda_i = d \text{ for } i \gg 0\Big\}. \tag{6.1}$$

We define $P_{\overline{Y}}^d$ by changing $I(m|\mathbb{N})$ in the definition of P_Y^d to $I(m|\frac{1}{2}+\mathbb{N})$ and changing $\lambda_i = d$ to $\lambda_i = -d$ for $i \gg 0$, and let $P_{\overline{Y}} = \cup_{d\in\mathbb{Z}} P_{\overline{Y}}^d$.

Then, for $\lambda \in P_Y^d$, $\lambda^+ := (\lambda_1 - d, \lambda_2 - d, \dots)$ is a partition. Also, for $\lambda \in P_{\overline{Y}}^d$, $\lambda^+ := (\lambda_{1/2} + d, \lambda_{3/2} + d, \dots)$ is a partition. We have a bijection of sets:

$$\natural : P_Y^d \longrightarrow P_{\overline{Y}}^d, \qquad \lambda \mapsto \lambda^\natural,$$

which is determined by the conditions

$$\lambda_i = \lambda_i^\natural, \quad \forall i \in I(m|0); \qquad (\lambda^\natural)^+ = (\lambda^+)'.$$

This further induces a bijection $\natural : P_Y \to P_{\overline{Y}}$.

We define another map

$$\theta : P_Y^d \to P_{\widetilde{Y}}^d, \quad \lambda \mapsto \lambda^\theta = \sum_{i\in I(m|0)} \lambda_i \epsilon_i + \sum_{i\in\mathbb{N}} \big((p_i - d)\epsilon_{i-\frac{1}{2}} + (q_i + d)\epsilon_i\big),$$

where the image λ^θ is determined by the conditions $\lambda_i = \lambda_i^\theta$ for all $i \in I(m|0)$ and $(p_i|q_i)_{i\geq 1}$ is the modified Frobenius coordinates of the partition $(\lambda^+)'$ (as defined in Example 3.16 (3)). The map θ is clearly injective, and so we obtain a bijection $\theta : P_Y^d \to P_{\widetilde{Y}}^d := \theta(P_Y^d)$ for each $d \in \mathbb{Z}$. Letting $P_{\widetilde{Y}} = \cup_{d\in\mathbb{Z}} P_{\widetilde{Y}}^d$ we also obtain a bijection $\theta : P_Y \to P_{\widetilde{Y}}$.

Following the construction of a parabolic category $\mathcal{O}$ in Section 6.1, we obtain the categories $\mathcal{O} := \mathcal{O}(\mathcal{G}, Y, P_Y), \overline{\mathcal{O}} := \mathcal{O}(\overline{\mathcal{G}}, \overline{Y}, P_{\overline{Y}})$, and $\widetilde{\mathcal{O}} := \mathcal{O}(\widetilde{\mathcal{G}}, \widetilde{Y}, P_{\widetilde{Y}})$. We observe the following direct sum decompositions of categories:

$$\mathcal{O} = \bigoplus_{d\in\mathbb{Z}} \mathcal{O}^d, \quad \overline{\mathcal{O}} = \bigoplus_{d\in\mathbb{Z}} \overline{\mathcal{O}}^d, \quad \widetilde{\mathcal{O}} = \bigoplus_{d\in\mathbb{Z}} \widetilde{\mathcal{O}}^d,$$

where the index d indicates additional weight constraints: $(\lambda, \epsilon_i) = d$ for $i \in \mathbb{Z}, i \gg 0$ for $\lambda \in P_Y$; $(\lambda, \epsilon_i) = -d$ for $i \in \frac{1}{2} + \mathbb{Z}, i \gg 0$ for $\lambda \in P_{\overline{Y}}$; $(\lambda, \epsilon_i) = (-1)^{2i} d$ for $i \in \frac{1}{2}\mathbb{Z}, i \gg 0$ for $\lambda \in P_{\widetilde{Y}}$, respectively.

Remark 6.2. A more conceptual formulation (see [CLW] for details) is to introduce central extensions of $\mathcal{G}, \overline{\mathcal{G}}, \widetilde{\mathcal{G}}$ respectively by a one-dimensional center, whose module categories at level d are equivalent to $\mathcal{O}^d, \overline{\mathcal{O}}^d, \widetilde{\mathcal{O}}^d$ above, respectively. In this way, the additional weight constraints above are transferred to be $(\lambda, \epsilon_i) = 0$ for $i \gg 0$, which is independent of d.

6.4 The equivalence of categories

We start with two simple observations:

(i) The Lie (super)algebras $\mathcal{G}$ and $\overline{\mathcal{G}}$ are naturally Lie subalgebras (though not Levi subalgebras) of $\widetilde{\mathcal{G}}$, and the standard triangular decompositions of the three Lie (super)algebras are compatible with the inclusions $\mathcal{G} \subset \widetilde{\mathcal{G}}$ and $\overline{\mathcal{G}} \subset \widetilde{\mathcal{G}}$. This follows by examining the simple roots of the three algebras.

(ii) We may naturally regard the lattices P and $\overline{P}$ as sublattices of $\widetilde{P}$, by definition of the lattices $P, \overline{P}, \widetilde{P}$.

Given a $\widetilde{\mathcal{G}}$-module with weight space decomposition $\widetilde{M} = \bigoplus_{\mu\in\widetilde{P}} \widetilde{M}_\mu$, we define

$$T(\widetilde{M}) := \bigoplus_{\mu\in P} \widetilde{M}_\mu, \qquad \text{and} \qquad \overline{T}(\widetilde{M}) := \bigoplus_{\mu\in\overline{P}} \widetilde{M}_\mu.$$

Clearly, $T(\widetilde{M})$ is a $\mathcal{G}$-module and $\overline{T}(\widetilde{M})$ is a $\overline{\mathcal{G}}$-module. We can check that T and $\overline{T}$ send objects in category $\widetilde{\mathcal{O}}$ to objects in category $\mathcal{O}$ and $\overline{\mathcal{O}}$, respectively, and hence we obtain functors

$$\begin{array}{ccc} \widetilde{\mathcal{O}} & \xrightarrow{T} & \mathcal{O} \\ \downarrow{\scriptstyle \overline{T}} & & \\ \overline{\mathcal{O}} & & \end{array}$$

More precisely, we have $T : \widetilde{\mathcal{O}}^d \to \mathcal{O}^d$ and $\overline{T} : \widetilde{\mathcal{O}}^d \to \overline{\mathcal{O}}^d$.

We will use $\overline{\Delta}(\lambda)$ and $\widetilde{\Delta}(\lambda)$ to denote the parabolic Verma modules in the categories $\overline{\mathcal{O}}$ and $\widetilde{\mathcal{O}}$ of highest weights λ, respectively. Similarly, we use $\overline{L}(\lambda)$ and $\widetilde{L}(\lambda)$ to denote the corresponding irreducible modules. We are ready to formulate the main result of [CLW].

Theorem 6.3. [CLW] *We have*

$$T(\widetilde{\Delta}(\lambda^\theta)) = \Delta(\lambda), \qquad \overline{T}(\widetilde{\Delta}(\lambda^\theta)) = \overline{\Delta}(\lambda^\natural),$$
$$T(\widetilde{L}(\lambda^\theta)) = L(\lambda), \qquad \overline{T}(\widetilde{L}(\lambda^\theta)) = \overline{L}(\lambda^\natural).$$

Moreover, $T : \widetilde{\mathcal{O}} \to \mathcal{O}$ *and* $\overline{T} : \widetilde{\mathcal{O}} \to \overline{\mathcal{O}}$ *are equivalences of categories.*

As a consequence, $T : \widetilde{\mathcal{O}}^d \to \mathcal{O}^d$ and $\overline{T} : \widetilde{\mathcal{O}}^d \to \overline{\mathcal{O}}^d$ are equivalences of categories.

Idea of a proof. The proof consists of several major steps:

(i) $T(L(\mathfrak{l}_{\widetilde{Y}}, \lambda^\theta)) = L(\mathfrak{l}_Y, \lambda)$. This follows by locating a highest weight λ for the module $L(\mathfrak{l}_{\widetilde{Y}}, \lambda^\theta)$ with respect to a new Borel of $\mathfrak{l}_{\widetilde{Y}}$ which is compatible with the imbedding $\mathfrak{l}_Y \subseteq \mathfrak{l}_{\widetilde{Y}}$. Then we check the equality on the character level (which boils down to the hook-Schur functions).

(ii) $T(\widetilde{\Delta}(\lambda^\theta))$ is a highest weight module of highest weight λ with respect to the standard Borel of $\mathcal{G}$. This follows from a weight argument.

(iii) $T(\widetilde{\Delta}(\lambda^\theta)) = \Delta(\lambda)$. Check on the character level, and use (i) and (ii).

(iv) $T(\widetilde{L}(\lambda^\theta)) = L(\lambda)$. Follows by showing that there is no singular vector in $T(\widetilde{L}(\lambda^\theta))$ of weight different from λ.

This proves the equality for T in the theorem. The proof of the equality for $\overline{T}$ is similar. More work is needed to establish the equivalence of categories. □

Corollary 6.4 (Super Duality). *The categories* $\mathcal{O}$ *and* $\overline{\mathcal{O}}$ *are equivalent.*

Remark 6.5 (History of Super Duality). First for type A.

(1) For $Y^0 = \Pi^{\mathfrak{a}}$, the super duality was conjectured in [CWZ, Conjecture 6.10]. This was motivated by Brundan's work [Br1] on finite-dimensional simple $\mathfrak{gl}(m|n)$-characters.
(2) For arbitrary Levi subalgebra associated to $Y^0 \subseteq \Pi^{\mathfrak{a}}$, the super duality conjecture was formulated in [CW4, Conjecture 4.18]. It is shown in [CWZ, CW4] that the Brundan-Kazhdan-Lusztig polynomials coincide with the usual Kazhdan-Lusztig polynomials in type A.
(3) The conjecture of [CWZ] is proved in [BrS], where the underlying algebras for the two categories were computed and shown to be isomorphic.
(4) The more general conjecture of [CW4] is independently proved in [CL2] essentially in the form of Theorem 6.3. In particular, this provides a new proof of the main result of [Br1].

Then for type $\mathfrak{osp}$.

(5) A first supporting evidence for the super duality was provided by the computation in [CKW] of the Kostant $\mathfrak{u}$-homology groups with coefficients in the modules of classical Lie superalgebras appearing in Howe duality decompositions.

(6) Theorem 6.3 is formulated and established in [CLW].

Note that very little has been known about representation theory of infinite-dimensional Lie superalgebras, such as affine superalgebras, with the exception of work of Kac and Wakimoto [KW].

(7) The super duality approach applies to and sheds new insight on more general Lie superalgebras, including affine superalgebras.

6.5 Irreducible $\overline{\mathcal{G}}$-characters

By Remark 6.1, the irreducible character problem for the category $\mathcal{O}$ is solved by the Kazhdan-Lusztig conjectures [KL] (theorem of Beilinson-Bernstein [BB] and Brylinski-Kashiwara [BK]). Hence by Theorem 6.3, the irreducible character problem for the module category $\overline{\mathcal{O}}$ for Lie superalgebras is solved by the same classical Kazhdan-Lusztig polynomials.

Recall for $\mathfrak{x} \in \{\mathfrak{a}, \mathfrak{b}, \mathfrak{c}, \mathfrak{d}\}$, we have defined a Lie superalgebra $\overline{\mathcal{G}} = \overline{\mathcal{G}}^{\mathfrak{x}}$ whose Dynkin diagram is obtained by connecting $\mathfrak{x}$ with the type $\mathfrak{gl}(1|\infty)$ tail diagram $\overline{\mathfrak{T}}$. For $n \in \mathbb{N}$, we will also consider a Lie superalgebra $\overline{\mathcal{G}}_n = \overline{\mathcal{G}}^{\mathfrak{x}}_n$ whose Dynkin diagram (see below) is obtained by connecting $\mathfrak{x}$ with a tail diagram $\overline{\mathfrak{T}}_n$ of type $\mathfrak{gl}(1|n)$:

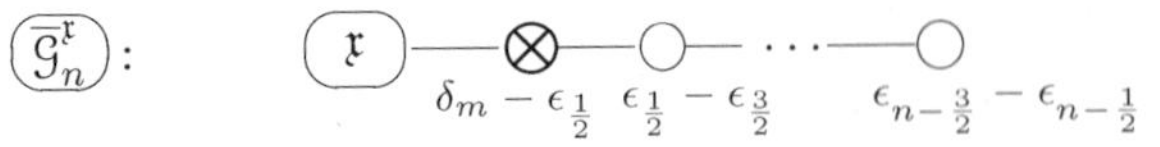

(6.2)

As $\mathfrak{x}$ is taken to be one of the four classical Dynkin diagrams, $\overline{\mathcal{G}}_n$ is a classical finite-dimensional Lie superalgebra of type either $\mathfrak{gl}$ or $\mathfrak{osp}$.

Let $\overline{\mathfrak{T}}^1_n$ be the subset of $\overline{\mathfrak{T}}_n$, which is obtained from $\overline{\mathfrak{T}}_n$ by the removal of the first simple root. Let $\overline{Y}_n = Y^0 \cup \overline{\mathfrak{T}}^1_n$, and let

$$P^d_{\overline{Y}_n} = \Big\{\lambda = \sum_{i \in I(m|n-\frac{1}{2})} \lambda_i \epsilon_i \mid \lambda \text{ is } \overline{Y}_n\text{-dominant and integral; } \lambda_{n-\frac{1}{2}} \geq -d\Big\}.$$

For $d \in \mathbb{Z}$, we introduce a category $\overline{\mathcal{O}}^d_n := \mathcal{O}(\overline{\mathcal{G}}_n, \overline{Y}_n, P^d_{\overline{Y}_n})$ of $\overline{\mathcal{G}}_n$-modules. Note that $\overline{\mathcal{O}}^d_n$ is a full subcategory of the (more standard) category $\overline{\mathcal{O}}_n := \mathcal{O}(\overline{\mathcal{G}}_n, \overline{Y}_n, P_{\overline{Y}_n})$, where

$$P_{\overline{Y}_n} = \Big\{\lambda = \sum_{i \in I(m|n-\frac{1}{2})} \lambda_i \epsilon_i \mid \lambda \text{ is } \overline{Y}_n\text{-dominant and integral}\Big\}.$$

Moreover, $\overline{\mathcal{O}}^d_n \subset \overline{\mathcal{O}}^{d+1}_n$ for all d, and $\overline{\mathcal{O}}_n = \cup_{d \in \mathbb{Z}} \overline{\mathcal{O}}^d_n$.

We also introduce truncation functors $\overline{\mathfrak{Tr}}_n : \overline{\mathcal{O}}^d \to \overline{\mathcal{O}}^d_n$ as follows: Let $M \in \overline{\mathcal{O}}^d$ such that $M = \bigoplus_\mu M_\mu$. We define

$$\overline{\mathfrak{Tr}}_n(M) = \bigoplus_{\{\mu|\mu_{n+\frac{1}{2}}=-d\}} M_\mu,$$

which is clearly a $\overline{\mathcal{G}}_n$-module. Note that $\mu_{n+\frac{1}{2}} = -d$ ensures $\mu_{k+\frac{1}{2}} = -d$ for all $k \geq n$. Such a truncation functor to a Levi subalgebra (except that we drop an abelian direct summand of the Levi) has its analogue in algebraic group setting studied by Donkin (cf. e.g. [CW4] for a formulation in type A). The functors $\overline{\mathfrak{Tr}}_n : \overline{\mathcal{O}}^d \to \overline{\mathcal{O}}^d_n$ have the following key property.

Proposition 6.6. *For $X = \Delta$ or L, we have*

$$\overline{\mathfrak{Tr}}_n(X(\lambda)) = \begin{cases} X_n(\lambda^{(n)}), & \text{if } \lambda_{n+\frac{1}{2}} = -d, \\ 0, & \text{otherwise}, \end{cases}$$

where $\lambda^{(n)}$ is obtained from λ by ignoring λ_j, for $j > n - \frac{1}{2}$.

Remark 6.7. By Proposition 6.6 and Theorem 6.3, the classical Kazhdan-Lusztig solution of the irreducible character problem of $\mathcal{O}$ induces a complete solution of the irreducible character problem of $\overline{\mathcal{O}}_n$ for each finite n. It is not hard to show [CLW] that every finite-dimensional irreducible modules of every ortho-symplectic Lie superalgebra appears in some $\overline{\mathcal{O}}_n$.

We end the paper with a list of symbols.

Symbol	Meaning
$\lvert v\rvert$	$\mathbb{Z}_2$-parity of a homogeneous vector v in a vector superspace
$\mathbb{C}^{m\|n}$	complex vector superspace of super dimension $m\|n$
$\mathfrak{gl}(m\|n)$	the general Lie superalgebra
$I(m\|n)$	the totally order set $\{\overline{1} < \overline{2} < \cdots < \overline{m} < 1 < \cdots < n\}$
$\mathfrak{sl}(m\|n)$	the special linear Lie superalgebra
$\mathfrak{h}$	Cartan subalgebra (of diagonal matrices)
δ_i	dual basis element corresponding to $E_{\overline{ii}}$
ϵ_j	dual basis element corresponding to E_{jj}
$\Phi, \Phi_{\overline{0}}$, or $\Phi_{\overline{1}}$	sets of all, even, or odd roots
Π	set of simple roots
$\Phi^{\pm}_{\epsilon}$	set of positive/negative roots of parity $\epsilon = \overline{0}, \overline{1}$
$\mathfrak{osp}(m\|n)$	ortho-symplectic Lie superalgebra preserving a non-degenerate supersymmetric bilinear form on $\mathbb{C}^{m\|n}$
$\mathfrak{spo}(m\|n)$	symplectic-orthogonal Lie superalgebra preserving a non-degenerate skew-supersymmetric bilinear form on $\mathbb{C}^{m\|n}$
$U(\mathfrak{g})$	universal enveloping algebra of a Lie (super)algebra $\mathfrak{g}$
$K(\lambda)$	Kac module of highest weight λ
$L(\mathfrak{g}, \lambda)$ or $L(\lambda)$	irreducible $\mathfrak{g}$-module of highest weight λ

Continued

Symbol	Meaning
$\mathrm{ch}M$	character of the $\mathfrak{h}$-semisimple module M
ρ_0, ρ_1	half sums of positive even and odd roots, respectively
ρ	$\rho_0 - \rho_1$
$\mathfrak{S}_d$	symmetric group in d letters
S^λ	Specht module corresponding to the partition λ
$\Delta(\lambda)$	(parabolic) Verma $\mathfrak{g}$-module of highest weight λ
μ'	conjugate partition of the partition μ
$\mathcal{P}(d, m\|n)$	set of $(m\|n)$-hook partitions of size d (Definition 2.10)
$\mathcal{P}(m\|n)$	$\bigcup_{d\geq 0} \mathcal{P}(d, m\|n)$
$s_\lambda(\underline{x})$	Schur function in $\underline{x}$ for the partition λ
$hs_\lambda(\underline{x}, \underline{y})$	hook Schur function for the hook partition λ (4.1)
$\diamondsuit_r, \diamondsuit_{k,r}$	row determinants given in (4.10) and (4.11), respectively
$\mathfrak{a}^+_\infty$	(one-sided) infinite-rank Kac-Moody Lie algebras of type A
$\mathfrak{b}_\infty, \mathfrak{c}_\infty, \mathfrak{d}_\infty$	infinite-rank Kac-Moody Lie algebras of types B, C, D
$\Lambda^{\mathfrak{c}}_i$	ith fundamental weight of $\mathfrak{c}_\infty$
λ_w	see (5.6)
$\mathcal{G}^{\mathfrak{x}}, \overline{\mathcal{G}}^{\mathfrak{x}}, \widetilde{\mathcal{G}}^{\mathfrak{x}}$	Lie (super)algebras corresponding to master diagrams of Section 6.2 of type $\mathfrak{x}$; also denoted by $\mathcal{G}, \overline{\mathcal{G}}, \widetilde{\mathcal{G}}$, respectively
$\mathcal{O}, \overline{\mathcal{O}}, \widetilde{\mathcal{O}}$	parabolic categories of $\mathcal{G}$-, $\overline{\mathcal{G}}$- and $\widetilde{\mathcal{G}}$-modules, respectively (Sections 6.1 and 6.3)
$\overline{\Delta}(\lambda), \widetilde{\Delta}(\lambda)$	parabolic Verma modules in $\overline{\mathcal{O}}, \widetilde{\mathcal{O}}$ (Section 6.4)
$\overline{L}(\lambda), \widetilde{L}(\lambda)$	irreducible modules in $\overline{\mathcal{O}}, \widetilde{\mathcal{O}}$ (Section 6.4)
$\boxed{\mathfrak{x}}$	head Dynkin diagrams of types $\mathfrak{x} = \mathfrak{a}, \mathfrak{b}, \mathfrak{c}, \mathfrak{d}$ (Section 6.2)
$P^d_Y, P^d_{\overline{Y}}, P^d_{\widetilde{Y}}$	set of dominant weights for $\mathcal{G}, \overline{\mathcal{G}}, \widetilde{\mathcal{G}}$, respectively (Section 6.3)
$\natural, \theta$	bijections from P^d_Y to $P^d_{\overline{Y}}$ and $P^d_{\widetilde{Y}}$ (Section 6.3), respectively
T, or $\overline{T}$	functors from $\widetilde{\mathcal{O}}$ to $\mathcal{O}$, or to $\overline{\mathcal{O}}$, respectively (Section 6.4)
$\overline{\mathcal{O}}_n$	parabolic category of a finite-dimensional Lie (super)algebra (Section 6.5)
$\mathfrak{Tr}_n$	truncation functor from $\overline{\mathcal{O}}$ to $\overline{\mathcal{O}}_n$ (Section 6.5)

References

[BB] A. Beilinson and J. Bernstein, Localisation de $\mathfrak{g}$-modules, C.R. Acad. Sci. Paris Ser. I Math. **292** (1981), 15–18.

[BeR] A. Berele and A. Regev, Hook Young diagrams with applications to combinatorics and to representations of Lie superalgebras, Adv. in Math. **64** (1987), 118–175.

[BL] I.N. Bernstein and D.A. Leites, A formula for the characters of the irreducible finite-dimensional representations of Lie superalgebras of series gl and sl, C.R. Aca. Bulg. Sci. **33** (1980), 1049–51 [in Russian].

[BKN] B. Boe, J. Kujawa and D. Nakano, Cohomology and support varieties for Lie superalgebras, Trans. AMS **362** (2010), 6551–6590.

[BPT] A. Brini, A. Palareti, and A. Teolis, Gordan-Capelli series in superalgebras, Proc. Natl. Acad. Sci. USA **85** (1988), 1330–1333.

[Br1] J. Brundan, Kazhdan-Lusztig polynomials and character formulae for the Lie superalgebra $\mathfrak{gl}(m|n)$, J. Amer. Math. Soc. **16** (2003), 185–231.

[Br2] J. Brundan, Kazhdan-Lusztig polynomials and character formulae for the Lie superalgebra $\mathfrak{q}(n)$, Adv. Math. **182** (2004), 28–77.

[BrK] J. Brundan and J. Kujawa, A new proof of the Mullineux conjecture, J. Alg. Combin. **18** (2003), 13–39.

[BrS] J. Brundan and C. Stroppel, Highest Weight Categories Arising from Khovanov's Diagram Algebras IV: the general linear supergroup, arXiv:0907.2543.

[BK] J.L. Brylinski and M. Kashiwara, Kazhdan-Lusztig conjecture and holonomic systerms, Invent. Math. **64** (1981), 387–410.

[CK] S.-J. Cheng and J.-H. Kwon, Howe duality and Kostant's homology formula for infinite-dimensional Lie superalgebras, IMRN (2008), Art. ID rnn 085, 52 pp.

[CKL] S.-J. Cheng, J.-H. Kwon and N. Lam, A BGG-type resolution for tensor modules over general linear superalgebra, Lett. Math. Phys. **84** (2008), 75–87.

[CKW] S.-J. Cheng, J.-H. Kwon and W. Wang, Kostant homology formulas for oscillator modules of Lie superalgebras, Adv. Math. **224** (2010), 1548–1588.

[CL1] S.-J. Cheng and N. Lam, Infinite-dimensional Lie superalgebras and hook Schur functions, Commun. Math. Phys. **238** (2003), 95–118.

[CL2] S.-J. Cheng and N. Lam, Irreducible characters of general linear superalgebra and super duality, Commun. Math. Phys. **298** (2010), 645–672.

[CLW] S.-J. Cheng, N. Lam and W. Wang, Super duality and irreducible characters of ortho-symplectic Lie superalgebras, Invent. Math. **183** (2011), 189–224.

[CLZ] S.-J. Cheng, N. Lam and R.B. Zhang, Character formula for infinite dimensional unitarizable modules of the general linear superalgebra, J. Algebra **273** (2004), 780–805.

[CW1] S.-J. Cheng and W. Wang, Howe duality for Lie superalgebras, Compositio Math. **128** (2001), 55–94.

[CW2] S.-J. Cheng and W. Wang, Remarks on Schur-Howe-Sergeev duality, Lett. Math. Phys. **52** (2000), 143–153.

[CW3] S.-J. Cheng and W. Wang, Lie subalgebras of differential operators on the super circle, Publ. Res. Inst. Math. Sci. **39** (2003), 545–600.

[CW4] S.-J. Cheng and W. Wang, Brundan-Kazhdan-Lusztig and super duality conjectures, Publ. Res. Inst. Math. Sci. **44** (2008), 1219–1272.

[CW5] S.-J. Cheng and W. Wang, Remarks on modules of the ortho-symplectic Lie superalgebras, Bull. Inst. Math. Acad. Sin. (N.S.) **3** (2008), 353–372.

[CWZ] S.-J. Cheng, W. Wang and R.B. Zhang, Super duality and Kazhdan-Lusztig polynomials, Trans. Amer. Math. Soc. **360** (2008), 5883–5924.

[CWZ2] S.-J. Cheng, W. Wang and R.B. Zhang, A Fock space approach to representation theory of $\mathrm{osp}(2|2n)$, Transformation Groups **12** (2007), 209–225.

[CZ1] S.-J. Cheng and R.B. Zhang, Analogue of Kostant's $\mathfrak{u}$-homology formula for general linear Superalgebras, International Mathematics Research Notices **2004**, no. 1, 31–53.

[CZ2] S.-J. Cheng and R.B. Zhang, Howe duality and combinatorial character formula for orthosymplectic Lie superalgebras, Adv. Math. **182** (2004), 124–172.

[J] J. Van der Jeugt, Character formulae for Lie superalgebra $C(n)$, Comm. Algebra **19** (1991), 199–222.

[JHKT] J. Van der Jeugt, J.W.B. Hughes, R. C. King and J. Thierry-Mieg, Character formulas for irreducible modules of the Lie superalgebras $\mathfrak{sl}(m/n)$, J. Math. Phys. **31** (1990), 2278–2304.

[DP] V. Dobrev and V. Petkova, Group-theoretical approach to extended conformal supersymmetry: function space realization and invariant differential operators, Fortschr. Phys. **35** (1987), 537–572.

[DS] M. Duflo and V. Serganova, On associated variety for Lie superalgebras, Preprint 2005, arXiv:math/0507198.

[Ger] J. Germoni, Indecomposable representations of $\mathrm{osp}(3,2)$, $D(2,1;\alpha)$ *and* $G(3)$, Colloquium on Homology and Representation Theory (Spanish) (Vaquerias, 1998), Bol. Acad. Nac. Cienc. (Cordoba) **65** (2000), 147–163.

[GW] R. Goodman and N. Wallach, Representations and invariants of the classical groups, Encyclopedia of Math. and Appl. **68**, Cambridge Univ. Press, 1998.

[Gor] M. Gorelik, Annihilation theorem and separation theorem for basic classical Lie superalgebras, J. AMS **15** (2001), 113–165.

[GS] C. Gruson and V. Serganova, Cohomology of generalized supergrassmannians and character formulae for basic classical Lie superalgebras, Proc. London Math. Soc. **101** (2010), 852–892.

[H1] R. Howe, Remarks on classical invariant theory, Trans. Amer. Math. Soc. **313** (1989), 539–570.

[H2] R. Howe, Perspectives on invariant theory: Schur duality, multiplicity-free actions and beyond, The Schur Lectures, Israel Math. Conf. Proc. **8**, Tel Aviv (1992), 1–182.

[K1] V. Kac, Lie superalgebras, Adv. in Math. **26** (1977), 8–96.

[K2] V. Kac, Representations of classical Lie superalgebras, Lect. Notes in Math. **676**, 597–626, Springer, Berlin, 1978.

[KW] V. Kac and M. Wakimoto, Integrable highest weight modules over affine superalgebras and Appell's function, Comm. Math. Phys. **215** (2001), 631–682.

[KL] D. Kazhdan and G. Lusztig, Representations of Coxeter groups and Hecke algebras, Invent. Math. **53** (1979), 165–184.

[LZ] N. Lam and R.B. Zhang, Quasi-finite modules for Lie superalgebras of infinite rank, Trans. Amer. Math. Soc. **358** (2006), 403–439.

[LSS] D. Leites, M. Saveliev and V. Serganova, Embeddings of $\mathrm{osp}(N/2)$ and the associated nonlinear supersymmetric equations, Group theoretical methods in physics, Vol. I (Yurmala, 1985), 255–297, VNU Sci. Press, Utrecht, 1986.

[Mac] I.G. Macdonald, Symmetric functions and Hall polynomials, Second Edition, Clarendon Press, Oxford, 1995.

[Ma] Yu. Manin, Gauge field theory and complex geometry, Grundlehren der mathematischen Wissenschaften **289**, Second Edition, Springer, Berlin, 1997.

[Mu] I. Musson, Lie superalgebras, Clifford algebras, induced modules and nilpotent orbits, Adv. Math. **207** (2006), 39–72.

[Naz] M. Nazarov, Capelli identities for Lie superalgebras, Ann. Sci. École Norm. Sup. (4) **30** (1997), 847–872.

[Pe] I. Penkov, Borel-Weil-Bott theory for classical Lie supergroups, J. Soviet Math. **51** (1990), 2108–2140.

[PS] I. Penkov and V. Serganova, Representations of classical Lie superalgebras of type I, Indag. Math. **3** (1992), 419–466.

[San] J. Santos, Foncteurs de Zuckermann pour les superalgébres de Lie, J. Lie Theory **9** (1999), 69–112.

[SNR] M. Scheunert, W. Nahm, and V. Rittenberg, Classification of all simple graded Lie algebras whose Lie algebra is reductive I, II. Construction of the exceptional algebras, J. Math. Phys. **17** (1976), 1626–1639, & 1640–1644.

[Sva] V. Serganova, Kazhdan-Lusztig polynomials and character formula for the Lie superalgebra $\mathfrak{gl}(m|n)$, Selecta Math. (N.S.) **2** (1996), 607–651.

[Sv1] A. Sergeev, The tensor algebra of the identity representation as a module over the Lie superalgebras $gl(n,m)$ and $Q(n)$, Math. USSR Sbornik **51** (1985), 419–427.

[Sv2] A. Sergeev, An Analog of the Classical Invariant Theory, I, II, Michigan J. Math. **49** (2001), 113–146, & 147–168.

[SW] B. Shu and W. Wang, Modular representations of the ortho-symplectic supergroups, Proc. London Math. Soc. **96** (2008), 251–271.

[Su] Y. Su, Composition factors of Kac modules for the general linear Lie superalgebras, Math. Z. **252** (2006), 731–754.

[SZ] Y. Su and R.B. Zhang, Generalised Verma modules for the orthosymplectic Lie superalgebra $\mathfrak{osp}(k|2)$, arXiv:0911.0735.

[Ta] T. Tanisaki, D-modules and Representation Theory, Lie Theory and Representation Theory, Surveys of Modern Mathematics **2**, 177–219, Higher Education Press and International Press, Beijing–Boston, 2010.

[Vo] D. Vogan, Irreducible characters of semisimple Lie Groups II: The Kazhdan-Lusztig Conjectures, Duke Math. J. **46** (1979), 805–859.

[Wa] W. Wang, Duality in infinite dimensional Fock representations, Commun. Contem. Math. **1** (1999), 155–199.

[WZ] W. Wang and L. Zhao, Representations of Lie superalgebras in prime characteristic I, Proc. London Math. Soc. **99** (2009), 145–167.

[We] H. Weyl, The Classical Groups: Their Invariants and Representations, Second Edition (15th Printing), Princeton University Press, Princeton, 1997.

[Zou] Y. Zou, Categories of finite-dimensional weight modules over type I classical Lie superalgebras, J. Algebra **180** (1996), 459–482.

Lie Theory and
Representation Theory
SMM 2, pp. 47–149

Combinatorial and Geometric Aspects of the Representation Theory of Finite Group Schemes

Rolf Farnsteiner*

Abstract

This paper provides an introduction to the aspects of the representation theory of finite group schemes that are related to the determination of their representation type.

2000 Mathematics Subject Classification: 14L15, 16G60, 16T05, 17B50.

Keywords and Phrases: Support spaces, representation type, McKay quivers.

0 Introduction

These notes grew out of a series of lectures given at East China Normal University (Shanghai) in July of 2006. The primary objective of my talks was two-fold: to provide a first introduction to the theories of Hopf algebras and algebraic groups on the one hand, and to delineate the role of certain combinatorial and geometric techniques within their representation theory on the other. Accordingly, our exposition emphasizes the discussion of illustrative examples and plausibility considerations highlighting the salient points of the theory. Proofs have been included primarily when they serve this purpose, and they often only sketch the main ideas rather than embarking on detailed discussions.

These notes are exclusively concerned with the "classical" theory of cocommutative Hopf algebras, leaving the very active field of quantum groups out of the account. Our Hopf algebras will also appear in a geometric guise, as affine algebraic group schemes. Since our point of view veils the historical sources of the subject matter, let me briefly mention that the axioms of our algebras first

*Mathematisches Seminar, Christian-Albrechts-Universität zu Kiel, Ludewig-Meyn Str. 4, 24098 Kiel, Germany. E-mail:rolf@math.uni-kiel.de

occurred in the work of the topologist Heinz Hopf [72], defining what are now called graded Hopf algebras.

Using broad brushstrokes the first section provides a quick tour of Hopf algebras, their associated group schemes, and restricted Lie algebras. Since the selection of the topics is dictated by what is relevant for the later sections, many interesting aspects have been left out of the account. The reader who is interested in a more in depth introduction may consult [81, 89] for the abstract theory of Hopf algebras, and [80, 123] for group schemes. The techniques involved in the representation theory of infinitesimal groups are often inductive, that is, one reduces problems to the consideration of "small" groups, whose structure is well enough understood to be amenable to the methods from the abstract representation theory of Artin algebras. The reduction process involves subgroups as well as quotients, and Section 1.5 provides a brief introduction to the latter. For almost all of our purposes, it suffices to know that the Hopf algebras associated to quotients are in fact quotient algebras. Accordingly, the notion of faithfully flat ring extensions, that is necessary for a more conceptual understanding of quotients, has been omitted.

In Section 2 we introduce two important tools that are defined in the wider context of self-injective algebras. The complexity of a module measures the polynomial rate of growth of its minimal projective resolution. We explore its relationship with the growth of certain Ext-groups and exhibit special features for Hopf algebras. The Heller operator is shown to induce a bijection on the set of isoclasses of non-projective indecomposable modules, leading to a necessary cohomological condition for an algebra to have finite representation type. The study of the algebras belonging to this class was the origin of many interesting developments in modern representation theory, such as the Auslander-Reiten theory of almost split sequences or tilting theory. Representation-finite algebras are one constituent of a trichotomy, whose other parts are the tame and wild algebras, which we briefly discuss in the final section.

In the context of finite group schemes, the complexity can also be interpreted as the dimension of a certain affine variety, the cohomological support variety of a module. These varieties have been available for modules over group algebras of finite groups for quite some time. The extension to arbitrary cocommutative Hopf algebras is based on the fundamental work by Friedlander-Suslin [63], who showed that the cohomology ring of a finite group scheme is finitely generated. We discuss representation theoretic support spaces, which, via the notion of a p-point, provide a non-cohomological interpretations of support varieties. Generalizing features from rank varieties, p-points are flat homomorphisms $k[X]/(X^p) \longrightarrow k\mathcal{G}$, whose associated pull-back functors $\operatorname{mod} k\mathcal{G} \longrightarrow \operatorname{mod} k[X]/(X^p)$ are used to study $\mathcal{G}$-modules.

As a first application, we show that modules over finite group schemes of tame representation type have complexity at most 2. For Frobenius kernels of smooth groups, the combination of rank varieties with basic results on nilpotent orbits yields the determination of the representation-finite and tame blocks. In fact, the latter are Morita equivalent to tame blocks of the first Frobenius kernel of SL(2). In particular, these blocks are special biserial and of domestic representation type.

Section 5 is devoted to a summary of recent progress concerning support spaces. The notion of a π-point, set forth by Friedlander-Pevtsova in [61], allows the geometric study of group schemes over arbitrary fields and also includes infinite-dimensional modules.

In Section 4 we introduce another geometric tool, given by varieties of diagonalizable subgroups. These are essential to understanding the structural impact of conditions on rank varieties of restricted Lie algebras. For Lie algebras of smooth groups, much information is derived from the so-called root space decomposition associated to a maximal torus. Since any two such tori are mapped onto each other by an automorphism of the ambient algebra, any maximal torus will do. By contrast, for arbitrary restricted Lie algebras the information encoded in the root space decomposition is highly sensitive to the choice of the torus. Schemes of tori (diagonalizable groups of height ≤ 1) are introduced to study all tori simultaneously and to consider algebraic families of Lie algebras.

Section 5 is concerned with the combinatorial tool given by the representation theory of bound quivers. We quickly review the fundamental results of Gabriel and Nazarova, Donovan-Freislich concerning basic algebras and the representation type of quivers. Representations of quivers correspond to modules over hereditary algebras, a class that substantially differs from algebras of measures. The notion of a trivial extension affords the passage from hereditary algebras to self-injective algebras. The concomitant doubling process involving the corresponding Gabriel quivers turns out to match the formation of the so-called McKay quivers, which we introduce in Section 5.5.

The final section combines the aforementioned techniques to study those finite group schemes, whose principal blocks have finite or tame representation type. We begin by reviewing the classical case concerning semi-simple Hopf algebras, which is covered by Nagata's Theorem on linearly reductive group schemes. In Section 6.2 we collect basic results on tensor products and explain the connection between the Ext-quivers and the McKay quivers of certain algebras. The latter arise in conjunction with the classification of the linearly reductive finite subgroup schemes of SL(2), the so-called binary polyhedral group schemes.

These notes are preliminary in the sense that they do not address the recent developments on π-points and Auslander-Reiten theory. Nevertheless, I feel that the current exposition may provide the interested reader with the background necessary to follow these emerging directions.

Finally, it is a pleasure to thank Department of Mathematics East China Normal University in general and Professor Bin Shu in particular for their hospitality during my stay in Shanghai.

1 Finite group schemes

1.1 Basic properties of Hopf algebras

Throughout these notes k will denote a field. The standard example of a Hopf algebra is the group algebra kG of some abstract group G. Recall that the G-modules correspond to the kG-modules. Accordingly, the module categories of group algebras enjoy special features that those of ordinary algebras do not have: If M and N are G-modules, then the vector spaces $M \otimes_k N$ and $\mathrm{Hom}_k(M,N)$ carry the structure of a G-module by setting

$$g.(m \otimes n) := g.m \otimes g.n \quad \text{and} \quad (g.\varphi)(m) := g\varphi(g^{-1}m)$$

for every $g \in G$, $m \in M$, $n \in N$, $\varphi \in \mathrm{Hom}_k(M,N)$, respectively.

Since these structures extend to kG, we should try to understand them without reference to G. Let's first look at tensor products: The spaces M and N are kG-modules, so $M \otimes_k N$ carries the structure of a $kG \otimes_k kG$-module by defining

$$(a \otimes b).(m \otimes n) := a.m \otimes b.n \quad \forall\, a,b \in kG,\ m \in M,\ n \in N.$$

Note that the map

$$G \times G \longrightarrow kG \otimes_k kG; \quad (g,h) \mapsto g \otimes h$$

induces an isomorphism $k(G \times G) \cong kG \otimes_k kG$. Hence the diagonal map $g \mapsto (g,g)$ gives rise to an algebra homomorphism $\Delta : kG \longrightarrow kG \otimes_k kG$. By definition, the kG-module structure on $M \otimes_k N$ is the pull-back of the $kG \otimes_k kG$-structure along Δ.

In order to understand the kG-structure of $\mathrm{Hom}_k(M,N)$, we observe that the map $g \mapsto g^{-1}$ induces an isomorphism $\eta : kG \longrightarrow kG^{\mathrm{op}}$ from kG to its *opposite algebra* kG^{op}. The space $\mathrm{Hom}_k(M,N)$ obtains the structure of a $kG \otimes_k kG^{\mathrm{op}}$-module via

$$((a \otimes b).\varphi)(m) := a\varphi(bm) \quad \forall\, a,b \in kG,\ m \in M,\ \varphi \in \mathrm{Hom}_k(M,N).$$

Our G-structure corresponds to the pull-back of this structure along the algebra homomorphism $(\mathrm{id}_{kG} \otimes \eta) \circ \Delta$.

The maps Δ and η have various properties that will be listed in the definition below. One obvious relation is $(\Delta \otimes \mathrm{id}_{kG}) \circ \Delta = (\mathrm{id}_{kG} \otimes \Delta) \circ \Delta$, ensuring that the natural k-linear identification $(X \otimes_k Y) \otimes_k Z \cong X \otimes_k (Y \otimes_k Z)$ is an isomorphism of kG-modules.

Unless mentioned otherwise, a *k-algebra* Λ is meant to be associative with an identity element that acts on all (left) modules via the identity operator. We will occasionally write $m : \Lambda \otimes_k \Lambda \longrightarrow \Lambda$ for the multiplication map. Given a k-algebra Λ, a Λ-module M, and k-linear maps $\varphi : V \longrightarrow \Lambda$, $\psi : W \longrightarrow M$ originating in some k-spaces V, W, we denote by $\varphi \hat{\otimes} \psi : V \otimes_k W \longrightarrow M$ the linear map given by $(\varphi \hat{\otimes} \psi)(v \otimes w) := \varphi(v)\psi(w)\ \ \forall\, v \in V,\ w \in W$.

Definition. Let H be a k-algebra, $\Delta : H \longrightarrow H \otimes_k H$, $\varepsilon : H \longrightarrow k$, and $\eta : H \longrightarrow H$ be k-linear maps. We say that (H, Δ, ε) is a *bialgebra* if

(1) Δ and ε are homomorphisms of k-algebras,

(2) $(\Delta \otimes \mathrm{id}_H) \circ \Delta = (\mathrm{id}_H \otimes \Delta) \circ \Delta$ (co-associativity),

(3) $(\mathrm{id}_H \hat{\otimes} \varepsilon) \circ \Delta = \mathrm{id}_H = (\varepsilon \hat{\otimes} \mathrm{id}_H) \circ \Delta$ (counit).

If, in addition, we have

(4) $(\eta \hat{\otimes} \mathrm{id}_H) \circ \Delta = \varepsilon 1 = (\mathrm{id}_H \hat{\otimes} \eta) \circ \Delta$,

then $(H, \Delta, \varepsilon, \eta)$ is referred to as a *Hopf algebra*.

Remarks. (i) If $H = kG$ is the group algebra of a group G, then ε is the unique homomorphism such that $\varepsilon(g) = 1 \ \forall\, g \in G$.

(ii) If one writes down the axioms for a k-algebra H as commutative diagrams involving the multiplication m and the canonical map $k \longrightarrow H$, then (2) and (3) follow by dualizing these diagrams, that is, by making all arrows point in the opposite direction.

(iii) When dealing with the *comultiplication* Δ it is convenient to use the so-called Heyneman-Sweedler notation

$$\Delta(h) = \sum_{(h)} h_{(1)} \otimes h_{(2)}.$$

For instance, (3) and (4) read

$$\sum_{(h)} h_{(1)}\varepsilon(h_{(2)}) = h = \sum_{(h)} \varepsilon(h_{(1)})h_{(2)}$$

and

$$\sum_{(h)} \eta(h_{(1)})h_{(2)} = \varepsilon(h) = \sum_{(h)} h_{(1)}\eta(h_{(2)}),$$

respectively. The main advantage lies in the treatment of iterated coproducts:

$$((\Delta \otimes \mathrm{id}_H) \circ \Delta)(h) = \sum_{(h)} h_{(1)} \otimes h_{(2)} \otimes h_{(3)} = ((\mathrm{id}_H \otimes \Delta) \circ \Delta)(h).$$

The identity $((\varepsilon \hat{\otimes} \mathrm{id}_H) \otimes \mathrm{id}_H) \circ (\Delta \otimes \mathrm{id}_H) \circ \Delta = \Delta$ then simply reads

$$\sum_{(h)} \varepsilon(h_{(1)})h_{(2)} \otimes h_{(3)} = \sum_{(h)} h_{(1)} \otimes h_{(2)}$$

for every $h \in H$.

(iv) The maps ε and η are called the *counit* and the *antipode* of the Hopf algebra H, respectively. The antipode is a homomorphism $H \longrightarrow H^{\mathrm{op}}$. It is bijective whenever H is finite-dimensional (cf. [102]).

(v) The ideal $H^\dagger := \ker \varepsilon$ is customarily referred to as the *augmentation ideal* of the Hopf algebra H.

An algebra homomorphism $f : H \longrightarrow H'$ between two bialgebras is a *bialgebra homomorphism* if $(f \otimes f) \circ \Delta = \Delta' \circ f$ and $\varepsilon' \circ f = \varepsilon$. If H and H' are Hopf algebras with antipodes η and η', respectively, then a *Hopf algebra homomorphism* additionally satisfies $\eta' \circ f = f \circ \eta$.

An element $g \neq 0$ of a Hopf algebra H is called *group-like* if $\Delta(g) = g \otimes g$. The set $G(H)$ of group-like elements of H is a subgroup of the group of units of H, the inverse of $g \in G(H)$ being given by $\eta(g)$. Moreover, $G(H) \subseteq H$ is a linearly independent subset, so that $G(H)$ is finite whenever H has finite dimension.

We say that $x \in H$ is *primitive*, provided $\Delta(x) = x \otimes 1 + 1 \otimes x$. The set $\mathrm{Lie}(H)$ of primitive elements is a subspace, that is closed under the Lie bracket

$[x, y] := xy - yx$ of H. In other words, Lie(H) is a Lie subalgebra of the commutator algebra H^- of the associative algebra H.

A subalgebra $H' \subseteq H$ of a Hopf algebra H is referred to as a *Hopf subalgebra* provided $\Delta(H') \subseteq H' \otimes_k H'$ and $\eta(H') \subseteq H'$. An ideal I of a bialgebra H is called a *bi-ideal* if $\Delta(I) \subseteq H \otimes_k I + I \otimes_k H$ and $\varepsilon(I) = (0)$. In that case, the factor space H/I canonically obtains the structure of a bialgebra. If, in addition, H is a Hopf algebra and $\eta(I) \subseteq I$, then I is called a *Hopf ideal* and H/I inherits the Hopf algebra structure from H.

Examples. (1) Let T be an indeterminate over k. We can endow the polynomial ring $k[T]$ and its localization $k[T]_T$ at T with the following Hopf algebra structures:

(a) $\Delta(T) = T \otimes 1 + 1 \otimes T$, $\varepsilon(T) = 0$, $\eta(T) = -T$.

(b) $\Delta(T) = T \otimes_k T$, $\varepsilon(T) = 1$, $\eta(T) = T^{-1}$.

Note that the Hopf algebra $k[T]_T$ is the group algebra $k\mathbb{Z}$ of the group $\mathbb{Z}$ of integers.

(2) Let $n \in \mathbb{N}$, and consider the bialgebra $k[\mathrm{Mat}_n] := k[X_{ij}, 1 \leq i, j \leq n]$, whose comultiplication and counit are given by

$$\Delta(X_{ij}) := \sum_{\ell=1}^{n} X_{i\ell} \otimes X_{\ell j}; \quad \varepsilon(X_{ij}) = \delta_{ij},$$

respectively. Note that the polynomial $\det((X_{ij}))$ belongs to $G(k[\mathrm{Mat}_n])$.

(3) Given $q \in k \setminus \{0\}$, we consider the k-algebra $k_q[k^2] := k\langle x, y\rangle/(yx - qxy)$. Then $k_q[k^2]$ has a bialgebra structure given by

$$\Delta(x) = x \otimes 1 + y \otimes x, \quad \Delta(y) = y \otimes y, \quad \varepsilon(x) = 0, \quad \varepsilon(y) = 1.$$

This bialgebra is often referred to as *Manin's quantum plane.*

Let H be a bialgebra, Λ be any k-algebra. Given linear maps φ, $\psi : H \longrightarrow \Lambda$, we define the *convolution* $\varphi * \psi : H \longrightarrow \Lambda$ of φ and ψ via

$$(\varphi * \psi)(h) := \sum_{(h)} \varphi(h_{(1)})\psi(h_{(2)}) \qquad \forall\, h \in H.$$

Lemma 1.1. *Let H be a Hopf algebra. Then the following statements hold:*

(1) *The dual space H^* is a k-algebra with multiplication given by convolution.*

(2) *If H is finite-dimensional, then H^* is a Hopf algebra with comultiplication*

$$\Delta^*(f) = \sum_{(f)} f_{(1)} \otimes f_{(2)} :\Leftrightarrow f(ab) = \sum_{(f)} f_{(1)}(a) f_{(2)}(b) \quad \forall\, a, b \in H,$$

and counit and antipode defined via $\varepsilon^*(f) = f(1)$ and $\eta^*(f) = f \circ \eta$, respectively.

Proof. We illustrate the existence of an identity element: Note that

$$(\varepsilon * f)(h) = \sum_{(h)} \varepsilon(h_{(1)}) f(h_{(2)}) = \sum_{(h)} f(\varepsilon(h_{(1)}) h_{(2)}) = f(h) \qquad \forall\, h \in H.$$

Consequently, ε is the identity element of H^*. □

Suppose that $\dim_k H < \infty$. The group-like elements of H^* are the *characters* of H, and the group $X(H) := G(H^*)$ of k-algebra homomorphisms $H \longrightarrow k$ is referred to as the *character group* of H. The space $\mathrm{Lie}(H^*)$ consists of the *derivations* $H \longrightarrow k$, that is, of all linear maps ψ such that $\psi(ab) = \psi(a)\varepsilon(b) + \varepsilon(a)\psi(b) \ \forall\, a, b \in H$.

Example. Consider the Hopf algebra $k[T]$ with

$$\Delta(T) = T \otimes 1 + 1 \otimes T, \quad \varepsilon(T) = 0, \quad \eta(T) = -T.$$

Suppose that $\mathrm{char}(k) = p > 0$. Given $r \in \mathbb{N}$, the binomial formula yields $\Delta(T^{p^r}) = T^{p^r} \otimes 1 + 1 \otimes T^{p^r}$. Accordingly, (T^{p^r}) is a Hopf ideal, and $k[\mathbb{G}_{a(r)}] := k[T]/(T^{p^r})$ has the structure of a Hopf algebra. We put $t := T + (T^{p^r})$, so that $k[\mathbb{G}_{a(r)}]$ has basis $\{t^i;\ 0 \leq i \leq p^r - 1\}$. Let $\{\delta_i;\ 0 \leq i \leq p^r - 1\}$ be the dual basis within $k\mathbb{G}_{a(r)} := k[\mathbb{G}_{a(r)}]^*$. Then we have

$$(\delta_i * \delta_j)(t^m) = \begin{cases} \binom{i+j}{i} & i + j = m, \\ 0 & \text{otherwise.} \end{cases}$$

Consequently, $\delta_i * \delta_j = \binom{i+j}{i}\delta_{i+j}$, where the right-hand term is understood to be zero whenever $i + j \notin \{0, \ldots, p^r - 1\}$. This readily yields $\delta_i^p = 0$ for $i \geq 1$, and it follows that the map $X_i \mapsto \delta_{p^i}$ induces an isomorphism

$$k[X_0, \ldots, X_{r-1}]/(X_0^p, \ldots, X_{r-1}^p) \longrightarrow k\mathbb{G}_{a(r)}$$

of k-algebras. Thus, as an algebra, $k\mathbb{G}_{a(r)}$ is isomorphic to the group algebra of the p-elementary abelian group $(\mathbb{Z}/(p))^r$. However, the Hopf algebras $k\mathbb{G}_{a(r)}$ and $k(\mathbb{Z}/(p))^r$ are not isomorphic: The dual of $k\mathbb{G}_{a(r)}$ is local (its group scheme is *infinitesimal*), while the dual of $k(\mathbb{Z}/(p))^r$ is separable (its group scheme is *étale*).

Let M be a k-vector space. The linear map

$$\tau_M : M \otimes_k M \longrightarrow M \otimes_k M; \quad m \otimes n \mapsto n \otimes m$$

is called the *flip*.

Definition. A bialgebra H is called *cocommutative* if $\tau_H \circ \Delta = \Delta$.

Note that a bialgebra is cocommutative exactly when its dual algebra is commutative.

Examples. (1) The Hopf algebras kG, $k[T]$, $k[T]_T$, and $k\mathbb{G}_{a(r)}$ are cocommutative.

(2) The bialgebras $k_q[k^2]$ and $k[\mathrm{Mat}_n]$ are not cocommutative.

Lemma 1.2. *Let H be a Hopf algebra. Then the following statements hold:*

(1) *If $H' \subseteq H$ is a finite-dimensional subalgebra such that $\Delta(H') \subseteq H' \otimes_k H'$, then H' is a Hopf subalgebra of H.*

(2) *If H is finite-dimensional and $I \subsetneq H$ is an ideal such that $\Delta(I) \subseteq I \otimes_k H + H \otimes_k I$, then I is a Hopf ideal of H.*

Proof. (1) Let $f \in \mathrm{End}_k(H')$ be a k-linear map. By assumption, we have $(f * \mathrm{id}_{H'})(h) = \sum_{(h)} f(h_{(1)})h_{(2)} \in H'$ for every $h \in H'$. Hence

$$\psi : \mathrm{End}_k(H') \longrightarrow \mathrm{End}_k(H'); \quad f \mapsto f * \mathrm{id}_{H'}$$

is a well-defined linear map. Given $f \in \ker \psi$ and $h \in H'$, we obtain, observing $\Delta(H') \subseteq H' \otimes_k H'$,

$$f(h) = \sum_{(h)} f(h_{(1)})\varepsilon(h_{(2)}) = \sum_{(h)} f(h_{(1)})h_{(2)}\eta(h_{(3)}) = \sum_{(h)} \psi(f)(h_{(1)})\eta(h_{(2)}) = 0,$$

so that ψ is injective. Since H' is finite-dimensional, there exists an element $g \in \mathrm{End}_k(H')$ such that

$$g * \mathrm{id}_{H'} = \varepsilon|_{H'}1.$$

A computation similar to the one above then implies $g(h) = \eta(h)$ for every $h \in H'$, so that $\eta(H') \subseteq H'$. Consequently, H' is a Hopf subalgebra.

(2) We consider the subspace $U := \{f \in \mathrm{End}_k(H);\ f(I) \subseteq I\} \subseteq \mathrm{End}_k(H)$. Since I is an ideal satisfying $\Delta(I) \subseteq I \otimes_k H + H \otimes_k I$, we have $U * U \subseteq U$, so that $A := k\varepsilon 1 + U$ is a subalgebra of $\mathrm{End}_k(H)$ containing id_H. As $\mathrm{id}_H \in \mathrm{End}_k(H)$ is invertible, the map

$$\psi : A \longrightarrow A; \quad f \mapsto f * \mathrm{id}_H$$

is injective, and, in view of A being finite-dimensional, also surjective. Hence there exist $g \in U$ and $\alpha \in k$ such that $\varepsilon 1 = (g + \alpha\varepsilon 1) * \mathrm{id}_H = (g * \mathrm{id}_H) + \alpha\,\mathrm{id}_H$. Given $x \in I$, we obtain

$$\varepsilon(x)1 = (g * \mathrm{id}_H)(x) + \alpha x \in I,$$

and our assumption $I \subsetneq H$ forces $\varepsilon(x) = 0$. As a result, $\varepsilon 1 \in U$ and the above arguments provide $f \in U$ with $f * \mathrm{id}_H = \varepsilon 1$. Consequently, $\eta = f \in U$, so that $\eta(I) \subseteq I$. □

We continue by recording an important homological property of Hopf algebras. Recall that a finite-dimensional k-algebra Λ is referred to as a *Frobenius algebra* if it admits a nondegenerate bilinear form $(\ ,\ ) : \Lambda \times \Lambda \longrightarrow k$ such that

$$(ab, c) = (a, bc) \qquad \forall\ a, b, c \in \Lambda.$$

Forms with the latter property are referred to as *associative*. Associative forms correspond to linear maps $\lambda : H \longrightarrow k$, the correspondence being given by

$$(a, b)_\lambda := \lambda(ab) \qquad \forall\ a, b \in \Lambda.$$

The associative form $(\,,\,)$ is usually not symmetric. Its departure from symmetry is measured by the *Nakayama automorphism* $\mu : \Lambda \longrightarrow \Lambda$ that is given by

$$(b, a) = (\mu(a), b) \qquad \forall\ a, b \in \Lambda.$$

Our next results show that Hopf algebras are Frobenius algebras affording a Nakayama automorphism of finite order.

Given a Hopf algebra H, we let $\int_H^\ell := \{x \in H;\ hx = \varepsilon(h)x\ \ \forall\ h \in H\}$ be the space of *left integrals* of H. Analogously, $\int_H^r := \{x \in H;\ xh = \varepsilon(h)x\ \ \forall\ h \in H\}$ is the space of *right integrals* of H.

Theorem 1.3 ([118, 85]). *Let H be a finite-dimensional Hopf algebra.*

(1) *We have* $\dim_k \int_H^\ell = 1$.

(2) *If $\lambda \in \int_{H^*}^\ell \setminus\{0\}$, then $(\,,\,)_\lambda$ is a nondegenerate, associative form.* □

Owing to (1), there exists an algebra homomorphism $\zeta_\ell : H \longrightarrow k$ such that $xh = \zeta(h)x$ for every $h \in H$ and $x \in \int_H^\ell$. The function ζ_ℓ is called the *left modular function* of H. The unique algebra homomorphism $\zeta_r : H \longrightarrow k$ with $hx = \zeta(h)x$ for every $h \in H$ and $x \in \int_H^r$ is referred to as the *right modular function* of H. Note that

$$\zeta_\ell = \zeta_r \circ \eta^{-1}.$$

Given an algebra homomorphism (a character) $\gamma : H \longrightarrow k$, the map $\mathrm{id}_H * \gamma$ is readily seen to be an automorphism of the algebra H. In fact,

$$X(H) \longrightarrow \mathrm{Aut}_k(H); \quad \gamma \mapsto \mathrm{id}_H * \gamma$$

is group a homomorphism from the character group $X(H)$ to the automorphism group $\mathrm{Aut}_k(H)$ of the algebra H. In particular, the automorphisms $\mathrm{id}_H * \gamma$ have finite order.

Proposition 1.4 ([92, 56]). *Let H be a finite-dimensional Hopf algebra with antipode η and left modular function ζ_ℓ. Then $\eta^{-2} \circ (\mathrm{id}_H * \zeta_\ell)$ is a Nakayama automorphism of H.*

Proof. Let $\pi \in \int_{H^*}^\ell$ be a non-zero left integral of H^*, so that

$$(x, y)_\pi = \pi(xy)$$

endows H with the structure of a Frobenius algebra. Directly from the defining property of π, we obtain

$$\pi(h)1 = \sum_{(h)} \pi(h_{(2)})h_{(1)} \qquad \forall\ h \in H. \tag{$*$}$$

Since the map $H \longrightarrow H^*;\ h \mapsto \pi.h$ is an isomorphism of right H-modules, there exists a unique element $u_\pi \in H$ such that

$$\pi.u_\pi = \varepsilon.$$

In other words, we have

$$\pi(u_\pi h) = \varepsilon(h) \qquad \forall\, h \in H.$$

In view of $\pi((u_\pi h - \varepsilon(h)u_\pi)x) = 0$ for all $h, x \in H$, we see that $u := u_\pi$ is a non-zero right integral of H. Given $x \in H$, we now obtain, observing $(*)$,

$$\begin{aligned}\sum_{(u)} \eta(u_{(1)})\pi(u_{(2)}x) &= \sum_{(u),(x)} \eta(u_{(1)})u_{(2)}x_{(1)}\pi(u_{(3)}x_{(2)}) = \sum_{(u),(x)} \varepsilon(u_{(1)})x_{(1)}\pi(u_{(2)}x_{(2)})\\ &= \sum_{(x)} x_{(1)}\pi(ux_{(2)}) = \sum_{(x)} x_{(1)}\varepsilon(x_{(2)}) = x.\end{aligned}$$

Thus, letting μ be the Nakayama automorphism relative to $(\ ,\ )_\pi$, we have

$$\mu^{-1}(x) = \sum_{(u)} \eta(u_{(1)})\pi(u_{(2)}\mu^{-1}(x)) = \sum_{(u)} \pi(xu_{(2)})\eta(u_{(1)}).$$

Using $(*)$ again, we compute $\eta^{-2} \circ \mu^{-1}$:

$$\begin{aligned}(\eta^{-2} \circ \mu^{-1})(x) &= \sum_{(u)} \pi(xu_{(2)})\eta^{-1}(u_{(1)}) = \sum_{(u),(x)} \pi(x_{(2)}u_{(3)})x_{(1)}u_{(2)}\eta^{-1}(u_{(1)})\\ &= \sum_{(u),(x)} \pi(x_{(2)}u_{(2)})x_{(1)}\varepsilon(u_{(1)}) = \sum_{(x)} \pi(x_{(2)}u)x_{(1)} = \sum_{(x)} \zeta_r(x_{(2)})\pi(u)x_{(1)}\\ &= \sum_{(x)} \zeta_r(x_{(2)})x_{(1)} = (\mathrm{id}_H * \zeta_r)(x).\end{aligned}$$

It follows that

$$\mu = (\mathrm{id}_H * \zeta_r)^{-1} \circ \eta^{-2} = (\mathrm{id}_H * (\zeta_r \circ \eta)) \circ \eta^{-2} = \eta^{-2} \circ (\mathrm{id}_H * (\zeta_r \circ \eta^{-1})) = \eta^{-2} \circ (\mathrm{id}_H * \zeta_\ell),$$

as desired. □

If H is cocommutative, then $\eta^2 = \mathrm{id}_H$, so that $\mathrm{id}_H * \zeta$ is a Nakayama automorphism of finite order. Since the antipode has finite order, this result also holds in the general case [102]. Integrals and modular functions are usually hard to compute. If $H = kG$ is the group algebra of a finite group G, then the element $\sum_{g \in G} g$ is an integral, and ε is the modular function. In particular, group algebras are *symmetric* algebras (i.e., they afford a symmetric, nondegenerate associative form). By contrast, the Hopf algebras associated to arbitrary finite group schemes are usually not symmetric.

In our prefatory remarks we emphasized the fact that tensor products of modules over a Hopf algebra H are also H-modules. Given H-modules M and N, we define H-module structures on $M \otimes_k N$ and $\mathrm{Hom}_k(M, N)$ in analogy with the special case kG:

$$h.(m \otimes n) := \sum_{(h)} h_{(1)}m \otimes h_{(2)}n \qquad \forall\, m \in M,\ n \in N,\ h \in H$$

and

$$(h.\varphi)(m) := \sum_{(h)} h_{(1)}\varphi(\eta(h_{(2)})m) \qquad \forall\, m \in M,\ \varphi \in \mathrm{Hom}_k(M,N),\ h \in H.$$

Let me indicate the utility of these concepts by giving a result that illustrates the way in which tensor products can be exploited.

Lemma 1.5. *Let H be a finite-dimensional Hopf algebra, P be a projective H-module, M be an arbitrary H-module. Then $P \otimes_k M$ is projective and injective.*

Proof. Recall that the functors $\mathrm{Hom}_k(P \otimes_k M, -)$ and $\mathrm{Hom}_k(P, \mathrm{Hom}_k(M, -))$ are naturally equivalent. Direct computation shows that this equivalence induces an equivalence

$$\mathrm{Hom}_H(P \otimes_k M, -) \cong \mathrm{Hom}_H(P, \mathrm{Hom}_k(M, -)).$$

Consequently, the left-hand functor is, as the composite of two exact functors, exact. Thus, $P \otimes_k M$ is projective, and since H is a Frobenius algebra, $P \otimes_k M$ is also injective. □

Using Lemma 1.5 one obtains identities for Ext-groups such as

$$\mathrm{Ext}^n_H(M,N) \cong \mathrm{Ext}^n_H(k, \mathrm{Hom}_k(M,N)) \qquad \forall\, n \geq 0,$$

where the right-hand groups coincide with the Hochschild cohomology groups of the augmented algebra (H, ε) with coefficients in the H-module $\mathrm{Hom}_k(M,N)$.

We conclude our survey of basic properties by quoting a Hopf algebra freeness theorem that was first verified in [92] for cocommutative Hopf algebras.

Theorem 1.6 ([91]). *Let K be a Hopf subalgebra of the finite-dimensional Hopf algebra H. Then H is a free left and right K-module.* □

1.2 Group schemes

In this section we introduce the geometric interpretation of the theory of cocommutative Hopf algebras. For a more thorough discussion we refer to [80] and [123]. Throughout, M_k and Ens will denote the categories of commutative k-algebras and sets, respectively. A functor $\mathcal{X} : M_k \longrightarrow \mathrm{Ens}$ is called a *k-functor*. The k-functors we will primarily be interested in are the so-called *affine schemes*: given a commutative k-algebra A, we consider the k-functor

$$\mathrm{Spec}_k(A) : M_k \longrightarrow \mathrm{Ens}; \quad R \mapsto \mathrm{Alg}_k(A, R),$$

where $\mathrm{Alg}_k(A,R)$ is the set of k-algebra homomorphisms from A to R. An affine scheme is called *algebraic* if A is a finitely generated k-algebra. Accordingly, a k-functor $\mathcal{X}$ is affine algebraic if and only if there exist polynomials $f_1, \dots, f_m \in k[X_1, \dots, X_n]$ such that

$$\mathcal{X}(R) = \{(x_1, \dots, x_n) \in R^n;\ f_i(x_1, \dots, x_n) = 0,\ 1 \leq i \leq m\},$$

for every commutative k-algebra R. Thus, the *k-rational points* $\mathcal{X}(k)$ of the affine algebraic schemes are the affines varieties from classical algebraic geometry. Note, however, that the *coordinate ring* $k[\mathcal{X}]$ of an affine scheme $\mathcal{X} = \mathrm{Spec}_k(k[\mathcal{X}])$ may have nilpotent elements.

Group functors take values in the category Gr of groups. In order to see the connection with Hopf algebras, we require the following basic result:

Theorem 1.7 (Yoneda's Lemma). *Let A, B be two commutative k-algebras. The assignment $\Phi \mapsto \Phi_A(\mathrm{id}_A)$ is a bijection between the set of natural transformations $\mathrm{Spec}_k(A) \longrightarrow \mathrm{Spec}_k(B)$ and $\mathrm{Alg}_k(B, A)$. In other words, for every natural transformation $\Phi : \mathrm{Spec}_k(A) \longrightarrow \mathrm{Spec}_k(B)$ there exists a unique homomorphism $\varphi : B \longrightarrow A$ of k-algebras such that $\Phi_R(\lambda) = \lambda \circ \varphi$ for every $\lambda \in \mathrm{Alg}_k(A, R)$ and $R \in M_k$.*

Proof. Each natural transformation $\Phi : \mathrm{Spec}_k(A) \longrightarrow \mathrm{Spec}_k(B)$ is determined by $\Phi_A(\mathrm{id}_A) \in \mathrm{Spec}_k(B)(A)$: If $\lambda : A \longrightarrow R$ is a homomorphism of commutative k-algebras, then

$$\begin{aligned}\Phi_R(\lambda) &= \Phi_R(\lambda \circ \mathrm{id}_A) = \Phi_R(\mathrm{Spec}_k(A)(\lambda)(\mathrm{id}_A)) \\ &= \mathrm{Spec}_k(B)(\lambda)(\Phi_A(\mathrm{id}_A)) = \lambda \circ \Phi_A(\mathrm{id}_A).\end{aligned}$$

Since the above equation defines a natural transformation, our result follows. □

Let $\Phi : \mathrm{Spec}_k(A) \longrightarrow \mathrm{Spec}_k(B)$ be a natural transformation. The corresponding homomorphism $B \longrightarrow A$ is often denoted Φ^* and referred to as the *comorphism* associated to Φ. In view of Theorem 1.7 Φ^* is uniquely determined by

$$\Phi_R(\lambda) = \lambda \circ \Phi^* \qquad \forall\, \lambda \in \mathrm{Spec}_k(A)(R),\ R \in M_k.$$

Let A and B be commutative k-algebras. Then there is a natural equivalence

$$\Pi : \mathrm{Spec}_k(A) \times \mathrm{Spec}_k(B) \longrightarrow \mathrm{Spec}_k(A \otimes_k B)$$

sending a pair (x, y) of algebra homomorphisms with values in R to the unique map

$$x \hat{\otimes} y : A \otimes_k B \longrightarrow R; \quad a \otimes b \mapsto x(a)y(b).$$

Definition. A k-functor $\mathcal{G} : M_k \longrightarrow \mathrm{Gr}$ from M_k to the category Gr of groups is called a *k-group functor*. We say that $\mathcal{G}$ is an *affine group scheme* if the k-functor $\mathcal{G}$ is affine. If the representing algebra $k[\mathcal{G}]$ is finitely generated, then the affine group scheme is called *algebraic*. In that case $\mathcal{G}$ is often referred to as an (*affine*) *algebraic k-group*.

Directly from the definition we obtain that

(a) the multiplication $(m_R : \mathcal{G}(R) \times \mathcal{G}(R) \longrightarrow \mathcal{G}(R))_{R \in M_k}$, and

(b) the inverse map $(\iota_R : \mathcal{G}(R) \longrightarrow \mathcal{G}(R))_{R \in M_k}$

are natural transformations.

Examples. (1) Consider the k-group functor $\mathrm{GL}_n : M_k \longrightarrow \mathrm{Gr}$

$$\mathrm{GL}_n(R) := \{(x_{ij}) \in \mathrm{Mat}_n(R);\ \det((x_{ij})) \text{ is invertible}\}, \qquad R \in M_k.$$

Observe that $\mathrm{GL}_n = \mathrm{Spec}_k(k[\mathrm{Mat}_n]_{\det(X_{ij})})$ is an affine algebraic group, the so-called *general linear group.*

(2) The *additive group* $\mathbb{G}_a$ is defined by means of $\mathbb{G}_a : M_k \longrightarrow \mathrm{Gr}$; $\mathbb{G}_a(R) := (R, +)$ for every $R \in M_k$.

(3) Consider $\mu : M_k \longrightarrow \mathrm{Gr}$; $\mu(R) := (R^\times, \cdot)$ for every $R \in M_k$. The k-group functor $\mu = \mathrm{GL}_1$ is represented by the Hopf algebra $k[T]_T$.

(4) Suppose that $\mathrm{char}(k) = p > 0$. Given $r \in \mathbb{N}$, we consider the group $\mathbb{G}_{a(r)}$ that is given by

$$\mathbb{G}_{a(r)}(R) := \{x \in \mathbb{G}_a(R);\ x^{p^r} = 0\}.$$

Note that $\mathbb{G}_a$ and $\mathbb{G}_{a(r)}$ are represented by the commutative Hopf algebras $k[T]$ and $k[T]/(T^{p^r})$, respectively. For $\mathbb{G}_a$ the points of R are identified with the values of $x(T)$, where $x \in \mathrm{Spec}_k(k[T])(R)$. The operation of $\mathbb{G}_a$ is induced by the comultiplication of $k[T]$:

$$x(T) + y(T) = (x\hat{\otimes}y)(T \otimes 1 + 1 \otimes T) = ((x\hat{\otimes}y) \circ \Delta)(T).$$

Inverses are given by the antipode:

$$-x(T) = x(-T) = (x \circ \eta)(T).$$

For every $R \in M_k$ the group $\mathbb{G}_{a(r)}(R) = \mathrm{Spec}_k(k[T]/(T^{p^r}))(R)$ is contained in $\mathbb{G}_a(R)$, and the inclusion is induced by the surjective map $k[T] \longrightarrow k[T]/(T^{p^r})$. This is an example of a closed subgroup of an algebraic group.

The group GL_n is an example of a reduced group scheme. Recall the bialgebra $k[\mathrm{Mat}_n]$. This algebra represents the monoid functor Mat_n that associates to every commutative k-algebra R the multiplicative monoid $\mathrm{Mat}_n(R)$ of $(n \times n)$-matrices with coefficients in R. Since this algebra is *reduced* (i.e., zero is the only nilpotent element), its localization $k[\mathrm{Mat}_n]_{\det(X_{ij})}$ has the same property. The algebraic groups represented by reduced Hopf algebras, are called *reduced group schemes.* They are determined by their k-rational points in case k is algebraically closed. By contrast, $\mathbb{G}_{a(r)}(k) = \{0\}$, so the k-rational points don't provide any information in this case.

Suppose that $(A, \Delta, \varepsilon, \eta)$ is a commutative Hopf algebra. Given $R \in M_k$, we define a multiplication on $\mathrm{Spec}_k(A)(R)$ via convolution:

$$(x * y)(a) = \sum_{(a)} x(a_{(1)})y(a_{(2)}) \tag{1}$$

for $x, y \in \mathrm{Spec}_k(A)$ and $a \in A$. (Observe that we need the commutativity of R to ensure that $x * y$ is indeed a homomorphism of k-algebras.) Then $(\mathrm{Spec}_k(A)(R), *)$ is a group with identity element ε and inverse $x^{-1} = x \circ \eta$. These operations endow $\mathrm{Spec}_k(A)$ with the structure of an affine group scheme.

The following result shows that all affine group schemes arise in this fashion.

Proposition 1.8. *Let A be a commutative k-algebra such that* $\mathrm{Spec}_k(A)$ *is a group scheme. Then A has the structure of a Hopf algebra such that the group structure on* $\mathrm{Spec}_k(A)$ *is given by* (1).

Proof. Let m be the multiplication on $\mathcal{G} := \mathrm{Spec}_k(A)$. Then $m : \mathcal{G} \times \mathcal{G} \longrightarrow \mathcal{G}$ is a natural transformation. We compose this transformation with the natural equivalence $\Pi^{-1} : \mathrm{Spec}_k(A \otimes_k A) \longrightarrow \mathcal{G} \times \mathcal{G}$. By Yoneda's Lemma, there exists an algebra homomorphism $\Delta : A \longrightarrow A \otimes_k A$ such that

$$(m \circ \Pi^{-1})(x) = x \circ \Delta \qquad \forall\, x \in \mathrm{Spec}_k(A \otimes_k A)(R),\ R \in M_k.$$

Let $g, h \in \mathcal{G}(R)$. Since $\Pi^{-1}(g\hat{\otimes}h) = (g, h)$, we have

$$(g \cdot h)(a) = ((g\hat{\otimes}h) \circ \Delta)(a) = \sum_{(a)} g(a_{(1)})h(a_{(2)}) = (g * h)(a)$$

for every $a \in A$. Yoneda's Lemma also provides an algebra homomorphism $\eta : A \longrightarrow A$ such that $g^{-1} = g \circ \eta$ for every element $g \in \mathcal{G}(R)$. The Hopf algebra axioms are now readily seen to correspond to the group axioms. □

Remark. Since $(g^{-1})^{-1} = g$ for every $g \in \mathcal{G}(R)$, we conclude that the antipode η of a commutative Hopf algebra H satisfies $\eta^2 = \mathrm{id}_H$. In view of Lemma 1.1 the antipode of a finite-dimensional cocommutative Hopf algebra satisfies the same identity. By the same token, an algebra homomorphism $\varphi : H \longrightarrow H'$ between two finite-dimensional cocommutative Hopf algebras satisfying $\Delta' \circ \varphi = (\varphi \otimes \varphi) \circ \Delta$ is a homomorphism of Hopf algebras.

A natural transformation $\varphi : \mathcal{G} \longrightarrow \mathcal{H}$ between two k-group functors is a *homomorphism of k-group functors*. Yoneda's Lemma then shows that the homomorphisms between affine group schemes correspond to the Hopf algebra homomorphisms $k[\mathcal{H}] \longrightarrow k[\mathcal{G}]$. More precisely, we have:

Proposition 1.9. *The categories of affine group schemes and commutative Hopf algebras are anti-equivalent.*

Proof. We only have to understand how to retrieve A from $\mathrm{Spec}_k(A)$. For any scheme $\mathcal{X}$, we define $k[\mathcal{X}]$ to be the set of natural transformations $\mathcal{X} \longrightarrow \mathrm{Spec}_k(k[T])$. This set naturally has the structure of a k-algebra. In case $\mathcal{X} = \mathrm{Spec}_k(A)$, Yoneda's Lemma provides an identification $k[\mathcal{X}] \cong \mathrm{Spec}_k(k[T])(A) \cong A$. □

Definition. Let $\mathcal{X}$ be a k-functor. Then the k-algebra $k[\mathcal{X}]$ of all natural transformations $\mathcal{X} \longrightarrow \mathrm{Spec}_k(k[T])$ is called the *function algebra* of $\mathcal{X}$. If $\mathcal{X}$ is an affine scheme, then $k[\mathcal{X}]$ is the coordinate ring of $\mathcal{X}$.

Given a homomorphism $\varphi : \mathcal{G} \longrightarrow \mathcal{H}$ of affine group schemes, we consider the functor $\ker \varphi$ that is defined by $\ker \varphi\,(R) := \ker \varphi_R$ for every $R \in M_k$. Let $\varphi^* : k[\mathcal{H}] \longrightarrow k[\mathcal{G}]$ be the Hopf algebra homomorphism corresponding to φ, and let $I := k[\mathcal{G}]\varphi^*(k[\mathcal{H}]^\dagger)$ be the ideal generated by the image of the augmentation ideal $k[\mathcal{H}]^\dagger$ of $k[\mathcal{H}]$. Then $x \in \ker\varphi\,(R)$ if and only if $x(I) = (0)$. In other words, the functor $\ker\varphi$ is represented by the Hopf algebra $k[\mathcal{G}]/I$.

Suppose that k is algebraically closed, and let A be a finitely generated, commutative k-algebra. Every finite-dimensional semi-simple subalgebra $S \subseteq A$ gives rise to a finite subset of mutually orthogonal idempotents of A. Since A is noetherian, it follows that A possesses a unique maximal finite-dimensional semi-simple subalgebra $\pi_0(A)$. Moreover, if A is a Hopf algebra, then $\pi_0(A)$ is a Hopf subalgebra of A. (For arbitrary fields, one considers separable subalgebras instead of semi-simple ones.)

Now let $\mathcal{G} = \mathrm{Spec}_k(A)$ be an affine algebraic k-group. We put $\pi_0(\mathcal{G}) := \mathrm{Spec}_k(\pi_0(A))$, and consider the homomorphism $\pi : \mathcal{G} \longrightarrow \pi_0(\mathcal{G})$ that is given by restriction. The subgroup $\mathcal{G}^0 := \ker\pi$ is called the *connected component* of $\mathcal{G}$.

Definition. An affine group scheme $\mathcal{G} := \mathrm{Spec}_k(k[\mathcal{G}])$ is *connected* if $k[\mathcal{G}]$ possesses exactly one idempotent.

Note that $\mathcal{G}^0$ is a connected, affine algebraic group scheme. An arbitrary affine algebraic group $\mathcal{G} = \mathrm{Spec}_k(A)$ is connected if and only if the prime ideal spectrum of A is connected.

1.3 Algebras of measures

In this section we interpret cocommutative Hopf algebras as the algebras of measures on the finite algebraic k-groups. *For simplicity we assume throughout that the ground field k is algebraically closed.*

Definition. An affine group scheme $\mathcal{G}$ is said to be *finite* if its coordinate ring $k[\mathcal{G}]$ is finite-dimensional. Given such a scheme $\mathcal{G}$, we call $k\mathcal{G} := k[\mathcal{G}]^*$ the *algebra of measures* on $\mathcal{G}$. The number $\mathrm{ord}(\mathcal{G}) := \dim_k k[\mathcal{G}]$ is referred to as the *order* of the finite algebraic group $\mathcal{G}$.

By definition, $k\mathcal{G}$ is a finite-dimensional cocommutative Hopf algebra. Our previous results now show that the category of finite-dimensional cocommutative Hopf algebras is equivalent to the category of finite group schemes. Indeed, if H is such a Hopf algebra, then $\mathcal{G}_H := \mathrm{Spec}_k(H^*)$ is a finite group scheme such that $H \cong k\mathcal{G}_H$.

Algebras of measures can be viewed as "group algebras" of finite group schemes: Given a k-vector space V, we consider the k-functor $V_a : M_k \longrightarrow \mathrm{Ens}$; $V_a(R) := V \otimes_k R$. In particular, we can consider $V := k\mathcal{G}$. Note that $k\mathcal{G} \otimes_k R$ has the structure of a Hopf algebra over R with comultiplication

$$\Delta_R : k\mathcal{G}\otimes_k R \longrightarrow k\mathcal{G}\otimes_k k\mathcal{G}\otimes_k R \cong (k\mathcal{G}\otimes_k R)\otimes_R(k\mathcal{G}\otimes_k R);\ \ h\otimes x \mapsto \sum_{(h)} h_{(1)}\otimes h_{(2)}\otimes x.$$

There is an embedding $\iota_\mathcal{G} : \mathcal{G} \hookrightarrow k\mathcal{G}_a$, which interprets an element $g : k[\mathcal{G}] \longrightarrow R$ of $\mathcal{G}(R)$ as a homomorphism $k[\mathcal{G}] \otimes_k R \longrightarrow R$ of R-algebras. Since $\mathrm{Hom}_R(k[\mathcal{G}] \otimes_k$

$R, R) \cong \mathrm{Hom}_k(k[\mathcal{G}], R) \cong k\mathcal{G}\otimes_k R$, this amounts to identifying $\mathcal{G}(R)$ with the group $G(k\mathcal{G} \otimes_k R)$ of group-like elements of $k\mathcal{G} \otimes_k R$. Given any morphism $f : \mathcal{G} \longrightarrow V_a$, there exists a unique k-linear map $\hat{f} : k\mathcal{G} \longrightarrow V$ such that

$$f_R = (\hat{f} \otimes \mathrm{id}_R) \circ (\iota_{\mathcal{G}})_R \qquad \forall\ R \in M_k.$$

The interested reader may consult [122] for more details.

Definition. Let $\mathcal{G}$ be a k-group, V be a k-vector space. We say that V is a $\mathcal{G}$-*module* if there exists a natural transformation $\mathcal{G} \times V_a \longrightarrow V_a$ such that, for every $R \in M_k$, the map

$$\mathcal{G}(R) \times (V \otimes_k R) \longrightarrow V \otimes_k R$$

is an action of the group $\mathcal{G}(R)$ on $V \otimes_k R$ by R-linear transformations.

Suppose that $\mathcal{G}$ is a finite algebraic group. The aforementioned universal property of $k\mathcal{G}$ entails that the notions "$\mathcal{G}$-module" and "$k\mathcal{G}$-module" coincide: every $\mathcal{G}$-module possesses a unique $k\mathcal{G}$-module structure and vice versa. We shall therefore use the two notions interchangeably.

Examples. (1) Let $\mathrm{char}(k) = p > 0$ and consider the group $\mathbb{G}_{a(r)}$. We have observed before that $k\mathbb{G}_{a(r)} \cong k[X_0, \ldots, X_{r-1}]/(X_0^p, \ldots, X_{r-1}^p)$ as an algebra. Recall that the generator X_i corresponds to the functional δ_{p^i}, where $\{\delta_0, \ldots, \delta_{p^r-1}\}$ is the basis dual to the canonical basis $\{1, t, \ldots, t^{p^r-1}\}$ of the coordinate ring $k[T]/(T^{p^r})$ of $\mathbb{G}_{a(r)}$. Direct computation shows that

$$\Delta(\delta_i) = \sum_{j=0}^{i} \delta_j \otimes \delta_{i-j}; \quad \varepsilon(\delta_i) = \delta_{i,0}; \quad \eta(\delta_i) = (-1)^i \delta_i.$$

Thus, $\mathrm{Lie}(k\mathbb{G}_{a(r)}) = k\delta_1$, while $G(k\mathbb{G}_{a(r)}) = \{\delta_0\}$. In particular, the algebra $k\mathbb{G}_{a(r)}$ is generated by $\mathrm{Lie}(k\mathbb{G}_{a(r)})$ if and only if $r = 1$.

(2) Given $r > 0$, we consider the finite group scheme $\mu_{(r)} : M_k \longrightarrow \mathrm{Gr}$,

$$\mu_{(r)}(R) := \{x \in R;\ x^r = 1\}.$$

Note that $\mu_{(r)} = \mathrm{Spec}_k(k[T]/(T^r - 1))$. If $t := T + (T^r - 1)$, then $\Delta(t) = t \otimes t$, $\varepsilon(t) = 1$ and $\eta(t) = t^{r-1}$. Given $g, h \in k\mu_{(r)}$, we have

$$(g * h)(t^i) = g(t^i)h(t^i),$$

whence $k\mu_{(r)} \cong k^r$ is semi-simple. If $\{\delta_0, \ldots, \delta_{r-1}\}$ is the basis dual to $\{1, t, \ldots, t^{r-1}\}$, then the δ_i are the primitive idempotents of $k\mu_{(r)}$. Note that

$$\Delta(\delta_i) = \sum_{j=0}^{r-1} \delta_j \otimes \delta_{i-j}; \quad \varepsilon(\delta_i) = \delta_{i,0}; \quad \eta(\delta_i) = \delta_{-i},$$

where the subscripts are considered elements of $\mathbb{Z}/(r)$. Moreover, $\mathrm{Lie}(k\mu_{(r)}) = (0)$ for $p \nmid r$, and $\dim_k \mathrm{Lie}(k\mu_{(r)}) = 1$, otherwise.

So far, the characteristic of the underlying base field k has not played a major role. The following fundamental result shows how the classical characteristic zero theory differs from the modular theory. It says that algebraic groups in characteristic zero are reduced and hence are completely determined by their k-rational points. We require the following result from commutative algebra.

Theorem 1.10 (Krull's Intersection Theorem,[35]). *Let R be a commutative noetherian ring,* $\operatorname{Max}(R)$ *be the set of maximal ideals of R. Then*

$$\bigcap_{\mathfrak{M}\in\operatorname{Max}(R)} \bigcap_{n\in\mathbb{N}} \mathfrak{M}^n = (0).$$

Let A be a commutative k-algebra. A k-linear map, $d : A \longrightarrow M$ with values in an A-module M is called a *derivation* of A into M, provided

$$d(xy) = x.d(y) + y.d(x) \qquad \forall\ x, y \in A.$$

Given a derivation $d : A \longrightarrow M$, the map

$$\operatorname{id}_A \otimes d : A \otimes_k A \longrightarrow A \otimes_k M; \quad a \otimes b \mapsto a \otimes d(b)$$

is a derivation of the k-algebra $A \otimes_k A$ with values in the $(A \otimes_k A)$-module $A \otimes_k M$, whose structure is defined by

$$(a \otimes b).(a' \otimes m) := aa' \otimes b.m.$$

Let H be a commutative Hopf algebra with augmentation ideal $I := \ker \varepsilon$. We consider the map

$$\pi_H : H \longrightarrow I/I^2; \quad h \mapsto h - \varepsilon(h)1 + I^2,$$

which is readily seen to be a derivation. Hence $\operatorname{id}_H \otimes \pi_H$ is a derivation of $H \otimes_k H$ with values in $H \otimes_k (I/I^2)$. Consequently,

$$d_H := (\operatorname{id}_H \otimes \pi_H) \circ \Delta$$

is a derivation of H into $H \otimes_k (I/I^2)$, where the latter space is the tensor product of the H-modules H and I/I^2.

Theorem 1.11 (Cartier). *Suppose that* $\operatorname{char}(k) = 0$. *Then every finitely generated commutative Hopf algebra H over k is reduced.*

Proof. Since the k-algebra H is finitely generated, the vector space I/I^2 is finite-dimensional. Hence there are elements $x_1, \ldots, x_r \in I$ such that their residue classes form a basis of I/I^2. Using multi-index notation, we write

$$x^m := x_1^{m_1} \cdots x_r^{m_r} \text{ and } |m| := \sum_{i=1}^{r} m_i$$

for $m \in \mathbb{N}_0^r$. The following claim says that the graded algebra, defined by the powers of the augmentation ideal I, is a polynomial ring in r variables.

(i) *Let $n \geq 1$. The residue classes $\{x^m + I^{n+1};\ |m| = n\}$ form a basis of I^n/I^{n+1}.*

Given $i \in \{1, \ldots, r\}$, we consider the k-linear map

$$f_i : I/I^2 \longrightarrow k; \quad \bar{x}_j \mapsto \delta_{ij}$$

as well as

$$\tilde{f}_i : H \otimes_k I/I^2 \longrightarrow H; \quad a \otimes v \mapsto f_i(v)a.$$

For $a, b \in H$ and $v \in I/I^2$, we obtain

$$a.(b \otimes v) = \sum_{(a)} a_{(1)}b \otimes a_{(2)}v = \sum_{(a)} a_{(1)}b \otimes \varepsilon(a_{(2)})v = (\sum_{(a)} a_{(1)}\varepsilon(a_{(2)}))b \otimes v = ab \otimes v,$$

so that the map $\tilde{f}_i$ is H-linear.

Consequently, the map

$$d_i : H \longrightarrow H; \quad h \mapsto \tilde{f}_i \circ d_H(h)$$

is a derivation of H with $d_i(h) = \sum_{(h)} f_i(\pi_H(h_{(2)}))h_{(1)}$ for all $h \in H$. We thus obtain

$$(\varepsilon \circ d_i)(h) = \sum_{(h)} \varepsilon(h_{(1)})f_i(\pi_H(h_{(2)})) = f_i(\sum_{(h)} \varepsilon(h_{(1)})\pi_H(h_{(2)})) = (f_i \circ \pi_H)(h),$$

whence

$$d_i(x_j) \equiv \delta_{ij} \quad \mod I \qquad (*)$$

for $1 \leq i \leq r$. Let $\underline{1}_i \in \mathbb{N}_0^r$ be the element with coordinates δ_{ij}. Since d_i is a derivation, $(*)$ implies

$$d_i(x^m) \equiv m_i x^{m-\underline{1}_i} \quad \mod I^{|m|}.$$

In particular, $d_i(I^n) \subseteq I^{n-1}$ for all $n \geq 1$. Thus, if

$$\sum_{|m|=n} \alpha_m x^m \equiv 0 \quad \mod I^{n+1},$$

then, applying d_i, we obtain

$$0 = \sum_{|m|=n} m_i \alpha_m x^{m-\underline{1}_i} \quad \mod I^n,$$

so that induction implies $m_i\alpha_m = 0$. Since $\mathrm{char}(k) = 0$, we conclude that $\alpha_m = 0$. ◇

(ii) *If $h^2 = 0$, then $h \in \bigcap_{n\in\mathbb{N}} I^n$.*

We have $h \in I$, and if $h \notin \bigcap_{n\in\mathbb{N}} I^n$, then there exists $n \in \mathbb{N}$ with $h \in I^n \setminus I^{n+1}$. We write

$$h = \sum_{|m|=n} \alpha_m x^m + z,$$

with $\alpha_m \in k$ and $z \in I^{n+1}$. Our assumption in conjunction with (i) implies

$$\sum_{m+m'=t} \alpha_m \alpha_{m'} = 0$$

for all $t \in \mathbb{N}_0^r$ with $|t| = 2n$. Upon ordering the elements of $\mathbb{N}_0^r$ lexicographically we obtain $0 = \alpha_{\tilde{m}}$, where $\tilde{m} = \max_{|m|=n}\{\alpha_m \neq 0\}$, a contradiction. ◇

If $\mathfrak{M} \trianglelefteq H$ is a maximal ideal of H, then Hilbert's Nullstellensatz provides an algebra homomorphism $\lambda : H \longrightarrow k$ such that $\mathfrak{M} = \ker \lambda$. Direct computation shows that λ induces an automorphism

$$\psi_\lambda : H \longrightarrow H; \quad h \mapsto \sum_{(h)} \lambda(h_{(1)}) h_{(2)}$$

of H, whose inverse is $\psi_{\lambda\circ\eta}$. Since

$$(\varepsilon \circ \psi_\lambda)(h) = \sum_{(h)} \lambda(h_{(1)})\varepsilon(h_{(2)}) = \lambda(\sum_{(h)} h_{(1)}\varepsilon(h_{(2)})) = \lambda(h)$$

for every $h \in H$, it follows that $\Psi_\lambda(\mathfrak{M}) = I$.

Let $h \in H$ be nilpotent. Without loss of generality, we may assume that $h^2 = 0$. Given a maximal ideal $\mathfrak{M} = \ker \lambda$, (ii) implies that $\psi_\lambda(h) \in \bigcap_{n\in\mathbb{N}} I^n$, whence $h \in \bigcap_{n\in\mathbb{N}} \mathfrak{M}^n$. Krull's Intersection Theorem now yields $h = 0$, as desired. □

Corollary 1.12. *If H is a finite-dimensional, cocommutative Hopf algebra such that H^* is reduced, then there exists a finite group G such that $H \cong kG$. In particular, all finite-dimensional cocommutative Hopf algebras of characteristic zero are semi-simple.*

Proof. By assumption, H^* is a finite-dimensional reduced algebra and thus is a product of copies of k. We let $G := X(H^*)$ be the character group of H^*, endowed with the convolution product. Then $\operatorname{ord}(G) = \dim_k H$, and $G \subseteq H$ is linearly independent. Consequently, the canonical map $kG \longrightarrow H$ is an isomorphism of Hopf algebras.

If $\operatorname{char}(k) = 0$, then Cartier's Theorem ensures that H^* is reduced, so that there exists a finite group G with $H \cong kG$. Owing to Maschke's Theorem, the latter algebra is semi-simple. □

Since we will be mainly interested in questions related to the representation type of an algebra, Corollary 1.12 shows that we will ultimately be studying Hopf algebras that are defined over fields of positive characteristic.

Let H be a finite-dimensional Hopf algebra with counit ε. As an algebra, H decomposes into its blocks

$$H = \mathcal{B}_0 \oplus \mathcal{B}_1 \oplus \cdots \oplus \mathcal{B}_n.$$

Here $\mathcal{B}_0$ is the block to which the *trivial H-module* k, with H acting via ε, belongs. In other words, $\mathcal{B}_0$ is determined by the property $\varepsilon(\mathcal{B}_0) \neq (0)$. This block is usually referred to as the *principal block* $\mathcal{B}_0(H)$ of H. It is not true in general that the principal block of a Hopf algebra is a Hopf subalgebra. In fact, if H is commutative, then $\mathcal{B}_0(H)$ is a Hopf subalgebra if and only if $\mathcal{B}_0(H) = H$ ($\mathcal{B}_0(H)$ is local with unique non-zero idempotent e_0. Thus, if $\mathcal{B}_0(H)$ is a Hopf subalgebra, then $\Delta(e_0) = e_0 \otimes e_0$, and e_0 is invertible, whence $e_0 = 1$.

Definitions. Let $\mathcal{G}$ be an affine group scheme. A subfunctor $\mathcal{H} \subseteq \mathcal{G}$ is called a (*closed*) *subgroup* if there exists a Hopf ideal $I \subseteq k[\mathcal{G}]$ such that $\mathcal{H}(R) = \mathcal{V}(I)(R) := \{g \in \mathcal{G}(R);\ g(I) = (0)\}$ for every commutative k-algebra R.

A homomorphism $\mathcal{G} \longrightarrow \mathcal{G}'$ of affine group schemes is called a *closed embedding* if the associated comorphism $k[\mathcal{G}'] \longrightarrow k[\mathcal{G}]$ of k-algebras is surjective.

Let $\mathcal{N} \subseteq \mathcal{G}$ be a subgroup. We say that $\mathcal{N}$ is a *normal* subgroup of $\mathcal{G}$ if $\mathcal{N}(R)$ is normal in $\mathcal{G}(R)$ for every $R \in M_k$.

Note that the comorphism $c^* : k[\mathcal{G}] \longrightarrow k[\mathcal{G}] \otimes_k k[\mathcal{G}]$ of the conjugation $c : \mathcal{G} \times \mathcal{G} \longrightarrow \mathcal{G};\ (g,h) \mapsto ghg^{-1}$ is given by $c^*(a) = \sum_{(a)} a_{(1)}\eta(a_{(3)}) \otimes a_{(2)}$. Hence $\mathcal{N} = \mathcal{V}(I)$ is normal if and only if $c^*(I) \subseteq k[\mathcal{G}] \otimes_k I$.

Suppose that $\operatorname{char}(k) = p > 0$, and consider the group $\mu_{(r)}$, where $r = p^s\ell$ with p not dividing ℓ. The map $x \mapsto (x^{p^s}, x^\ell)$ is an isomorphism $\mu_{(r)} \cong \mu_{(\ell)} \times \mu_{(p^s)}$. Note that the first factor is represented by $k[T]/(T^\ell - 1) \cong k^\ell$, while the second has local coordinate ring $k[T]/(T^{p^s} - 1)$. We now prove that finite algebraic groups always afford such a decomposition.

Theorem 1.13. *Let $\mathcal{G} = \operatorname{Spec}_k(A)$ be a finite algebraic group. Then $\mathcal{G}$ is a semidirect product*

$$\mathcal{G} = \mathcal{G}^0 \rtimes \mathcal{G}_{\mathrm{red}},$$

where $\mathcal{G}^0$ is normal and $\mathcal{G}_{\mathrm{red}} \cong \pi_0(\mathcal{G})$. As a k-functor, the connected component $\mathcal{G}^0$ is represented by the principal block $\mathcal{B}_0(A)$ of A.

Proof. We decompose $A = \bigoplus_{i=0}^n \mathcal{B}_i$ into its blocks, and denote the (central) primitive idempotents by $\{e_0, \dots, e_n\}$ with $\mathcal{B}_0 := \mathcal{B}_0(A) = Ae_0$. Since each block $\mathcal{B}_i = Ae_i$ is local and of the form $\mathcal{B}_i = ke_i \oplus \operatorname{Rad}(\mathcal{B}_i)$, the subalgebra $S := \sum_{i=0}^n ke_i$ is the largest semi-simple subalgebra of A, and we have $A = S \oplus \operatorname{Rad}(A)$.

As $\operatorname{Rad}(A)$ is the set of nilpotent elements of A, it is a Hopf ideal. Consider the closed subgroup $\mathcal{G}_{\mathrm{red}} := \mathcal{V}(\operatorname{Rad}(A))$ of $\mathcal{G}$.

Note that $\{e_i \otimes e_j;\ 0 \leq i, j \leq n\}$ is the set of orthogonal primitive idempotents of $A \otimes_k A$. Since $\Delta(e_i)$ is an idempotent of $A \otimes_k A$, we have $\Delta(e_i) \in S \otimes_k S$. Consequently, S is a Hopf subalgebra of A, and the composition $\lambda \circ \iota$ of the canonical projection $A \xrightarrow{\lambda} A/\operatorname{Rad}(A)$ with the inclusion $\iota : S \hookrightarrow A$ is an isomorphism $S \cong A/\operatorname{Rad}(A)$ of Hopf algebras. Accordingly, the corresponding isomorphism $\operatorname{Spec}_k(\lambda \circ \iota) : \mathcal{G}_{\mathrm{red}} \longrightarrow \pi_0(\mathcal{G})$ factors as

$$\mathcal{G}_{\mathrm{red}} \xrightarrow{\operatorname{Spec}_k(\lambda)} \mathcal{G} \xrightarrow{\pi} \pi_0(\mathcal{G}),$$

so that $\mathcal{G} = \mathcal{G}^0 \rtimes \mathcal{G}_{\mathrm{red}}$. By definition, the closed subgroup $\mathcal{G}^0 = \ker \pi$ is normal in $\mathcal{G}$. Since $\mathcal{G}^0 = \mathcal{V}(AS^\dagger)$ and $AS^\dagger = \bigoplus_{i=1}^n \mathcal{B}_i$, we see that the representing algebra $A/AS^\dagger$ is isomorphic to $\mathcal{B}_0(A)$. □

The foregoing result can also be interpreted at the level of Hopf algebras. Recall from Corollary 1.12 that, given a reduced finite algebraic group $\mathcal{G}$, we have $k\mathcal{G} \cong kG$, where $G = \operatorname{Alg}_k(k[\mathcal{G}], k) = \mathcal{G}(k)$ is the finite group of k-rational points of $\mathcal{G}$. Now let $\mathcal{G} = \mathcal{G}^0 \rtimes \mathcal{G}_{\mathrm{red}}$ be an arbitrary finite algebraic group. Since the connected component $\mathcal{G}^0$ is represented by a local algebra, we have

$$\mathcal{G}(k) = \mathcal{G}^0(k) \cdot \mathcal{G}_{\mathrm{red}}(k) = \mathcal{G}_{\mathrm{red}}(k).$$

It follows that

$$k\mathcal{G} \cong k\mathcal{G}^0 \# k\mathcal{G}_{\mathrm{red}} \cong k\mathcal{G}^0 \# k\mathcal{G}(k) \cong (k\mathcal{G}^0)\mathcal{G}(k)$$

is the *smash product* of $k\mathcal{G}^0$ with the group algebra of the group of k-rational points of $\mathcal{G}$. The right-hand term interprets the smash product as a *skew group algebra.*

Definition. A finite group scheme $\mathcal{G}$ is *infinitesimal* if its coordinate ring $k[\mathcal{G}]$ is local. In that case, $k\mathcal{G}$ is also called the *distribution algebra* of $\mathcal{G}$ and one also writes $k\mathcal{G} = \operatorname{Dist}(\mathcal{G})$ (see also the comments below).

Note that an affine algebraic k-group $\mathcal{G}$ is infinitesimal if and only if $\mathcal{G}(k) = \{1\}$. By Cartier's Theorem, any infinitesimal group of characteristic zero is trivial.

When studying the representations of a cocommutative Hopf algebra one thus has to understand the following disciplines:

(a) The modular representation theory of finite groups. By now, this field is rather well-understood.

(b) The representation theory of the infinitesimal group $\mathcal{G}^0$.

(c) The fusion of (a) and (b).

It has turned out that the methods figuring prominently in (a), such as the Mackey decomposition theorem and the defect theory of blocks, usually break down for infinitesimal group schemes. This has ultimately led to the approach via the geometric methods to be outlined in Sections 3 and 4.

For the remainder of this section we assume that $\operatorname{char}(k) = p > 0$.

Examples. (1) The groups $\mathbb{G}_{a(r)}$ and $\mu_{(p^r)}$ are infinitesimal for every $r > 0$.

(2) Given $r \in \mathbb{N}$, we consider the closed subgroup of $\operatorname{GL}(2)$ given by

$$\operatorname{GL}(2)_r(R) = \left\{ \begin{pmatrix} a & b \\ c & d \end{pmatrix} \in \operatorname{GL}(2)(R);\ a^{p^r} = 1 = d^{p^r},\ b^{p^r} = 0 = c^{p^r} \right\}$$

for every commutative k-algebra R. The group $\mathrm{GL}(2)_r$ is readily seen to be the kernel of the homomorphism

$$F^r : \mathrm{GL}(2) \longrightarrow \mathrm{GL}(2); \quad \begin{pmatrix} a & b \\ c & d \end{pmatrix} \mapsto \begin{pmatrix} a^{p^r} & b^{p^r} \\ c^{p^r} & d^{p^r} \end{pmatrix}.$$

Accordingly,

$$k[\mathrm{GL}(2)_r] \cong k[\mathrm{GL}(2)]/(\{X_{11}^{p^r} - 1, X_{22}^{p^r} - 1, X_{12}^{p^r}, X_{21}^{p^r}\}),$$

so that $\mathrm{GL}(2)_r$ is finite. Since $\mathrm{GL}(2)_r(k) = \{1\}$, the group scheme $\mathrm{GL}(2)_r$ is infinitesimal.

(3) Let $\mathcal{L} : M_k \longrightarrow \mathrm{Gr}$ be the group scheme, given by

$$\mathcal{L}(R) := \left\{ \begin{pmatrix} a & b \\ c & d \end{pmatrix} \in \mathrm{SL}(2)(R);\ a^p = 1 = d^p,\ b^{p^2} = b^p,\ c^p = 0 \right\}$$

for every $R \in M_k$. Then $\mathcal{L}$ is a closed subgroup of $\mathrm{SL}(2)$ such that $\mathcal{L}(k) = \left\{ \begin{pmatrix} 1 & b \\ 0 & 1 \end{pmatrix};\ b \in \mathbb{F}_p \right\} \cong \mathbb{Z}/(p)$. Moreover, using Proposition 1.14 below, we see that $\mathcal{L}^0 = \mathrm{SL}(2)_1 := \ker F|_{\mathrm{SL}(2)}$. Consequently,

$$\mathcal{L} = \mathrm{SL}(2)_1 \rtimes \mathcal{U},$$

where, for every commutative k-algebra R, the R-points of $\mathcal{U}$ are the group

$$\mathcal{U}(R) = \left\{ \begin{pmatrix} 1 & b \\ 0 & 1 \end{pmatrix};\ b^p = b \right\}.$$

The second example above shows how one can generate infinitesimal groups. The relevant notion in this context is that of the Frobenius homomorphism. If V is a k-vector space, we denote by $V^{(r)}$ the k-vector space with underlying abelian group V and action given by

$$\alpha \cdot v := \alpha^{p^{-r}} v \qquad \forall\, \alpha \in k,\ v \in V.$$

Given an affine k-group $\mathcal{G}$, we let $\mathcal{G}^{(r)} := \mathrm{Spec}_k(k[\mathcal{G}]^{(r)})$ be the affine group scheme defined by the twisted coordinate ring of $\mathcal{G}$.

Definition. Let $\mathcal{G}$ be an affine algebraic group scheme over the algebraically closed field k of characteristic $p > 0$. The homomorphism $F : \mathcal{G} \longrightarrow \mathcal{G}^{(1)}$ satisfying

$$F_R : \mathcal{G}(R) \longrightarrow \mathcal{G}^{(1)}(R); \quad F_R(\lambda)(x) = \lambda(x)^p \qquad \forall\, \lambda \in \mathcal{G}(R),\ x \in k[\mathcal{G}],\ R \in M_k$$

is called the *Frobenius homomorphism* of $\mathcal{G}$. The kernel $\mathcal{G}_r$ of its iterate $F^r : \mathcal{G} \longrightarrow \mathcal{G}^{(r)}$ is referred to as the *r-th Frobenius kernel* of $\mathcal{G}$.

Remarks. (1) Note that $F_R(\lambda)$ is indeed a k-linear map:

$$F_R(\lambda)(\alpha \cdot x) = \lambda(\alpha^{\frac{1}{p}} x)^p = \alpha F_R(\lambda)(x)$$

for every $x \in k[\mathcal{G}]^{(1)}$ and $\alpha \in k$.

(2) We have $F^r = \mathrm{Spec}_k(\varphi_r)$, where

$$\varphi_r : k[\mathcal{G}]^{(r)} \longrightarrow k[\mathcal{G}]; \quad x \mapsto x^{p^r}$$

is the comorphism of F^r. It follows that $k[\mathcal{G}_r] = k[\mathcal{G}]/k[\mathcal{G}]\varphi_r(k[\mathcal{G}]^\dagger) = k[\mathcal{G}]/k[\mathcal{G}]\{x^{p^r}; x \in k[\mathcal{G}]^\dagger\}$. Consequently, $\mathcal{G}_r$ is an infinitesimal k-group whenever $\mathcal{G}$ is algebraic.

(3) Certain problems on representations of connected algebraic groups can already be decided on sufficiently large Frobenius kernels. For instance, two finite-dimensional $\mathcal{G}$-modules are isomorphic if and only if their restrictions to a suitable Frobenius kernel $\mathcal{G}_r$ enjoy this property (see [80, (I.9.8)]).

Examples. (1) The group $\mathbb{G}_{a(r)}$ is the r-th Frobenius kernel of $\mathbb{G}_a$.

(2) For $r \geq 1$, let

$$\mathcal{A}_{[r]}(R) := \left\{ \begin{pmatrix} a & b \\ c & d \end{pmatrix} \in \mathrm{GL}(2)(R);\ a^{p^r} = 1 = d^{p^r},\ b^{p^2} = 0 = c^p \right\}, \quad R \in M_k.$$

Then $\mathcal{A}_{[r]} \subseteq \mathrm{GL}(2)_r$ is an infinitesimal subgroup of $\mathrm{GL}(2)$ containing $\mathrm{GL}(2)_1$, which is not a Frobenius kernel of $\mathrm{GL}(2)$.

Proposition 1.14. *Let $\mathcal{G}$ be a finite algebraic k-group. Then $\mathcal{G}$ is infinitesimal if and only if $\mathcal{G} = \mathcal{G}_r$ for some $r \geq 0$.*

Proof. Suppose that $\mathcal{G}$ is infinitesimal, and put $I := k[\mathcal{G}]^\dagger$. By assumption, the ideal I is nilpotent, so there exists an integer $r \geq 0$ such that $x^{p^r} = 0$ for every $x \in I$. By our remark above, this implies that $\mathcal{G}_r = \mathcal{V}(k[\mathcal{G}]\{x^{p^r}; x \in I\}) = \mathcal{V}((0)) = \mathcal{G}$. □

Definition. Let $\mathcal{G}$ be an infinitesimal k-group. Then $\mathrm{ht}(\mathcal{G}) := \min\{r \in \mathbb{N}_0;\ \mathcal{G} = \mathcal{G}_r\}$ is called the *height* of $\mathcal{G}$.

The distribution algebra of an infinitesimal group scheme is a special case of a more general construction that applies to arbitrary affine group schemes. We briefly indicate the definition; a thorough account can be found in Jantzen's book [80, I.§7]. Let $\mathcal{G} = \mathrm{Spec}_k(A)$ be an affine group scheme. Then

$$\mathrm{Dist}(\mathcal{G}) := \{h \in A^*;\ h((A^\dagger)^r) = (0) \text{ for some } r \in \mathbb{N}\}$$

is a subalgebra of A^*. If $\mathcal{G}$ is algebraic, then the definition of the comultiplication for finite-dimensional A still works, and $\mathrm{Dist}(\mathcal{G})$ has the structure of a Hopf algebra. This Hopf algebra is called the *distribution algebra* of $\mathcal{G}$. If $\mathcal{G}$ is a finite algebraic group, then $\mathrm{Dist}(\mathcal{G}) \subseteq k\mathcal{G}$ with equality holding if and only if $\mathcal{G}$ is infinitesimal.

Definition. Let $\mathcal{G}$ be an affine algebraic k-group. Then $\mathrm{Lie}(\mathcal{G}) := \mathrm{Lie}(\mathrm{Dist}(\mathcal{G}))$ is called the *Lie algebra of* $\mathcal{G}$.

Let $x \in \mathrm{Lie}(\mathcal{G})$. Then we have $x(ab) = x(a)\varepsilon(b) + \varepsilon(a)x(b)$ for $a, b \in k[\mathcal{G}]$, so that $x((k[\mathcal{G}]^\dagger)^2) = (0)$. Consequently, the natural map $\mathrm{Dist}(\mathcal{G}_r) \hookrightarrow \mathrm{Dist}(\mathcal{G})$ induces an isomorphism $\mathrm{Lie}(\mathcal{G}_r) \cong \mathrm{Lie}(\mathcal{G})$.

Let $\mathcal{G}$ be an affine algebraic k-group. Since $e_k = \mathcal{V}(k[\mathcal{G}]^\dagger)$ is a normal subgroup, the comorphism $c^* : k[\mathcal{G}] \longrightarrow k[\mathcal{G}] \otimes_k k[\mathcal{G}]$ of the conjugation action satisfies $c^*(k[\mathcal{G}]^\dagger) \subseteq k[\mathcal{G}] \otimes_k k[\mathcal{G}]^\dagger$. Hence we have an action, defined by

$$k[\mathcal{G}]^* \otimes_k \mathrm{Dist}(\mathcal{G}) \longrightarrow \mathrm{Dist}(\mathcal{G});\ (\varphi \cdot \psi)(a) := ((\varphi \hat{\otimes} \psi) \circ c^*)(a) = \sum_{(a)} \varphi(a_{(1)}\eta(a_{(3)}))\psi(a_{(2)})$$

for every $a \in k[\mathcal{G}]$. Direct computation shows that $\varphi \cdot \psi = \sum_{(\varphi)} \varphi_{(1)} * \psi * \eta^*(\varphi_{(2)})$ for $\varphi,\ \psi \in \mathrm{Dist}(\mathcal{G})$. Thus, our action specializes to the (left) adjoint representation of the Hopf algebra $\mathrm{Dist}(\mathcal{G})$. One verifies that

$$\varphi \cdot \psi \in \mathrm{Lie}(\mathcal{G}) \qquad \forall\ \varphi \in k[\mathcal{G}]^*,\ \psi \in \mathrm{Lie}(\mathcal{G}).$$

In particular, the group $\mathcal{G}(k)$ acts on $\mathrm{Lie}(\mathcal{G})$ via the *adjoint representation*:

$$g \cdot \psi := g * \psi * g^{-1} \qquad \forall\ g \in \mathcal{G}(k),\ \psi \in \mathrm{Lie}(\mathcal{G}).$$

Finally, we have $\varphi \cdot \psi = \varphi * \psi - \psi * \varphi,\ \forall\ \varphi,\ \psi \in \mathrm{Lie}(\mathcal{G})$.

1.4 Restricted Lie algebras

Given a Hopf algebra H, we have defined the associated Lie algebra $\mathrm{Lie}(H)$ via

$$\mathrm{Lie}(H) := \{x \in H;\ \Delta(x) = x \otimes 1 + 1 \otimes x\}.$$

The subspace $\mathrm{Lie}(H)$ is closed under the commutator product $[x, y] := xy - yx$ of H, that is, $\mathrm{Lie}(H)$ is a Lie subalgebra of the Lie algebra $(H, [\ ,\ ])$. If $\mathrm{char}(k) = p > 0$ we also have $x^p \in \mathrm{Lie}(H)$ for every $x \in \mathrm{Lie}(H)$. Lie algebras with the latter property are examples of restricted Lie algebras. The abstract notion of a restricted Lie algebra arose first in work by N. Jacobson [75, 76] concerning a Galois theory for purely inseparable field extensions of exponent 1.

Throughout this section we assume that k is a field of characteristic $\mathrm{char}(k) = p > 0$. Given a Lie algebra $\mathfrak{g}$, the left multiplication effected by the element $x \in \mathfrak{g}$ is customarily denoted $\mathrm{ad}\, x : \mathfrak{g} \longrightarrow \mathfrak{g};\ y \mapsto [x, y]$. If $(\mathfrak{g}, [\ ,\ ])$ is a Lie algebra over k, and R is a commutative k-algebra, then $\mathfrak{g} \otimes_k R$ obtains the structure of a Lie algebra over R via $[x \otimes r, y \otimes s] := [x, y] \otimes rs$ for all $x, y \in \mathfrak{g},\ r, s \in R$.

Definition. A *restricted Lie algebra* $(\mathfrak{g}, [p])$ is a pair consisting of a Lie algebra $\mathfrak{g}$ and a map $[p] : \mathfrak{g} \longrightarrow \mathfrak{g}\,;\ x \mapsto x^{[p]}$ such that

(1) $\mathrm{ad}\, x^{[p]} = (\mathrm{ad}\, x)^p \quad \forall\ x \in \mathfrak{g}$,

(2) $(\alpha x)^{[p]} = \alpha^p x^{[p]} \quad \forall\ \alpha \in k,\ x \in \mathfrak{g}$,

(3) $(x+y)^{[p]} = x^{[p]} + y^{[p]} + \sum_{i=1}^{p-1} s_i(x,y)$, where the s_i are given by the identity

$$(\mathrm{ad}\,(x \otimes T + y \otimes 1))^{p-1}(x \otimes 1) = \sum_{i=1}^{p-1} i\, s_i(x,y) \otimes T^{i-1} \text{ in } \mathfrak{g} \otimes_k k[T].$$

A map $[p] : \mathfrak{g} \longrightarrow \mathfrak{g}$ satisfying (1)–(3) is called a *p-map.*

If $(\mathfrak{g}, [p])$ is a restricted Lie algebra, a subalgebra $\mathfrak{h} \subseteq \mathfrak{g}$ (an ideal $\mathfrak{m} \subseteq \mathfrak{g}$) is called a *p-subalgebra* (*a p-ideal*) if $x^{[p]} \in \mathfrak{h}\ \forall\, x \in \mathfrak{h}$ ($x^{[p]} \in \mathfrak{m}\ \forall\, x \in \mathfrak{m}$). The notions of homomorphisms and factor algebras of restricted Lie algebras are defined in the canonical fashion.

Examples. (1) Given an associative k-algebra Λ, its *commutator algebra* $(\Lambda^-, [\,,\,])$ with product $[x,y] = xy - yx$ is restricted with respect to the ordinary p-th power operator $x \mapsto x^p$.

(2) Let $\Lambda := \mathrm{Mat}_n(k)$ be the algebra of $(n \times n)$-matrices over k. The corresponding restricted Lie algebra, which we denote by $\mathfrak{gl}(n)$, is called the *general linear Lie algebra.* Note that the *special linear Lie algebra*

$$\mathfrak{sl}(n) := \{x \in \mathfrak{gl}(n);\ \mathrm{tr}(x) = 0\}$$

is a p-ideal of $\mathfrak{gl}(n)$.

(3) For $n = 2$, we shall refer to the basis $\{e, h, f\}$ defined via

$$e := \begin{pmatrix} 0 & 1 \\ 0 & 0 \end{pmatrix};\ h := \begin{pmatrix} 1 & 0 \\ 0 & -1 \end{pmatrix};\ f := \begin{pmatrix} 0 & 0 \\ 1 & 0 \end{pmatrix}$$

as the *standard basis* of $\mathfrak{sl}(2)$.

Suppose that $\mathfrak{g}$ is an *abelian* Lie algebra, i.e., $[x,y] = 0$ for $x,\ y \in \mathfrak{g}$. Then the p-maps on $\mathfrak{g}$ are just the p-semilinear maps. These are determined by their values on a basis. To see how this generalizes to arbitrary restricted Lie algebras, we quote several basic results, whose proofs can be found in [77, 115].

Theorem 1.15. *Let $\mathfrak{g}$ be a Lie algebra with basis $(e_i)_{i \in I}$. Suppose there exist $x_i \in \mathfrak{g}$ $(i \in I)$ such that*

$$(\mathrm{ad}\, e_i)^p = \mathrm{ad}\, x_i \qquad \forall\, i \in I.$$

Then there exists a unique p-map $[p] : \mathfrak{g} \longrightarrow \mathfrak{g}$ such that $e_i^{[p]} = x_i$ for every $i \in I$. □

The foregoing result enables us to construct simple examples.

Examples. (1) We consider the $(2n+1)$-dimensional *Heisenberg algebra* $\mathfrak{h}_n$ with basis $\{x_1, \ldots, x_n, y_1, \ldots, y_n, z\}$. The Lie product is given by

$$[x_i, y_j] = \delta_{ij} z;\quad [x_i, x_j] = 0 = [y_i, y_j];\quad [z, \mathfrak{h}_n] = (0).$$

We endow $\mathfrak{h}_n$ with the following p-maps:

(a) $x_i^{[p]} = 0 = y_i^{[p]}$; $z^{[p]} = 0$.
(b) $x_i^{[p]} = 0 = y_i^{[p]}$; $z^{[p]} = z$.

(2) Let $\mathfrak{g} = kt \oplus kx$, $[t, x] = x$. Then $\mathfrak{g}$ possesses exactly one p-map, namely the one satisfying

$$t^{[p]} = t; \quad x^{[p]} = 0.$$

(3) We endow the vector space $\mathfrak{sl}(2) \oplus k$ with the structure of a Lie algebra by setting

$$[(x, \alpha), (y, \beta)] = ([x, y], 0]) \quad \forall\, x, y \in \mathfrak{sl}(2),\ \alpha, \beta \in k.$$

In other words, $\mathfrak{sl}(2) \oplus k$ is the direct sum of $\mathfrak{sl}(2)$ and the one-dimensional Lie algebra. This Lie algebra supports essentially three p-structures with $k^{[p]} = 0$, defining four-dimensional restricted Lie algebras $\mathfrak{sl}(2)_0$, $\mathfrak{sl}(2)_s$, and $\mathfrak{sl}(2)_n$. In the following formulae we consider the basis e, h, f, v_0 of $\mathfrak{sl}(2) \oplus k$, where $v_0 := 1$:
(a) $\mathfrak{sl}(2)_0$: $e^{[p]} = 0$; $h^{[p]} = h$; $f^{[p]} = 0$; $v_0^{[p]} = 0$.
(b) $\mathfrak{sl}(2)_s$: $e^{[p]} = 0$; $h^{[p]} = h + v_0$; $f^{[p]} = 0$; $v_0^{[p]} = 0$.
(c) $\mathfrak{sl}(2)_n$: $e^{[p]} = 0$; $h^{[p]} = h$; $f^{[p]} = v_0$; $v_0^{[p]} = 0$.
We shall see later that the representation theories of these three restricted Lie algebras are rather different.

(4) Let $(\mathfrak{g}, [p])$ be a restricted Lie algebra, R be a commutative k-algebra. The extended Lie algebra $\mathfrak{g} \otimes_k R$ has the structure of a restricted R-Lie algebra via

$$(x \otimes r)^{[p]} := x^{[p]} \otimes r^p \quad \forall\, x \in \mathfrak{g},\ r \in R.$$

The following subsidiary result explains how many different p-maps can be defined on a given Lie algebra. We denote by

$$C(\mathfrak{g}) := \{x \in \mathfrak{g};\ [x, y] = 0 \quad \forall\, y \in \mathfrak{g}\}$$

the *center* of the Lie algebra $\mathfrak{g}$.

Lemma 1.16. *Let $(\mathfrak{g}, [p])$ be a restricted Lie algebra, $\{p\} : \mathfrak{g} \longrightarrow \mathfrak{g}$ be a map. Then the following statements are equivalent:*
(1) *$\{p\}$ is a p-map.*
(2) *There exists a p-semilinear map $f : \mathfrak{g} \longrightarrow C(\mathfrak{g})$ such that $\{p\} = [p] + f$.* □

Let $\mathfrak{g}$ be a Lie algebra with *universal enveloping algebra* $(U(\mathfrak{g}), \iota)$. By definition, $U(\mathfrak{g})$ is an associative k-algebra and $\iota : \mathfrak{g} \longrightarrow U(\mathfrak{g})^-$ is a homomorphism of Lie algebras satisfying the following universal property: for any associative k-algebra Λ and any homomorphism $f : \mathfrak{g} \longrightarrow \Lambda^-$ of Lie algebras there exists a unique homomorphism $\varphi : U(\mathfrak{g}) \longrightarrow \Lambda$ of associative algebras such that $\varphi \circ \iota = f$.

It will be convenient to employ multi-index notation. Let Λ be a k-algebra, $a := (a_1, \ldots, a_\ell) \in \Lambda^\ell$, and $i = (i_1, \ldots, i_\ell) \in \mathbb{N}_0^\ell$. Then we put

$$a^i := \prod_{j=1}^{\ell} a_j^{i_j}.$$

Given ℓ-tuples $r = (r_1, \ldots, r_\ell)$; $s = (s_1, \ldots, s_\ell) \in \mathbb{N}_0^\ell$, we define

$$r \leq s :\Leftrightarrow r_i \leq s_i, \quad 1 \leq i \leq \ell.$$

We also put $\tau := (p-1, \ldots, p-1)$.

The following result is usually referred to as the PBW-Theorem:

Theorem 1.17 (Poincaré-Birkhoff-Witt). *Let $\mathfrak{g}$ be a Lie algebra with basis $\{x_1, \ldots, x_\ell\}$. Then $\{\iota(x)^n;\ n \in \mathbb{N}_0^\ell\}$ is a basis of $U(\mathfrak{g})$ over k.* □

In particular, $\iota : \mathfrak{g} \longrightarrow U(\mathfrak{g})$ is an embedding, and we will henceforth consider $\mathfrak{g}$ a subalgebra of $U(\mathfrak{g})^-$. Since $x \mapsto x \otimes 1 + 1 \otimes x$ is a homomorphism $\mathfrak{g} \longrightarrow (U(\mathfrak{g}) \otimes_k U(\mathfrak{g}))^-$ of Lie algebras, there is a unique extension $\Delta : U(\mathfrak{g}) \longrightarrow U(\mathfrak{g}) \otimes_k U(\mathfrak{g})$ to a homomorphism of associative k-algebras. By the same token, there exist unique homomorphisms $\eta : U(\mathfrak{g}) \longrightarrow U(\mathfrak{g})^{\mathrm{op}}$ and $\varepsilon : U(\mathfrak{g}) \longrightarrow k$ such that $\eta(x) = -x$ and $\varepsilon(x) = 0$ for every $x \in \mathfrak{g}$. Consequently, $U(\mathfrak{g})$ is a Hopf algebra that is generated by $\mathfrak{g}$.

Many features from the theory of finite-dimensional Hopf algebras lose their validity for $U(\mathfrak{g})$. For instance, the global dimension of $U(\mathfrak{g})$ coincides with $\dim_k \mathfrak{g}$. By contrast, Frobenius algebras never have finite non-zero global dimension. Moreover, $U(\mathfrak{g})$ is free of zero divisors. In view of Theorem 1.3, this shows that $k \cdot 1$ is the only finite-dimensional Hopf subalgebra of $U(\mathfrak{g})$. Thus, while $\mathrm{Dist}(\mathcal{G}) \cong U(\mathrm{Lie}(\mathcal{G}))$ in case $\mathrm{char}(k) = 0$ (cf. [31, II.§6]), it will follow from Theorem 1.20 below that these algebras are not isomorphic over fields of positive characteristic.

Definition. Let $(\mathfrak{g}, [p])$ be a restricted Lie algebra with universal enveloping algebra $U(\mathfrak{g})$. Let $I \subseteq U(\mathfrak{g})$ be the two-sided ideal generated by $\{x^p - x^{[p]};\ x \in \mathfrak{g}\}$. Then

$$U_0(\mathfrak{g}) := U(\mathfrak{g})/I$$

is called the *restricted enveloping algebra* of $\mathfrak{g}$.

If $(\mathfrak{g}, [p])$ is a restricted Lie algebra with basis $\{x_1, \ldots, x_\ell\}$, we define $z_i := x_i^p - x_i^{[p]} \in U(\mathfrak{g})$. Then one can modify the PBW-Theorem to show that $\{x^i z^j;\ 0 \leq i \leq \tau,\ j \in \mathbb{N}_0^\ell\}$ is a basis of $U(\mathfrak{g})$ over k (cf. [75]). Denoting by $\iota : \mathfrak{g} \longrightarrow U_0(\mathfrak{g})$ the natural map, we thus obtain:

Corollary 1.18. *The set $\{\iota(x)^r;\ 0 \leq r \leq \tau\}$ is a basis of $U_0(\mathfrak{g})$ over k. In particular, ι is injective, and $\dim_k U_0(\mathfrak{g}) = p^{\dim_k \mathfrak{g}}$.* □

Accordingly, we will henceforth consider $\mathfrak{g}$ a subalgebra of $U_0(\mathfrak{g})^-$. Since the ideal I is generated by primitive elements, I is a Hopf ideal, and $U_0(\mathfrak{g})$ inherits the Hopf algebra structure from $U(\mathfrak{g})$. As $U_0(\mathfrak{g})$ is generated by primitive elements, it is a cocommutative Hopf algebra. Note that $U_0(\mathfrak{g})$ has the following universal property: for any k-algebra Λ and any homomorphism $f : \mathfrak{g} \longrightarrow \Lambda^-$ of restricted Lie algebras, there exists a unique homomorphism $\varphi : U_0(\mathfrak{g}) \longrightarrow \Lambda$ of associative k-algebras such that $\varphi|_{\mathfrak{g}} = f$.

Example. For restricted enveloping algebras, modular functions and Nakayama automorphisms can be written down explicitly. Let $(\mathfrak{g}, [p])$ be a restricted Lie algebra with basis $\{e_1, \ldots, e_n\}$, and corresponding basis $\{e^r;\ 0 \leq r \leq \tau\}$ of $U_0(\mathfrak{g})$. We consider the linear form

$$\psi : U_0(\mathfrak{g}) \longrightarrow k; \quad \sum_{0 \leq r \leq \tau} \alpha_r e^r \mapsto \alpha_\tau.$$

Since each e_i is primitive, we have $\Delta(e^r) = \sum_{0 \leq s \leq r} \binom{r}{s} e^s \otimes e^{r-s}$, where $\binom{r}{s} := \prod_{i=1}^n \binom{r_i}{s_i}$. For $\varphi \in U_0(\mathfrak{g})^*$ we thus obtain

$$(\varphi * \psi)(e^r) = \sum_{0 \leq s \leq r} \binom{r}{s} \varphi(e^s)\psi(e^{r-s}) = \delta_{r,\tau}\varphi(1) = \varepsilon(\varphi)\psi(e^r).$$

Hence ψ is a left integral of the commutative Hopf algebra $U_0(\mathfrak{g})^*$. It was shown in [45] that $\psi(e^r x) = \psi((x + \operatorname{tr}(\operatorname{ad} x)1)e^r)$ for every $x \in \mathfrak{g}$. Owing to Proposition 1.4, the unique automorphism $\mu : U_0(\mathfrak{g}) \longrightarrow U_0(\mathfrak{g})$ satisfying $\mu(x) = x + \operatorname{tr}(\operatorname{ad} x)1$ for every $x \in \mathfrak{g}$ is a Nakayama automorphism of $U_0(\mathfrak{g})$. By the same token, the map $x \mapsto \operatorname{tr}(\operatorname{ad} x)1$ gives rise to the left modular function of $U_0(\mathfrak{g})$.

In contrast to group algebras of finite groups, restricted enveloping algebras are usually not symmetric. In fact, they are symmetric precisely when $\operatorname{tr}(\operatorname{ad} x) = 0$ for every $x \in \mathfrak{g}$, which was first observed by Schue [110].

Let $\mathcal{G}$ be a reduced affine algebraic k-group. Then $\mathcal{G}$ acts on $k\mathcal{G}_r$ via the adjoint representation Ad, and a combination of [80, (I.9.7)] and [80, (I.8.8)] shows that the character $g \mapsto \det(\operatorname{Ad}(g))$ defines a modular function of $k\mathcal{G}_r$. As noted in [80, (I.9.7)], these results are also valid for arbitrary infinitesimal groups of height 1. Our next result implies that this also yields the above formula for the Nakayama automorphism of $U_0(\mathfrak{g})$.

We will return to the structure of restricted Lie algebras in Section 4 when we study schemes of tori. Presently, we are interested in the interpretation of Lie algebras as infinitesimal groups of height ≤ 1. We record the following basic result from ring theory:

Proposition 1.19. *Let $\Lambda' \subseteq \Lambda$ be a subring of the artinian ring Λ such that $\Lambda = \Lambda' + \operatorname{Rad}(\Lambda)^2$. Then we have $\Lambda = \Lambda'$.* □

Theorem 1.20. *Let $\mathcal{G}$ be an infinitesimal k-group.*

(1) *There is an embedding $U_0(\operatorname{Lie}(\mathcal{G})) \hookrightarrow k\mathcal{G}$ of Hopf algebras.*
(2) *The group $\mathcal{G}$ has height ≤ 1 if and only if $U_0(\operatorname{Lie}(\mathcal{G})) \cong k\mathcal{G}$.*

Proof. Let $\mathfrak{g} := \operatorname{Lie}(\mathcal{G})$. The universal property of $U_0(\mathfrak{g})$ guarantees the existence of an algebra homomorphism $\psi : U_0(\mathfrak{g}) \longrightarrow k\mathcal{G}$ such that $\psi|_{\mathfrak{g}} = \operatorname{id}_{\mathfrak{g}}$. Thus, $((\psi \otimes \psi) \circ \Delta)|_{\mathfrak{g}} = (\Delta \circ \psi)|_{\mathfrak{g}}$, $\varepsilon_{\mathcal{G}} \circ \psi = \varepsilon_{\mathfrak{g}}$, and $(\eta_{\mathcal{G}} \circ \psi)|_{\mathfrak{g}} = (\psi \circ \eta_{\mathfrak{g}})|_{\mathfrak{g}}$, so that ψ is in fact a homomorphism of Hopf algebras.

Let $\{x_1, \ldots, x_n\}$ be a basis of $\mathfrak{g}$ over k. By Corollary 1.18, the set $\{x^r;\ 0 \leq r \leq \tau\}$ is a basis of $U_0(\mathfrak{g})$ over k. Let $\{\delta_r;\ 0 \leq r \leq \tau\}$ be the dual basis of the commutative Hopf algebra $k[\mathfrak{g}] := U_0(\mathfrak{g})^*$. Direct computation shows that $\delta_r * \delta_s = \binom{r+s}{r}\delta_{r+s}$. If ϵ_i denotes the n-tuple with i-th entry 1 and all other entries zero, then the map $X_i \mapsto \delta_{\epsilon_i}$ induces an isomorphism $k[X_1, \ldots, X_n]/(X_1^p, \ldots, X_n^p) \cong k[\mathfrak{g}]$ of k-algebras. Moreover, $k[\mathfrak{g}]^\dagger = (\delta_{\epsilon_1}, \ldots, \delta_{\epsilon_n})$.

Recall that $\mathfrak{g}$ is the space of derivations $k[\mathcal{G}] \longrightarrow k$, which is isomorphic to $k[\mathcal{G}]^\dagger/(k[\mathcal{G}]^\dagger)^2$. Application of this argument to $k[\mathfrak{g}]$ shows that $\dim_k \mathrm{Lie}(U_0(\mathfrak{g})) = n$, so that $\mathfrak{g} = \mathrm{Lie}(U_0(\mathfrak{g}))$.

(1) Consider the transpose map $\psi^t : k[\mathcal{G}] \longrightarrow k[\mathfrak{g}]$. Since $\psi|_{\mathfrak{g}} = \mathrm{id}_{\mathfrak{g}}$, we see that ψ^t induces an isomorphism $k[\mathcal{G}]^\dagger/(k[\mathcal{G}]^\dagger)^2 \cong k[\mathfrak{g}]^\dagger/(k[\mathfrak{g}]^\dagger)^2$. Consequently,

$$k[\mathfrak{g}] = \psi^t(k[\mathcal{G}]) + (k[\mathfrak{g}]^\dagger)^2 = \psi^t(k[\mathcal{G}]) + \mathrm{Rad}(k[\mathfrak{g}])^2,$$

so that Proposition 1.19 implies that ψ^t is surjective. Consequently, ψ is injective.

(2) Suppose that $\mathcal{G}$ is infinitesimal of height ≤ 1. Then $x^p = 0$ for every $x \in k[\mathcal{G}]^\dagger$. Since $\dim_k \mathfrak{g} = n$, the local algebra $k[\mathcal{G}]$ is generated by n elements of $k[\mathcal{G}]^\dagger$. Thus, the resulting surjective homomorphism $k[X_1, \ldots, X_n] \longrightarrow k[\mathcal{G}]$ factors through the truncated polynomial ring $k[X_1, \ldots, X_n]/(X_1^p, \ldots, X_n^p)$. Hence $\dim_k \mathcal{G} = \dim_k k[\mathcal{G}] \leq p^n = \dim_k U_0(\mathfrak{g})$, and the injection ψ is surjective. □

Remark. Thanks to Theorem 1.20, the functors $\mathcal{G} \mapsto \mathrm{Lie}(\mathcal{G})$ and $\mathfrak{g} \mapsto \mathrm{Spec}_k(U_0(\mathfrak{g})^*)$ induce equivalences between the categories of infinitesimal groups of height ≤ 1 and restricted Lie algebras, respectively.

Proposition 1.21. *Let $\mathcal{G}$ be an infinitesimal k-group. Then there exists $n \in \mathbb{N}_0$ with $\dim_k k\mathcal{G} = p^n$.*

Proof. Let A be a local, commutative Hopf algebra. We show that A has dimension a power of p. We proceed inductively, and consider the Hopf subalgebra $B := \{a^p;\ a \in A\}$. If $B = k1$, then $\mathcal{G} := \mathrm{Spec}_k(A)$ is an infinitesimal group of height ≤ 1 with function algebra A. Hence our assertion follows from a consecutive application of Corollary 1.18 and Theorem 1.20.

Alternatively, Theorem 1.6 implies that A is free over B. Moreover, $\mathrm{rk}_B(A) = \dim_k A/AB^\dagger$. Since $A/AB^\dagger$ is also a local Hopf algebra, the inductive hypothesis ensures that $\dim_k B$ and $\mathrm{rk}_B(A)$ are p-powers. Hence $\dim_k A = \mathrm{rk}_B(A)\dim_k B$ is also a p-power. □

1.5 Quotients

Let $\mathcal{G}$ be an affine algebraic k-group, $\mathcal{N} \trianglelefteq \mathcal{G}$ be a normal subgroup. We would like to define a quotient $\mathcal{G}/\mathcal{N}$. The naive approach, setting $\mathcal{G}/\mathcal{N}(R) := \mathcal{G}(R)/\mathcal{N}(R)$ does not work, as the resulting k-functor is not necessarily representable (see the example below). What we need is a *categorical quotient*, i.e., a map $\pi : \mathcal{G} \longrightarrow \mathcal{G}/\mathcal{N}$ with kernel $\mathcal{N}$ that satisfies the following universal property:

- If $\varphi : \mathcal{G} \longrightarrow \mathcal{H}$ is a homomorphism of affine group schemes such that $\mathcal{N} \subseteq \ker\varphi$, then there exists a unique homomorphism $\psi : \mathcal{G}/\mathcal{N} \longrightarrow \mathcal{H}$ such that $\psi \circ \pi = \varphi$.

Let us see what this condition amounts to: Let $\pi^* : k[\mathcal{G}/\mathcal{N}] \longrightarrow k[\mathcal{G}]$ be the associated Hopf algebra homomorphism. Setting $I := \ker \pi^*$, we consider the subgroup $\mathcal{H} := \mathcal{V}(I) \subseteq \mathcal{G}/\mathcal{N}$ as well as the canonical embedding $\iota : \mathcal{H} \longrightarrow \mathcal{G}/\mathcal{N}$. By construction, there is a homomorphism $\gamma^* : k[\mathcal{H}] \longrightarrow k[\mathcal{G}]$ of Hopf algebras such that $\gamma^* \circ \iota^* = \pi^*$. By the universal property, the corresponding homomorphism $\gamma : \mathcal{G} \longrightarrow \mathcal{H}$ induces a unique map $\omega : \mathcal{G}/\mathcal{N} \longrightarrow \mathcal{H}$ with $\omega \circ \pi = \gamma$. Accordingly, we have $\pi = \iota \circ \gamma = (\iota \circ \omega) \circ \pi$, and unicity implies $\iota \circ \omega = \mathrm{id}_{\mathcal{G}/\mathcal{N}}$. Thus, $\omega^* \circ \iota^* = \mathrm{id}_{k[\mathcal{G}/\mathcal{N}]}$, so that $I = \ker \iota^* = (0)$. We therefore make the following definition:

Definition. A homomorphism $\pi : \mathcal{G} \longrightarrow \mathcal{H}$ between two affine k-groups is a *quotient map* if the corresponding Hopf algebra map $\pi^* : k[\mathcal{H}] \longrightarrow k[\mathcal{G}]$ is injective.

It is by no means clear that for any normal subgroup $\mathcal{N} \trianglelefteq \mathcal{G}$ a quotient map with kernel $\mathcal{N}$ exists. It turns out that the algebra $k[\mathcal{G}]^{\mathcal{N}}$ of invariants gives rise to the quotient group. Thus, if $\mathcal{G}$ is algebraic, then the question as to whether $\mathcal{G}/\mathcal{N}$ inherits this property is related to Hilbert's fourteenth problem. Of course, for finite algebraic groups we don't need to worry about such issues.

Theorem 1.22. *Let $\mathcal{G}$ be an affine k-group, $\mathcal{N} \trianglelefteq \mathcal{G}$ be a normal subgroup. Then there exists a quotient map $\pi : \mathcal{G} \longrightarrow \mathcal{H}$ with kernel $\mathcal{N}$. If $\mathcal{G}$ is algebraic, so is $\mathcal{H}$.* □

Remarks. (1) By the universal property, the pair $(\mathcal{H}, \pi)$ is unique up to isomorphism. One thus writes $\mathcal{G}/\mathcal{N} := \mathcal{H}$ and calls $\mathcal{G}/\mathcal{N}$ the *factor group of $\mathcal{G}$ by $\mathcal{N}$*.

(2) Note that the quotient map $\pi : \mathcal{G} \longrightarrow \mathcal{G}/\mathcal{N}$ is usually not surjective at each point. Suppose it is, then there exists $\lambda \in \mathcal{G}(k[\mathcal{G}/\mathcal{N}])$ such that $\pi(\lambda) = \mathrm{id}_{k[\mathcal{G}/\mathcal{N}]}$. Accordingly, the comorphism $\pi^* : k[\mathcal{G}/\mathcal{N}] \longrightarrow k[\mathcal{G}]$ is split injective.

Example. Consider the group $\mathbb{G}_{a(2)}$ with coordinate ring $k[\mathbb{G}_{a(2)}] = k[T]/(T^{p^2})$. We write $t := T + (T^{p^2})$ and observe that the primitive element t^p generates a Hopf ideal $I \trianglelefteq k[\mathbb{G}_{a(2)}]$. Thus, the corresponding normal subgroup $\mathcal{N} := \mathcal{V}(I)$ is isomorphic to $\mathbb{G}_{a(1)}$. Consider the group scheme

$$\mathcal{X} : M_k \longrightarrow \mathrm{Gr}; \quad R \mapsto \mathcal{G}(R)/\mathcal{N}(R).$$

Then $\pi := (\pi_R : \mathcal{G}(R) \longrightarrow \mathcal{X}(R))_{R \in M_k}$ is a homomorphism of k-group functors. If $\mathcal{X} = \mathrm{Spec}_k(k[\mathcal{X}])$ is representable, then $\pi^* : k[\mathcal{X}] \longrightarrow k[\mathcal{G}]$ is split injective, and there exists an ideal $J \trianglelefteq k[\mathcal{G}]$ such that

$$k[\mathcal{G}] = k[\mathcal{X}] \oplus J.$$

Since $k[\mathcal{G}]$ is free over $k[\mathcal{X}]$, we have $\dim_k k[\mathcal{X}] = p$. Consequently, $x^p = 0$ for every $x \in k[\mathcal{X}]^\dagger$, so that $k[\mathcal{X}]^\dagger \subseteq \bigoplus_{i=p}^{p^2-1} kt^i$. The condition $\Delta(k[\mathcal{X}]) \subseteq k[\mathcal{X}] \otimes_k k[\mathcal{X}]$ then implies $k[\mathcal{X}] = k[t^p]$. Thus, $x^p \in k[\mathcal{X}]$ for every $x \in k[\mathcal{G}]$, and the p-th power map is trivial on J. Since $k[\mathcal{G}]^\dagger = k[\mathcal{X}]^\dagger \oplus J$, we obtain $x^p = 0$ for every $x \in k[\mathcal{G}]^\dagger$, a contradiction.

For finite algebraic groups, we could have defined quotients via the following argument: If $\mathcal{N} \trianglelefteq \mathcal{G}$ is a normal subgroup, then, as we shall see below, $k\mathcal{G}k\mathcal{N}^\dagger$ is a Hopf ideal of $k\mathcal{G}$. In view of the equivalence between the categories of finite group schemes and finite-dimensional cocommutative Hopf algebras, there exists a finite group scheme, $\mathcal{G}/\mathcal{N}$ say, such that

$$k(\mathcal{G}/\mathcal{N}) \cong k\mathcal{G}/(k\mathcal{G}k\mathcal{N}^\dagger).$$

We now show that this brute force definition agrees with the previous one. In this context, the following notion is convenient:

Let $\mathcal{G}$ be a finite algebraic k-group. If η denotes the antipode of $k\mathcal{G}$, then the action given by

$$h.x := \sum_{(h)} h_{(1)} x \, \eta(h_{(2)}) \qquad \forall \; h, x \in k\mathcal{G}$$

is called the *(left) adjoint representation of* $k\mathcal{G}$.

Proposition 1.23. *Let $\mathcal{G}$ be a finite algebraic k-group, $\mathcal{N} \trianglelefteq \mathcal{G}$ be a normal subgroup. Then $k\mathcal{G}k\mathcal{N}^\dagger$ is a Hopf ideal, and $k(\mathcal{G}/\mathcal{N}) \cong k\mathcal{G}/(k\mathcal{G}k\mathcal{N}^\dagger)$.*

Proof. Consider the natural map $\mathcal{G} \times \mathcal{N} \longrightarrow \mathcal{N}$; $(g, n) \mapsto gng^{-1}$. By the universal property, there exists a unique linear map $k(\mathcal{G} \times \mathcal{N}) \longrightarrow k\mathcal{N}$ extending the above action. Direct computation, using $k(\mathcal{G} \times \mathcal{N}) \cong k\mathcal{G} \otimes_k k\mathcal{N}$, shows that this map is given by the adjoint representation. Accordingly,

$$h.x \in k\mathcal{N} \quad \forall \; h \in k\mathcal{G}, \; x \in k\mathcal{N}.$$

Given $x \in k\mathcal{N}^\dagger$ and $h \in k\mathcal{G}$, we therefore have

$$xh = \sum_{(h)} \varepsilon(h_{(1)}) x h_{(2)} = \sum_{(h)} h_{(1)} \eta(h_{(2)}) x h_{(3)} = \sum_{(h)} h_{(1)} (\eta(h_{(2)}).x) \in k\mathcal{G}k\mathcal{N}^\dagger.$$

Thus, $k\mathcal{G}k\mathcal{N}^\dagger$ is a two-sided ideal, and it readily follows that it is also a Hopf ideal.

Let $\pi : \mathcal{G} \longrightarrow \mathcal{G}/\mathcal{N}$ and $\iota : \mathcal{N} \hookrightarrow \mathcal{G}$ be the quotient map and the canonical embedding, respectively. Since $\pi^* : k[\mathcal{G}/\mathcal{N}] \longrightarrow k[\mathcal{G}]$ is injective, the induced map $\hat{\pi} : k\mathcal{G} \longrightarrow k(\mathcal{G}/\mathcal{N})$ is surjective. By the same token, $\hat{\iota} : k\mathcal{N} \longrightarrow k\mathcal{G}$ is injective. Since $\pi \circ \iota \equiv 1$, we have $\hat{\pi} \circ \hat{\iota} = \varepsilon$. Consequently, $k\mathcal{G}k\mathcal{N}^\dagger \subseteq \ker \hat{\pi}$.

Consider the cocommutative Hopf algebra $H := k\mathcal{G}/(k\mathcal{G}k\mathcal{N}^\dagger)$ as well as $\mathcal{H} := \operatorname{Spec}_k(H^*)$. There results a factorization $\hat{\pi} = \hat{\zeta} \circ \hat{\gamma}$, with a surjective homomorphism $\hat{\gamma} : k\mathcal{G} \longrightarrow H$ of Hopf algebras. Note that $\hat{\gamma}$ corresponds to a homomorphism $\gamma : \mathcal{G} \longrightarrow \mathcal{H}$. Since $\hat{\gamma} \circ \hat{\iota} = \varepsilon$, we have $\gamma \circ \iota \equiv 1$, whence $\mathcal{N} \subseteq \ker \gamma$. The universal property now provides a homomorphism $\omega : \mathcal{G}/\mathcal{N} \longrightarrow \mathcal{H}$ such that $\omega \circ \pi = \gamma$. Consequently,

$$(\hat{\omega} \circ \hat{\zeta}) \circ \hat{\gamma} = \hat{\omega} \circ \hat{\pi} = \hat{\gamma},$$

so that the surjectivity of $\hat{\gamma}$ yields $\hat{\omega} \circ \hat{\zeta} = \mathrm{id}_H$. As a result, $\hat{\zeta}$ is injective, so that $\ker \hat{\pi} = \ker(\hat{\zeta} \circ \hat{\gamma}) = \ker \hat{\gamma} = k\mathcal{G}k\mathcal{N}^\dagger$. □

2 Complexity and representation type

Let Λ be a finite-dimensional k-algebra. By the theorem of Krull-Remak-Schmidt, each finite dimensional Λ-module M decomposes in an essentially unique fashion into a direct sum

$$M = M_1 \oplus M_2 \oplus \cdots \oplus M_n$$

of indecomposables. Thus, when studying the category $\operatorname{mod}\Lambda$ of finite-dimensional Λ-modules, attention first focusses on the classification of indecomposable Λ-modules. The classical theory of semi-simple algebras is concerned with the case, where all indecomposables are simple. Wedderburn's Theorem then shows that the block decomposition of Λ is particularly nice: There exist division algebras $\Delta_1, \ldots, \Delta_r$ over k with

$$\Lambda \cong \operatorname{Mat}_{n_1}(\Delta_1) \oplus \cdots \oplus \operatorname{Mat}_{n_r}(\Delta_r).$$

Of course, determining the simple Λ-modules may still be rather difficult.

By definition, semi-simple algebras have (up to isomorphism) only finitely many indecomposables. In the 1950's, people working on representations of finite groups started to look at algebras with the latter property. Later, the notion of tameness was developed and it turned out that for algebras that are neither representation-finite nor tame, the abovementioned classification of indecomposable modules is rather hopeless.

Aside from providing the fundamental notions, this Section introduces two important tools: The Heller operator and the complexity of a module. Our basic references are [8, 4, 12, 13].

2.1 The Heller operator

Let Λ be a finite-dimensional self-injective algebra, $M \in \operatorname{mod}\Lambda$. Given a minimal projective resolution $(P_i, \partial_i)_{i\geq 0}$ of M, the syzygies $\Omega^n_\Lambda(M) := \ker \partial_{n-1}$ $(n \geq 1)$ are uniquely determined up to isomorphism. Hence $\Omega_\Lambda := \Omega^1_\Lambda$ is a well-defined operator on the isoclasses of Λ-modules. This operator is customarily referred to as the *Heller operator* or *loop space operator*. Directly from the definition we obtain $\Omega^m_\Lambda \circ \Omega^n_\Lambda = \Omega^{m+n}_\Lambda$ for $m, n \geq 0$. Note that Ω_Λ induces a functor on the *stable module category* $\underline{\operatorname{mod}}\,\Lambda$. This category has the same objects as $\operatorname{mod}\Lambda$ and morphisms

$$\underline{\operatorname{Hom}}_\Lambda(M, N) := \operatorname{Hom}_\Lambda(M, N)/P(M, N),$$

where $P(M, N)$ is the subspace of those homomorphisms that factor through a projective module.

Dually, we can construct for $n \geq 1$ operators Ω^{-n}_Λ by setting $\Omega^{-n}_\Lambda(M) := \operatorname{coker} \partial^{n-1}$, where $(E_i, \partial^i)_{i\geq 0}$ is a minimal injective resolution of M.

Given $M \in \operatorname{mod}\Lambda$, the theorem of Krull-Remak-Schmidt yields a decomposition

$$M = M_{\text{pf}} \oplus (\text{proj.}),$$

in which the first summand is the sum of all non-projective indecomposable constituents of M. In the following, we shall consider the full subcategory $(\operatorname{mod}\Lambda)_{\text{pf}}$

of $\operatorname{mod}\Lambda$ consisting of those finite-dimensional Λ-modules M for which $M = M_{\mathrm{pf}}$. Since Λ is self-injective, each object of $(\operatorname{mod}\Lambda)_{\mathrm{pf}}$ is characterized by the property that it admits no non-zero projective submodules.

Since $\Omega^n_\Lambda(M)$ is defined via a minimal projective resolution of M, Schanuel's Lemma implies that, given an arbitrary projective resolution $(P_i, \partial_i)_{i\geq 0}$ of M, the module $\Omega^n_\Lambda(M)$ is a direct summand of $\ker \partial_{n-1}$. Moreover, $\Omega^n_\Lambda(M) \in (\operatorname{mod}\Lambda)_{\mathrm{pf}}$ for all $n \in \mathbb{Z} \setminus \{0\}$, so that $\Omega^n_\Lambda(M) \cong (\ker \partial_{n-1})_{\mathrm{pf}}$. The objects of $(\operatorname{mod}\Lambda)_{\mathrm{pf}}$ are referred to as *projective-free* modules.

Lemma 2.1. *Let $M, N \in (\operatorname{mod}\Lambda)_{\mathrm{pf}}$. Then the following statements hold:*

(1) $\Omega^{-1}_\Lambda(\Omega_\Lambda(M)) \cong M \cong \Omega_\Lambda(\Omega^{-1}_\Lambda(M))$

(2) $\Omega_\Lambda(M \oplus N) \cong \Omega_\Lambda(M) \oplus \Omega_\Lambda(N)$.

(3) $\Omega^{-1}_\Lambda(M \oplus N) \cong \Omega^{-1}_\Lambda(M) \oplus \Omega^{-1}_\Lambda(N)$.

(4) *M is indecomposable if and only if $\Omega_\Lambda(M)$ is indecomposable.*

(5) *M is indecomposable if and only if $\Omega^{-1}_\Lambda(M)$ is indecomposable.*

Proof. (1) We show that the projective cover (P_M, ε_M) of M is an injective hull of $\Omega_\Lambda(M)$. We denote this injective hull by I. Since Λ is self-injective, P_M is an injective Λ-module. Hence there exists a Λ-linear map $f : I \longrightarrow P_M$ such that $f|_{\Omega_\Lambda(M)}$ is the canonical inclusion $\Omega_\Lambda(M) \hookrightarrow P_M$. Thus, f is injective, and there exists a Λ-linear map $\sigma : P_M \longrightarrow I$ such that $\sigma \circ f = \mathrm{id}_I$. We thus obtain a decomposition $P_M = (\ker \sigma) \oplus (\operatorname{im} f)$. Since $\ker \varepsilon_M = \Omega_\Lambda(M) \subseteq \operatorname{im} f$, the map $\varepsilon|_{\ker\sigma} : \ker \sigma \longrightarrow M$ is injective and M contains the injective module $\ker \sigma$. Hence $\ker \sigma$ is a direct summand of M and our assumption $M \in (\operatorname{mod}\Lambda)_{\mathrm{pf}}$ yields $\ker \sigma = (0)$. Consequently, $P_M \cong I$ is the injective hull of $\Omega_\Lambda(M)$, so that $\Omega^{-1}_\Lambda(\Omega_\Lambda(M)) \cong M$. The other assertion is proved analogously.

(4) Suppose M to be indecomposable, and write $\Omega_\Lambda(M) = X \oplus Y$. Since $\Omega^{-1}_\Lambda(\Omega_\Lambda(M)) \cong M$, (3) yields

$$M \cong \Omega^{-1}_\Lambda(X) \oplus \Omega^{-1}_\Lambda(Y),$$

so that we may assume without loss of generality that $\Omega^{-1}_\Lambda(X) = (0)$. As a result, $X = X_{\mathrm{pf}} \cong \Omega_\Lambda(\Omega^{-1}_\Lambda(X)) = (0)$, proving that $\Omega_\Lambda(M)$ is indecomposable. $\square$

Remark. The condition of self-injectivity is essential for the validity of (4). Consider for instance the three-dimensional local algebra $\Lambda := k[X, Y]/(X^2, XY, Y^2)$. If we define the Heller operator as above, then $\Omega_\Lambda(k) \cong \operatorname{Rad}(\Lambda)$ is isomorphic to the semi-simple Λ-module $k \oplus k$.

In the sequel, we shall write $M \cong N \oplus (\mathrm{proj.})$ to indicate that M is isomorphic to the direct sum of N and an unspecified projective module.

Lemma 2.2. *Let Λ be a finite-dimensional Hopf algebra, M, N be finite-dimensional Λ-modules. Then we have*

$$\Omega^r_\Lambda(M) \otimes_k \Omega^s_\Lambda(N) \cong \Omega^{r+s}_\Lambda(M \otimes_k N) \oplus (\mathrm{proj.})$$

for every $r, s \geq 0$.

Proof. We first consider the case, where $s = 0$. Let $(P_n, \partial_n)_{n\geq 0}$ be a minimal projective resolution of M. Thanks to Lemma 1.5, $(P_n \otimes_k N, \partial_n \otimes \mathrm{id}_N)_{n\geq 0}$ is a projective resolution of $M \otimes_k N$, so that our observations above imply

$$\Omega^r_\Lambda(M) \otimes_k N \cong (\ker \partial_{r-1}) \otimes_k N \cong \ker(\partial_{r-1} \otimes \mathrm{id}_N) \cong \Omega^r_\Lambda(M \otimes_k N) \oplus (\text{proj.}).$$

Applying the analogous result for $r = 0$, we obtain

$$\begin{aligned}\Omega^r_\Lambda(M) \otimes_k \Omega^s_\Lambda(N) &\cong \Omega^r_\Lambda(M \otimes_k \Omega^s_\Lambda(N)) \\ &\oplus(\text{proj.}) \cong \Omega^r_\Lambda(\Omega^s_\Lambda(M \otimes_k N) \oplus (\text{proj.})) \oplus (\text{proj.}) \\ &\cong \Omega^{r+s}_\Lambda(M \otimes_k N) \oplus (\text{proj.}),\end{aligned}$$

as desired. □

2.2 Complexity

The notion of the complexity of a module, first introduced by Alperin for group algebras of finite groups [1], has proven to be an effective tool in representation theory. Like the Heller operator, it makes methods from homological algebra amenable to applications.

Let $(a_i)_{i\geq 0}$ be a sequence of natural numbers. We call

$$\mathrm{gr}((a_i)_{i\geq 0}) := \min\{s \in \mathbb{N} \cup \{\infty\};\ \exists\ \lambda > 0 \text{ such that } a_n \leq \lambda n^{s-1} \quad \forall\ n \geq 1\}$$

the *polynomial rate of growth* of the sequence $(a_i)_{i\geq 0}$. If $\mathcal{V} := (V_i)_{i\geq 0}$ is a sequence of finite-dimensional k-vector spaces, then we write $\mathrm{gr}(\mathcal{V}) := \mathrm{gr}((\dim_k V_i)_{i\geq 0})$.

Throughout this section, Λ is assumed to be a finite-dimensional k-algebra.

Definition. Let M be a finite-dimensional Λ-module, $\mathcal{P} := (P_i, \partial_i)_{i\geq 0}$ be a minimal projective resolution of M. Then $\mathrm{cx}_\Lambda(M) := \mathrm{gr}(\mathcal{P})$ is called the *complexity* of M.

Remarks. (1) Since any two minimal projective resolutions are isomorphic, the complexity of a module is well-defined.

(2) The modules of complexity zero are precisely the modules of finite projective dimension. Thus, if Λ is self-injective, then a Λ-module M has complexity zero if and only if it is projective.

By our last observation, semi-simple algebras are characterized by the property that they are self-injective with all their modules having complexity zero. In the following sections we want to provide similar characterizations for algebras of finite- and tame representation types. We begin with an interpretation of the complexity in terms of extension groups. Let $\mathcal{S}$ be a complete set of representatives for the isomorphism classes of the simple Λ-modules. The projective cover of the simple Λ-module S will be denoted $P(S)$.

The following result, due to Alperin-Evens [3], relates the complexity of a module to the growth of certain Ext-groups.

Proposition 2.3. *Let M be a finite-dimensional Λ-module. Then*

$$\mathrm{cx}_\Lambda(M) = \max_{S\in\mathcal{S}} \mathrm{gr}((\mathrm{Ext}^n_\Lambda(M,S))_{n\geq 0}).$$

Proof. Given a minimal projective resolution $(P_n)_{n\geq 0}$ of M, we decompose each P_n into its indecomposable constituents and write $P_n \cong \bigoplus_{T\in\mathcal{S}} \ell_{n,T} P(T)$. Basic properties of Ext yield

$$\dim_k \mathrm{Ext}^n_\Lambda(M,S) = \sum_{T\in\mathcal{S}} \ell_{n,T} \dim_k \mathrm{Hom}_\Lambda(P(T),S) = \ell_{n,S} \dim_k \mathrm{Hom}_\Lambda(S,S)$$

for every $S \in \mathcal{S}$. Consequently,

$$\mathrm{cx}_\Lambda(M) = \max_{S\in\mathcal{S}} \mathrm{gr}((\ell_{n,S})_{n\geq 0}) = \max_{S\in\mathcal{S}} \mathrm{gr}((\mathrm{Ext}^n_\Lambda(M,S))_{n\geq 0}),$$

as desired. □

In certain cases, the rate of growth of a sequence $\mathcal{V} = (V_i)_{i\geq 0}$ of finite-dimensional k-vector spaces can also be characterized via power series. The *Poincaré series* $P_\mathcal{V}(t) \in k[[t]]$ is defined via

$$P_\mathcal{V}(t) = \sum_{i\geq 0} (\dim_k V_i)\, t^i.$$

If $\bigoplus_{i\geq 0} V_i$ is a finitely generated module, over a finitely generated, commutative, graded k-algebra $A = \bigoplus_{i\geq 0} A_i$, then the Hilbert-Serre Theorem provides a polynomial $f(t) \in k[t]$ such that

$$P_\mathcal{V}(t) = \frac{f(t)}{\prod_{j=1}^s (1 - t^{k_j})},$$

where $k_1, \ldots, k_s$ are the degrees homogeneous generators of A as an A_0-algebra. It then follows that $\mathrm{gr}(\mathcal{V})$ is the order of the pole of $P_\mathcal{V}(t)$ at $t = 1$, cf. [13, (5.3)] for more details.

We continue by collecting a few basic properties of the complexity of modules.

Proposition 2.4. *Let $\Gamma \subseteq \Lambda$ be a subalgebra, $M \in \mathrm{mod}\,\Lambda$, $N \in \mathrm{mod}\,\Gamma$.*

(1) *If $S_1, \ldots, S_n$ are the composition factors of M, then*

$$\mathrm{cx}_\Lambda(M) \leq \max\nolimits_{1\leq i\leq n} \mathrm{cx}_\Lambda(S_i).$$

(2) *If Λ is a projective left Γ-module, then $\mathrm{cx}_\Gamma(M) \leq \mathrm{cx}_\Lambda(M)$.*

(3) *If Λ is a projective right Γ-module, then $\mathrm{cx}_\Lambda(\Lambda \otimes_\Gamma N) \leq \mathrm{cx}_\Gamma(N)$.*

(4) *If Λ is a Hopf algebra, then $\mathrm{cx}_\Lambda(M) \leq \mathrm{cx}_\Lambda(k)$.*

Proof. (2),(3) Let $\mathcal{P} := (P_i)_{i\geq 0}$ be a minimal projective resolution of the Λ-module M. By assumption, the functor $\mathrm{Hom}_\Gamma(P_i, -) \cong \mathrm{Hom}_\Lambda(P_i, \mathrm{Hom}_\Gamma(\Lambda, -))$ is exact. Hence each P_i is a projective Γ-module, and $\mathrm{cx}_\Gamma(M) \leq \mathrm{gr}(\mathcal{P}) = \mathrm{cx}_\Lambda(M)$.

If $\mathcal{Q} := (Q_i)_{i\geq 0}$ is a minimal projective resolution of N, then the adjoint isomorphism

$$\mathrm{Hom}_\Lambda(\Lambda \otimes_\Gamma Q_i, -) \cong \mathrm{Hom}_\Gamma(Q_i, -)$$

yields the exactness of the left-hand functor, so that $\Lambda \otimes_\Gamma \mathcal{Q} := (\Lambda \otimes_\Gamma Q_i)_{i\geq 0}$ is a projective resolution of the induced module $\Lambda \otimes_\Gamma N$. Consequently, $\mathrm{cx}_\Lambda(\Lambda \otimes_\Gamma N) \leq \mathrm{gr}(\Lambda \otimes_\Gamma \mathcal{Q}) \leq \mathrm{gr}((\Lambda \otimes_k Q_i)_{i\geq 0}) = \mathrm{cx}_\Gamma(N)$.

(4) Let $\mathcal{P} := (P_i)_{i\geq 0}$ be a minimal projective resolution of the trivial Λ-module k. Thanks to Lemma 1.5, the complex $\mathcal{P} \otimes_k M := (P_i \otimes_k M)_{i\geq 0}$ is a projective resolution of M. Hence $\mathrm{cx}_\Lambda(M) \leq \mathrm{gr}(\mathcal{P} \otimes_k M) = \mathrm{cx}_\Lambda(k)$. □

Example. Suppose that $\mathrm{char}(k) = p > 0$, and consider the Hopf algebra

$$k\mathbb{G}_{a(2)} \cong k[X, Y]/(X^p, Y^p) \cong k[X]/(X^p) \otimes_k k[Y]/(Y^p).$$

Let $\mathcal{P} = (P_i)_{i\geq 0}$ be a minimal projective resolution of the trivial $k[X]/(X^p)$-module k, i.e., $P_i = k[X]/(X^p)$ for every $i \geq 0$. Setting $Q_i := \sum_{j=0}^{i} P_j \otimes_k P_{i-j}$, we obtain a minimal projective resolution $\mathcal{Q} := (Q_i)_{i\geq 0}$ of the $k\mathbb{G}_{a(2)}$-module $k \otimes_k k \cong k$. Since $\dim_k Q_i = (i+1)p^2$, we have $\mathrm{cx}_{k\mathbb{G}_{a(2)}}(k) = 2$. One can iterate this process to see that $\mathrm{cx}_{k\mathbb{G}_{a(r)}}(k) = r$.

In view of Proposition 2.3, we also have

$$\mathrm{cx}_{k\mathbb{G}_{a(r)}}(k) = \mathrm{gr}(\mathrm{H}^*(\mathbb{G}_{a(r)}, k)).$$

The structure of the cohomology ring $\mathrm{H}^*(\mathbb{G}_{a(r)}, k)$ is well-understood: For $p > 2$ there exists an isomorphism

$$\begin{aligned} \mathrm{H}^*(\mathbb{G}_{a(r)}, k) &\cong k[X_1, \ldots X_r, Y_1, \ldots, Y_r]/(Y_1^2, \ldots, Y_r^2) \\ &\cong \bigotimes_{i=1}^{r} k[X_i] \otimes_k \bigotimes_{j=1}^{r} k[Y_j]/(Y_j^2), \end{aligned}$$

of graded vector spaces, where $\deg(X_i) = 2$ and $\deg(Y_i) = 1$, see [22, (7.6)]. As a result, the Poincaré series of $\mathrm{H}^*(\mathbb{G}_{a(r)}, k)$ has the form

$$P(t) = \prod_{i=1}^{r} P_{k[X_i]}(t) \prod_{j=1}^{r} P_{k[Y_j]/(Y_j^2)}(t) = \prod_{i=1}^{r} \frac{1}{1-t^2} \prod_{j=1}^{r} (1+t) = \frac{1}{(1-t)^r},$$

so that $\mathrm{gr}(\mathrm{H}^*(\mathbb{G}_{a(r)}, k)) = r$.

As noted earlier, the algebra $k\mathbb{G}_{a(r)}$ is isomorphic to the group algebra $k(\mathbb{Z}/(p))^r$ of the p-elementary abelian group of rank r. Such groups play an important role in the representation theory of finite groups.

Theorem 2.5 ([3]). *Suppose that* $\mathrm{char}(k) = p > 0$. *Let G be a finite group, M be a finite-dimensional G-module. Then*

$$\mathrm{cx}_{kG}(M) = \max_{E \in \mathfrak{E}} \mathrm{cx}_{kE}(M),$$

where $\mathfrak{E}$ is the set of p-elementary abelian subgroups of G. □

Definition. A finite-dimensional non-projective indecomposable Λ-module M is said to be *periodic* if there exists $n > 0$ such that $\Omega^n_\Lambda(M) \cong M$.

Examples. (1) Suppose that $\mathrm{char}(k) = p > 0$. Let $\Lambda = k\mathbb{G}_{a(1)} = k[X]/(X^p)$ and write $x := X + (X^p)$. Since

$$(0) \longrightarrow kx^{p-1} \longrightarrow k[X]/(X^p) \xrightarrow{x\cdot} k[X]/(X^p) \xrightarrow{\varepsilon} k \longrightarrow (0)$$

defines the initial terms of a minimal projective resolution of the trivial module k, we have $\Omega^2_{k\mathbb{G}_{a(1)}}(k) \cong k$.

(2) Let $\mathfrak{b} := kt \oplus kx$ be the two-dimensional non-abelian restricted Lie algebra with product and p-map given by

$$[t, x] = x \quad \text{and} \quad t^{[p]} = t; \quad x^{[p]} = 0,$$

respectively. The simple $U_0(\mathfrak{b})$-modules are of the form k_λ, where the character $\lambda : U_0(\mathfrak{b}) \longrightarrow k$ is defined via

$$\lambda(x) = 0; \quad \lambda(t) \in \mathbb{F}_p.$$

We therefore identify the character group of $U_0(\mathfrak{g})$ with $\mathbb{Z}/(p)$, and write the convolution as addition. Let $\underline{1} : U_0(\mathfrak{b}) \longrightarrow k$ be the character defined by $\underline{1}(x) = 0$; $\underline{1}(t) = 1$. Since $P(\lambda) := U_0(\mathfrak{g}) \otimes_{U_0(kt)} k_\lambda$ is indecomposable projective, with $\mathrm{Rad}(P(\lambda))/\mathrm{Rad}^2(P(\lambda)) \cong k_{\lambda+\underline{1}}$ and $\mathrm{Soc}(P(\lambda)) \cong k_{\lambda-\underline{1}}$, there results an exact sequence

$$(0) \longrightarrow k_\lambda \longrightarrow P(\lambda + \underline{1}) \longrightarrow P(\lambda) \longrightarrow k_\lambda \longrightarrow (0),$$

whence $\Omega^2_{U_0(\mathfrak{b})}(k_\lambda) \cong k_\lambda$.

(3) Assume that $p \geq 3$, and consider the restricted Lie algebra $\mathfrak{g} = \mathfrak{sl}(2)$. The p-subalgebra $\mathfrak{b} := kh \oplus ke$ is isomorphic to the algebra of (2). Given $\lambda : U_0(\mathfrak{b}) \longrightarrow k$, we let

$$Z(\lambda) := U_0(\mathfrak{g}) \otimes_{U_0(\mathfrak{b})} k_\lambda$$

be the baby Verma module with highest weight λ. Since $U_0(\mathfrak{g})$ is a projective right $U_0(\mathfrak{b})$-module, the exact sequence of (2) yields a short exact sequence

$$\begin{aligned}(0) \longrightarrow Z(\lambda) \longrightarrow\ & U_0(\mathfrak{g}) \otimes_{U_0(\mathfrak{b})} P(\lambda + \underline{1}) \\ \longrightarrow\ & U_0(\mathfrak{g}) \otimes_{U_0(\mathfrak{b})} P(\lambda) \longrightarrow Z(\lambda) \longrightarrow (0),\end{aligned}$$

whose two middle terms are projective. Consequently,

$$\Omega^2_{U_0(\mathfrak{g})}(Z(\lambda)) \oplus (\text{proj.}) \cong Z(\lambda).$$

As $Z(\lambda)$ is indecomposable, it is thus either periodic or projective. It turns out that projectivity occurs precisely when $\lambda(t) = p - 1$. The corresponding Verma module is the *Steinberg module* St.

Remark. Note that every periodic Λ-module M has complexity $\mathrm{cx}_\Lambda(M) \leq 1$. Indeed, periodicity implies the existence of a minimal projective resolution $(P_i, \partial_i)_{i\geq 0}$ satisfying $P_{i+r} \cong P_i$ for some $r > 0$. Hence the dimensions of the P_i are bounded, and $\mathrm{cx}_\Lambda(M) \leq 1$. The converse is not true in general, as the example of the four-dimensional truncated quantum plane $\Lambda_q := k\langle x, y\rangle/(\{yx - qxy, x^2, y^2\})$ shows: The algebra Λ_q is a Frobenius algebra and the unique automorphism $\mu : \Lambda_q \longrightarrow \Lambda_q$ with $x \mapsto qx$ and $y \mapsto q^{-1}y$ is a Nakayama automorphism. Given $t \in k \setminus \{0\}$, we consider the two-dimensional indecomposable Λ_q-module $M_t := \Lambda_q(x+ty)$. Direct computation shows $\Omega_{\Lambda_q}(M_t) \cong M_{-q^{-1}t}$. Consequently, $\mathrm{cx}_{\Lambda_q}(M_t) = 1$, while M_t is not periodic whenever q is not a root of unity.

2.3 Representation type

In this section we subdivide the category of finite-dimensional associative algebras according to their representation type, a notion that provides a measure of how complicated the module category $\mathrm{mod}\,\Lambda$ of a k-algebra Λ is. For self-injective algebras the representation type turns out to be related to upper bounds on the complexity of modules.

Definition. An algebra Λ is *representation-finite* or of *finite representation type* if it admits only finitely many isoclasses of finite-dimensional indecomposable modules.

From linear algebra we know that the indecomposable modules of the truncated polynomial ring $k[X]/(X^n)$ are cyclic, so that such an algebra has finite representation type. In fact, the truncated polynomial rings in one variable belong to a particularly tractable class of representation-finite algebras, the so-called Nakayama algebras.

Given a finite-dimensional Λ-module M, the *Loewy length* $\ell\ell(M)$ is defined via

$$\ell\ell(M) := \min\{i \geq 0;\ \mathrm{Rad}(\Lambda)^i M = (0)\}.$$

We say that M is *uniserial* if it possesses exactly one composition series. Equivalently, the *Loewy series* $(\mathrm{Rad}(\Lambda)^i M)_{i\geq 0}$ is a composition series of M.

Definition. An algebra Λ is called a *Nakayama algebra* if all of its projective indecomposable modules and all of its injective indecomposable modules are uniserial.

The usual duality between injective left Λ-modules and projective right Λ-modules implies that Λ is a Nakayama algebra if and only if the projective indecomposable modules for Λ and its opposite algebra Λ^{op} are uniserial. Consequently, factor algebras of Nakayama algebras by powers of their radicals are also Nakayama algebras.

By work of Kupisch [84], Nakayama algebras are well-understood. We will return to the description of their Morita equivalence classes once we have discussed path algebras of quivers. For the moment we record the following basic result.

Proposition 2.6. *Let Λ be a Nakayama algebra. Then every indecomposable Λ-module is uniserial, and Λ has finite representation type.*

Proof. Let J be the radical of Λ. We prove the first assertion by induction on the Loewy length $\ell\ell(\Lambda)$ of Λ, the case $\ell\ell(\Lambda) = 1$ being trivial. Assuming $\ell := \ell\ell(\Lambda) \geq 2$, we consider an indecomposable Λ-module M. If $J^{\ell-1}M = (0)$, then M is an indecomposable module for the Nakayama algebra $\Lambda/J^{\ell-1}$, and the inductive hypothesis yields the assertion. Alternatively, there exists a simple left ideal $S \subseteq J^{\ell-1}$ with $SM \neq (0)$. We can therefore find $m \in M \setminus \{0\}$ such that

$$\psi_m : S \longrightarrow M; \quad s \mapsto s.m$$

is injective. Hence there is a map $\hat{\psi}_m : M \longrightarrow E(S)$ to the injective envelope $E(S)$ of S, whose composite with ψ_m is the canonical inclusion $S \hookrightarrow E(S)$. As $E(S)$ is uniserial, we can find $i \geq 0$ with $\hat{\psi}_m(M) = J^iE(S)$. Consequently, $J^{\ell-i}M \subseteq \ker \hat{\psi}_m$, while $J^{\ell-1}M \not\subseteq \ker \hat{\psi}_m$. As a result, $i = 0$, so that $\hat{\psi}_m$ is surjective and $J^{\ell-1}E(S) \neq (0)$. Since the uniserial projective cover $\pi : P \longrightarrow E(S)$ of $E(S)$ satisfies $\ell(P) = \ell\ell(P) \leq \ell = \ell\ell(E(S)) = \ell(E(S))$, we have $P \cong E(S)$. As M is indecomposable, it now follows that $\hat{\psi}_m$ is an isomorphism. Thus, M is uniserial.

As an upshot of the above, every indecomposable Λ-module M has a simple top and is thus of the form

$$M \cong P(S)/J^iP(S); \quad 0 \leq i \leq \ell\ell(\Lambda),$$

for some simple Λ-module S. Consequently, Λ has finite representation type. □

Remark. Let Λ be a local algebra of finite representation type. Then $\dim_k \mathrm{Rad}(\Lambda)/\mathrm{Rad}^2(\Lambda) \leq 1$, so that Λ is a truncated polynomial ring $k[X]/(X^n)$. In particular, every commutative algebra of finite representation type is a direct sum of truncated polynomial rings: $\Lambda \cong \bigoplus_{i=1}^m k[X]/(X^{n_i})$.

Here is a simple necessary condition for a self-injective algebra to have finite representation type.

Theorem 2.7 (Heller [68]). *Let Λ be self-injective. If Λ has finite representation type, then every indecomposable non-projective Λ-module is periodic.*

Proof. Let $\mathfrak{X}$ be the set of isomorphism types of the non-projective indecomposable Λ-modules. Owing to Lemma 2.1, the map $\Omega_\Lambda|_\mathfrak{X}$ is bijective. Since $\mathfrak{X}$ is finite, there exists $n \geq 1$ such that $\Omega^n_\Lambda|_\mathfrak{X} = \mathrm{id}_\mathfrak{X}$. This implies our result. □

The converse of Heller's Theorem does not hold in general. For instance, the group algebra of the quaternion group

$$Q := \langle x, y; \ \ x^2 = y^2, \ yxy^{-1} = x^{-1}\rangle$$

of order 8 over a field of characteristic 2 is known to possess only periodic modules: One verifies $\Omega^4_{kQ}(k) \cong k$ (see [25, (XII.7)]) and applies Lemma 2.2 to obtain

$\Omega^4_{kQ}(M) \cong M$ for every non-projective indecomposable kQ-module. By the above remark, the non-commutative local algebra kQ does not have finite representation type (in fact, it is tame). In Section 6 we shall see that distribution algebras of infinitesimal groups afford the converse of Heller's Theorem.

Among the algebras of infinite representation type one distinguishes between those, whose indecomposable modules can be parametrized by one-dimensional varieties and those requiring more parameters.

Definition. A k-algebra Λ is said to be *tame* if it is not representation-finite, and if for every $d > 0$ there exist $(\Lambda, k[X])$-bimodules $M_1, \ldots, M_{n(d)}$ that are finitely generated and free over $k[X]$, so that all but finitely many isoclasses of the d-dimensional indecomposable Λ-modules are of the form $[M_i \otimes_{k[X]} k_\lambda]$ for some $i \in \{1, \ldots, n(d)\}$ and $\lambda \in \mathrm{Alg}_k(k[X], k)$.

Remark. The tame local algebras were classified by Ringel [106]. The list of the tame symmetric local algebras, which plays a prominent role in the classification of the tame blocks of group algebras of finite groups, can be found in [36, Chap.III].

An easily described class of tame algebras is given by the so-called biserial algebras. If M is a non-simple, indecomposable Λ-module, then $\mathrm{Soc}(M) \subseteq \mathrm{Rad}(M)$, and we let

$$\mathrm{Ht}(M) := \mathrm{Rad}(M)/\,\mathrm{Soc}(M)$$

be the *heart* of M.

Definition. A self-injective algebra Λ is *biserial* if for every non-simple projective indecomposable Λ-module P the heart $\mathrm{Ht}(P)$ is a direct sum of at most two uniserial modules.

We will frequently encounter biserial algebras when studying infinitesimal groups of tame representation type. In a more classical context, any Brauer tree algebra (that is, any representation-finite block of a group algebra kG) is biserial (cf. [2] for more details). For the moment we record the following:

Theorem 2.8 ([29]). *A biserial algebra is tame or representation-finite.* □

The notion of a wild algebra, introduced by Drozd [33], captures a phenomenon first observed by Corner and Brenner: Given any k-algebra Λ, there exists a full exact embedding

$$\mathfrak{F}_\Lambda : \mathrm{mod}\,\Lambda \longrightarrow \mathrm{mod}\,k\langle x, y\rangle$$

from the module category of Λ to the module category of the free algebra $k\langle x, y\rangle$ with two generators.

Definition. A k-algebra Λ is referred to as *wild* if there exists a $(\Lambda, k\langle x, y\rangle)$-bimodule M, that is finitely generated and free over $k\langle x, y\rangle$, such that the functor

$$\mathrm{mod}\,k\langle x, y\rangle \longrightarrow \mathrm{mod}\,\Lambda; \quad X \mapsto M \otimes_{k\langle x,y\rangle} X$$

preserves indecomposables and reflects isomorphisms.

Thus, if Λ is wild and Γ is any k-algebra, then the functor

$$\operatorname{mod}\Gamma \longrightarrow \operatorname{mod}\Lambda; \quad V \mapsto M \otimes_{k\langle x,y\rangle} \mathfrak{F}_\Gamma(V)$$

induces an injection between the isomorphism classes of indecomposable modules for Γ and Λ, respectively. Accordingly, a complete classification of the indecomposable modules of a wild algebra is a rather hopeless endeavour.

Theorem 2.9 (Drozd (cf. [33, 28])). *Suppose that k is algebraically closed. Then Λ is either representation-finite, tame or wild.* □

Since periodic modules have complexity 1, Theorem 2.7 implies that

$$\operatorname{cx}_\Lambda(M) \leq 1 \quad \forall\, M \in \operatorname{mod}\Lambda,$$

whenever Λ has finite representation type. There is a similar criterion for certain algebras of tame representation type. This result requires geometric techniques that will be discussed in the following section. Its proof is considerably harder as it employs deep results by Crawley-Boevey [28] concerning the structure of the Auslander-Reiten quiver of tame algebras. Crawley-Boevey's Theorem implies the following result, which will be sufficient for our purposes.

Theorem 2.10. *Let k be algebraically closed and suppose that Λ is self-injective and tame. Then, for any $d > 0$, all but finitely many isoclasses of the d-dimensional indecomposable modules have complexity* 1. □

Note that for algebras of finite global dimension the statement of Theorem 2.10 provides no information.

3 Support varieties and support spaces

In this section we outline the geometric approach to the representation theory of finite algebraic groups. As before, we will be working over a field k of characteristic $p > 0$.

Given a module M of a commutative ring R, we recall that the *support*

$$\operatorname{Supp}(M) := \{\mathfrak{M} \in \operatorname{Max}(R);\ M_\mathfrak{M} \neq (0)\}$$

of M is the set of those maximal ideals $\mathfrak{M} \trianglelefteq R$ for which the localization of M at $\mathfrak{M}$ is not trivial. If R is noetherian and M is finitely generated, then

$$\operatorname{Supp}(M) = Z(\operatorname{ann}_R(M)) := \{\mathfrak{M} \in \operatorname{Max}(R);\ \operatorname{ann}_R(M) \subseteq \mathfrak{M}\}$$

is the zero locus of the annihilator of M. Thus, if R is an affine k-algebra, then $\operatorname{Supp}(M)$ is an affine variety. In our context, the even cohomology ring $\mathrm{H}^\bullet(\mathcal{G}, k)$ of a finite group scheme $\mathcal{G}$ assumes the role of R, with M being given by the cohomology space $\mathrm{H}^*(\mathcal{G}, \operatorname{Hom}_k(N, N))$ associated to a $\mathcal{G}$-module N.

3.1 The Friedlander-Suslin Theorem

Let $\mathcal{G}$ be a finite algebraic k-group. If M is a $\mathcal{G}$-module, we denote by

$$\mathrm{H}^n(\mathcal{G}, M) := \mathrm{Ext}^n_{\mathcal{G}}(k, M) \qquad (n \geq 0)$$

the *n-th cohomology group of $\mathcal{G}$ with coefficients in M*. Note that these groups are just the Hochschild cohomology groups of the augmented algebra $(k\mathcal{G}, \varepsilon)$.

For infinitesimal groups, these cohomology groups were first studied by Hochschild [70] in the context of restricted Lie algebras. He related them to the Chevalley-Eilenberg cohomology of the underlying Lie algebra and provided interpretations for H^1 and H^2. Further early results can be found in [93].

Given three $\mathcal{G}$-modules X, Y, Z, we recall the *Yoneda product*

$$\mathrm{Ext}^m_{\mathcal{G}}(Y, Z) \times \mathrm{Ext}^n_{\mathcal{G}}(X, Y) \longrightarrow \mathrm{Ext}^{m+n}_{\mathcal{G}}(X, Z).$$

This product endows $\mathrm{Ext}^*_{\mathcal{G}}(X, X) := \bigoplus_{n\geq 0} \mathrm{Ext}^n_{\mathcal{G}}(X, X)$ with the structure of a $\mathbb{Z}$-graded k-algebra. Moreover, the spaces $\mathrm{Ext}^*_{\mathcal{G}}(Y, X)$ and $\mathrm{Ext}^*_{\mathcal{G}}(X, Y)$ are graded left and right $\mathrm{Ext}^*_{\mathcal{G}}(X, X)$-modules, respectively. In particular, $\mathrm{H}^*(\mathcal{G}, M)$ is a graded right module over the *cohomology ring* $\mathrm{H}^*(\mathcal{G}, k)$. This ring is known to be *graded commutative*, i.e., we have

$$yx = (-1)^{\deg(x)\deg(y)}xy$$

for any two homogeneous elements $x, y \in \mathrm{H}^*(\mathcal{G}, k)$. Consequently, the subring

$$\mathrm{H}^\bullet(\mathcal{G}, k) := \begin{cases} \bigoplus_{i\geq 0} \mathrm{H}^{2i}(\mathcal{G}, k) & \text{if } p > 2, \\ \bigoplus_{i\geq 0} \mathrm{H}^{i}(\mathcal{G}, k) & \text{if } p = 2 \end{cases}$$

is a commutative, $\mathbb{Z}$-graded k-algebra (see [22, §6] for details).

The following result by Friedlander and Suslin, which generalizes earlier work by Venkov [121] and Evens [37] for finite groups, and Friedlander-Parshall [59] for infinitesimal groups of height ≤ 1, is fundamental for everything that follows.

Theorem 3.1 (Friedlander-Suslin [63]). *Let $\mathcal{G}$ be a finite algebraic k-group, M be a finite-dimensional $\mathcal{G}$-module. Then the following statements hold:*

(1) *$\mathrm{H}^\bullet(\mathcal{G}, k)$ is a finitely generated k-algebra.*

(2) *$\mathrm{H}^*(\mathcal{G}, M)$ is a finitely generated $\mathrm{H}^\bullet(\mathcal{G}, k)$-module.* □

In some cases, the cohomology ring $\mathrm{H}^\bullet(\mathcal{G}, k)$ can be computed explicitly. If $\mathcal{G}$ is smooth, semi-simple and simply connected and p exceeds the Coxeter number of $\mathcal{G}$, then $\mathrm{H}^\bullet(\mathcal{G}_1, k) \cong k[\mathcal{N}]$, where $\mathcal{N} := \{x \in \mathrm{Lie}(\mathcal{G});\ x^{[p]^n} = 0 \text{ for some } n \in \mathbb{N}\}$ is the *nilpotent cone* of $\mathrm{Lie}(\mathcal{G})$ (see [58]).

Let M be a finite-dimensional $\mathcal{G}$-module, $(P_i, \partial_i)_{i\geq 0}$ be a projective resolution of the trivial $\mathcal{G}$-module k. Since $(P_i \otimes_k M, \partial_i \otimes \mathrm{id}_M)_{i\geq 0}$ is a projective resolution of M, we obtain a homomorphism

$$\Phi_M : \mathrm{H}^\bullet(\mathcal{G}, k) \longrightarrow \mathrm{Ext}^*_{\mathcal{G}}(M, M);\quad [f] \mapsto [f \hat{\otimes} \mathrm{id}_M]$$

of graded k-algebras. In view of the standard isomorphism $\mathrm{Ext}^*_{\mathcal{G}}(M,M) \cong \mathrm{H}^*(\mathcal{G}, \mathrm{Hom}_k(M,M))$, Theorem 3.1 implies that Φ_M endows the Yoneda algebra with the structure of a finitely generated $\mathrm{H}^\bullet(\mathcal{G},k)$-module. We define the *cohomological support variety* of M via

$$\mathcal{V}_{\mathcal{G}}(M) := Z(\ker \Phi_M) \subseteq \mathrm{Max}(\mathrm{H}^\bullet(\mathcal{G},k)).$$

Since $\ker \Phi_M$ is a homogeneous ideal, the affine variety $\mathcal{V}_{\mathcal{G}}(M)$ is conical.

Our first result concerns the interpretation of the dimension of a support variety. Let

$$A = \bigoplus_{n\geq 0} A_n$$

be a finitely generated, commutative, graded k-algebra. We want to find the growth $\mathrm{gr}(A)$ of the sequence $(A_n)_{n\geq 0}$. By the Noether Normalization Lemma, there exists a graded subalgebra

$$R = \bigoplus_{n\geq 0} R_n$$

of A such that

(a) A is a finitely generated R-module, and

(b) $R \cong k[X_1, \dots, X_\ell]$, where $\deg(X_i) = d$ for some $d \geq 1$.

Owing to (a), we have $\mathrm{gr}(R) = \mathrm{gr}(A)$, while property (b) implies $\mathrm{gr}(R) = \ell$. The number ℓ is the *Krull dimension* of R, which, by the Cohen-Seidenberg Theorems coincides with the Krull dimension $\dim A$ of A. We therefore obtain

$$\mathrm{gr}(A) = \dim A.$$

In the sequel, we write $\mathrm{cx}_{\mathcal{G}}(M) := \mathrm{cx}_{k\mathcal{G}}(M)$ for any $\mathcal{G}$-module M.

Lemma 3.2. *Let $\mathcal{G}' \subseteq \mathcal{G}$ be a subgroup of the finite algebraic group $\mathcal{G}$, $M \in \mathrm{mod}\,\mathcal{G}$ be a $\mathcal{G}$-module.*

(1) $\dim \mathcal{V}_{\mathcal{G}}(M) = \mathrm{cx}_{\mathcal{G}}(M)$.

(2) $\dim \mathcal{V}_{\mathcal{G}'}(M) \leq \dim \mathcal{V}_{\mathcal{G}}(M)$.

Proof. (1) Let $\mathcal{S}$ be a complete set of representatives for the isoclasses of the simple $\mathcal{G}$-modules. Since the algebra $\mathrm{H}^\bullet(\mathcal{G},k)$ is finitely generated, we have

$$\begin{aligned}\dim \mathcal{V}_{\mathcal{G}}(M) &= \dim \mathrm{H}^\bullet(\mathcal{G},k)/\ker\Phi_M = \mathrm{gr}(\mathrm{H}^\bullet(\mathcal{G},k)/\ker\Phi_M)\\ &\leq \mathrm{gr}(\mathrm{Ext}^*_{\mathcal{G}}(M,M)) \leq \max_{S\in\mathcal{S}} \mathrm{gr}(\mathrm{Ext}^*_{\mathcal{G}}(M,S)),\end{aligned}$$

where the last inequality follows from the long exact sequence in cohomology. In view of Proposition 2.3, the latter number coincides with $\mathrm{cx}_{\mathcal{G}}(M)$.

To verify the reverse inequality, we let $S \in \mathcal{S}$ be a simple $\mathcal{G}$-module. Thanks to Theorem 3.1, the space $\mathrm{Ext}^*_{\mathcal{G}}(M,S) \cong \mathrm{H}^*(\mathcal{G}, \mathrm{Hom}_k(M,S))$ is a finitely generated $\mathrm{H}^\bullet(\mathcal{G},k)$-module. Since the action is induced by the scalar multiplication $\mathrm{Hom}_k(M,S) \times k \longrightarrow \mathrm{Hom}_k(M,S)$, it factors through the action $\mathrm{Hom}_k(M,S) \times$

$\mathrm{Hom}_k(M,M) \longrightarrow \mathrm{Hom}_k(M,S)$. As a result, $\ker \Phi_M$ annihilates $\mathrm{Ext}^*_{\mathcal{G}}(M,S)$. Consequently,

$$\mathrm{gr}(\mathrm{Ext}^*_{\mathcal{G}}(M,S)) \leq \mathrm{gr}(\mathrm{H}^\bullet(\mathcal{G},k)/\ker \Phi_M) = \dim \mathcal{V}_{\mathcal{G}}(M),$$

so that another application of Proposition 2.3 yields $\mathrm{cx}_{\mathcal{G}}(M) \leq \dim \mathcal{V}_{\mathcal{G}}(M)$, as desired.

(2). This follows directly from (1), Theorem 1.6 and Proposition 2.4. □

As an immediate consequence, we record the following basic criteria for blocks of finite representation type.

Corollary 3.3. *Let $\mathcal{B} \subseteq k\mathcal{G}$ be a representation-finite block. Then we have $\dim \mathcal{V}_{\mathcal{G}}(M) \leq 1$ for every $M \in \mathrm{mod}\,\mathcal{B}$.* □

Example. Suppose that $p \geq 3$. As noted before, the Künneth formula furnishes an isomorphism

$$\mathrm{H}^*(\mathbb{G}_{a(r)},k) \cong k[X_1,\ldots,X_r] \otimes_k \Lambda(Y_1,\ldots,Y_r),$$

where the X_i and Y_i have degrees 2 and 1, respectively (cf. [22, (7.6)]). Consequently, $k[X_1,\ldots,X_r]$ is a Noether normalization of $\mathrm{H}^\bullet(\mathbb{G}_{a(r)},k)$ and $\dim \mathcal{V}_{\mathbb{G}_{a(r)}}(k) = r$. Note that this agrees with our earlier observations.

3.2 The space of p-points

We have just seen that the dimension $\dim \mathcal{V}_{\mathcal{G}}(M)$ of the support variety of a $\mathcal{G}$-module M is an important invariant. Unfortunately, support varieties are usually hard to compute. It is therefore desirable to find representation-theoretic realizations of these varieties that do not resort to the computation of cohomology groups. The so-called *rank varieties*, that were first defined in the context of finite groups [20, 21], and later generalized successively to restricted Lie algebras [57, 59] and infinitesimal groups [116, 117], turn out to have all the requisite properties. In recent work [60, 61], Friedlander and Pevtsova have unified and extended the foregoing approaches by considering p-points and π-points of finite group schemes. Since their work is very much motivated by the historical origins pertaining to finite groups, we briefly review the by now classical definition of a rank variety.

Throughout this section, k is assumed to be a field of characteristic $\mathrm{char}(k) = p > 0$. Early work by Quillen [100, 101] showed that the support variety of the trivial module of a finite group G may be described as the union of the corresponding supports for the p-elementary abelian subgroups of G. In particular, the dimension of $\mathcal{V}_G(k)$ coincides with the maximal rank of all p-elementary abelian subgroups of G, the so-called *p-rank $r_p(G)$ of G*:

Theorem 3.4 (Quillen's Dimension Theorem,[100]).

$$\dim \mathcal{V}_G(k) = r_p(G).$$

Elementary abelian subgroups also arise in a different context, namely in connection with the detection of projectivity.

If M is a projective G-module and $G' \subseteq G$ is a subgroup, then the restriction $M|_{G'}$ is also projective. The question is, on which class of subgroups can projectivity effectively be detected. It is fairly easy to see that a module is projective, if its restriction $M|_P$ to a Sylow-p-subgroup enjoys the same property. The following refinement of this result is much harder to prove, see [3]:

Theorem 3.5 (Chouinard,[26]). *A G-module M is projective if and only if its restriction $M|_E$ to every p-elementary abelian subgroup $E \subseteq G$ is projective.* □

Let $E \cong \mathbb{Z}/(p)^r$ be p-elementary abelian with generators $g_1, \ldots, g_r$. Setting $x_i := g_i - 1$, the map $X_i \longrightarrow x_i$ induces an isomorphism

$$k[X_1, \ldots, X_r]/(X_1^p, \ldots, X_r^p) \xrightarrow{\sim} kE.$$

For $\lambda = (\lambda_1, \ldots, \lambda_r) \in k^r$, we put $u_\lambda := \sum_{i=1}^r \lambda_i x_i$. The following result, which requires the ground field to be algebraically closed, provides a criterion for the projectivity of an E-module:

Lemma 3.6 (Dade's Lemma,[30]). *Suppose that k is algebraically closed. The E-module M is projective if and only if $M|_{k[u_\lambda]}$ is projective for every $\lambda \in k^r$.* □

Example. Suppose that k is not algebraically closed and let $K : k$ be a finite field extension, $\alpha \in K \setminus k$. Consider the group $E := \mathbb{Z}/(p) \times \mathbb{Z}/(p)$. Then $kE \cong k[X, Y]/(X^p, Y^p)$, with canonical generators x, y. We consider the k-space $M := K^p$ and define a kE-action on M via the $(p \times p)$-matrices

$$x_M := \sum_{i=1}^{p-1} E_{i,i+1} \quad \text{and} \quad y_M := \alpha x_M.$$

If $(\lambda_1, \lambda_1) \in k^2 \setminus \{0\}$, then $\mathrm{rk}(\lambda_1 x_M + \lambda_2 y_M) = \frac{(p-1)\dim_k M}{p}$, so that $M|_{k[\lambda_1 x + \lambda_2 y]}$ is projective. On the other hand, comparing Loewy lengths, we get $\ell\ell(M) = p$, while $\ell\ell(kE) = 2p-1$, implying that M is not a projective kE-module.

In view of the above, we assume from now on that k is algebraically closed, and proceed as follows: Given a p-elementary abelian group E, we consider a subspace $V \subseteq kE$ with $\dim_k V = \mathrm{rk}(E)$, whose nonzero elements $u \in V \setminus \{0\}$ have the following property:

$u^p = 0$, and $k[u]$ is a local algebra of dimension p such that $kE|_{k[u]}$ is free.

For an E-module M, one defines

$$\widehat{\mathcal{V}}_E(M) := \{u \in V;\ M|_{k[u]} \text{ is not free}\} \cup \{0\}.$$

This is a closed subset of V and hence an affine variety. Avrunin and Scott [9] showed that $\widehat{\mathcal{V}}_E(M) \cong \mathcal{V}_E(M)$, and they thus obtained an intrinsic characterization of the cohomological support variety.

Let $\mathfrak{A}_p := k[T]/(T^p)$ be the p-dimensional truncated polynomial ring, with generator $t := T + (T^p)$. Given a finite group scheme $\mathcal{G}$, every element $u \in k\mathcal{G}$ satisfying $u^p = 0$ defines an algebra homomorphism

$$\alpha_u : \mathfrak{A}_p \longrightarrow k\mathcal{G}; \quad t \mapsto u.$$

The freeness requirement occurring for finite groups amounts to saying that α_u is left flat. In general, any homomorphism $\varphi : A \longrightarrow B$ of associative k-algebras, induces a pull-back functor $\varphi^* : \operatorname{mod} B \longrightarrow \operatorname{mod} A$, where $\varphi^*(M)$ is the A-module with underlying k-space M and action

$$a.m := \varphi(a)m \qquad \forall\, a \in A,\ m \in M.$$

We say that φ is *left flat* if $\varphi^*(B)$ is a flat A-module. In case A and B are finite-dimensional, this amounts to saying that $\varphi^*(B)$ is a projective A-module. If the algebra A is local, then $\varphi^*(B)$ is free, and φ is necessarily injective. Thus, if $\alpha_u : \mathfrak{A}_p \longrightarrow k\mathcal{G}$ is left flat, then $k[u] \subseteq k\mathcal{G}$ is a k-algebra of dimension p.

Elementary abelian p-groups are abelian p-groups. In the context of finite group schemes, the latter are the reduced abelian unipotent groups. A finite algebraic k-group $\mathcal{U}$ is referred to as *unipotent* if the trivial module k is the only simple $\mathcal{U}$-module. For example, the Frobenius kernels $\mathbb{G}_{a(r)}$ of the additive group $\mathbb{G}_a$ are unipotent.

Definition ([60]). Let $\mathcal{G}$ be a finite group scheme. A homomorphism $\alpha : \mathfrak{A}_p \longrightarrow k\mathcal{G}$ is a *p-point* if

(P1) α is left flat, and

(P2) there exists an abelian, unipotent subgroup $\mathcal{U} \subseteq \mathcal{G}$ such that $\operatorname{im} \alpha \subseteq k\mathcal{U}$.

Two p-points α and β are *equivalent* ($\alpha \sim \beta$) if, for every $M \in \operatorname{mod} \mathcal{G}$, we have

$$\alpha^*(M) \text{ is projective } \Leftrightarrow \beta^*(M) \text{ is projective.}$$

The set of equivalence classes of p-points will be denoted $P(\mathcal{G})$.

Remarks. (1) Condition (P2) looks somewhat contrived and mysterious, but it turns out to be of vital importance. In general, p-points are hard to come by and the equivalence relation $\sim$ is difficult to compute. However, using (P2) many problems can be reduced to the consideration of abelian unipotent group schemes. For such a group $\mathcal{U}$, general theory (cf. [123]) provides an isomorphism

$$k\mathcal{U} \cong k[X_1, \ldots, X_n]/(X_1^{p^{r_1}}, \ldots, X_n^{p^{r_n}}),$$

with the latter algebra being amenable to computations.

(2) By considering algebra homomorphisms rather than homomorphisms of Hopf algebras we disregard the coalgebra structure. It turns out, however, that at the level of abelian unipotent groups every equivalence class $x \in P(\mathcal{U})$ can be represented by a homomorphism of Hopf algebras (see [43] for more details).

The following important projectivity criterion, which provides some control over the equivalence relation $\sim$, underscores the importance of abelian unipotent subgroups.

Proposition 3.7 ([60]). *Let $n \geq 1$ and consider the algebra $A := k[X, Y, Z] / (X^p, Y^p, Z^n)$, whose canonical generators are denoted x, y, z. Let $M \in \operatorname{mod} A$ be an A-module. Then M is projective as a $k[x]$-module if and only if M is projective as a $k[x + yz]$-module.*

Proof. Observe that the subalgebras $k[x]$ and $k[x+yz]$ are both isomorphic to $\mathfrak{A}_p$. By decomposing an $\mathfrak{A}_p$-module N into its cyclic direct summands, we see that N is projective if and only if the endomorphism t_N associated to its canonical generator $t \in \mathfrak{A}_p$ satisfies the equivalent conditions $\ker t_N = \operatorname{im} t_N^{p-1}$ and $\ker t_N^{p-1} = \operatorname{im} t_N$.

Suppose that $M|_{k[x]}$ is projective and consider the A-module $N := \ker(x + yz)_M / \operatorname{im}(x+yz)_M^{p-1}$. A rather involved computation shows that the operator z_N is injective. As z is nilpotent, we obtain $N = (0)$, so that $M|_{k[x+yz]}$ is projective. □

We illustrate the relevance of Condition (P2) by considering the case of infinitesimal groups $\mathcal{G}$ of height 1. Thanks to Theorem 1.20, $k\mathcal{G} \cong U_0(\mathfrak{g})$ is the restricted enveloping algebra of the restricted Lie algebra $\mathfrak{g} := \operatorname{Lie}(\mathcal{G})$.

Let $\alpha : \mathfrak{A}_p \longrightarrow U_0(\mathfrak{g})$ be a p-point. Then α factors through the restricted enveloping algebra $U_0(\mathfrak{u})$ of a suitable abelian, p-unipotent p-subalgebra $\mathfrak{u} \subseteq \mathfrak{g}$. In this case, the generators $x_1, \ldots, x_n$ of the algebra $U_0(\mathfrak{u})$ are primitive elements and they satisfy $x_i^{p^{r_i}} = 0$. Repeated application of Proposition 3.7 shows that α is equivalent to the p-point $\alpha_0 : \mathfrak{A}_p \longrightarrow U_0(\mathfrak{u})$ given by

$$\alpha_0(t) = \sum_{i=1}^{n} \zeta_i x_i^{p^{r_i - 1}},$$

where $(\zeta_1, \ldots, \zeta_n) \in k^n \setminus \{0\}$. Thus, we may assume that $u := \alpha(t)$ belongs to $\mathfrak{u} \subseteq \mathfrak{g}$ and satisfies the identity $u^{[p]} = 0$. Consequently, the element u belongs to the *nullcone*

$$\widehat{\mathcal{V}}_{\mathfrak{g}} := \{x \in \mathfrak{g};\ x^{[p]} = 0\}$$

of $\mathfrak{g}$. In view of Corollary 1.18, given $v, w \in \widehat{\mathcal{V}}_{\mathfrak{g}} \setminus \{0\}$, the $U_0(\mathfrak{g})$-module $M_v := U_0(\mathfrak{g}) \otimes_{k[v]} k$ is $k[w]$-projective if and only if $w \notin kv$. Thus, the line $k\alpha_0(t)$ only depends on the equivalence class $[\alpha]$, and we obtain a bijection

$$P(\mathcal{G}) \xrightarrow{\sim} \operatorname{Proj}(\widehat{\mathcal{V}}_{\mathfrak{g}}); \quad [\alpha] \mapsto k\alpha_0(t).$$

We will refine this result at the end of this section.

Returning to the general situation, we observe that every algebra homomorphism $\alpha : \mathfrak{A}_p \longrightarrow k\mathcal{G}$ induces a homomorphism

$$\alpha^\bullet : \mathrm{H}^\bullet(\mathcal{G}, k) \longrightarrow \mathrm{H}^\bullet(\mathfrak{A}_p, k)$$

of commutative, graded k-algebras. The even cohomology ring of $\mathfrak{A}_p$ is known to be isomorphic to the polynomial ring $k[X]$, where X has degree 2.

Let $\mathcal{V}$ be a conical affine variety, whose coordinate ring is a finitely generated, commutative, graded k-algebra

$$R = \bigoplus_{n\geq 0} R_n; \quad R_0 = k.$$

Recall that

$$\mathrm{Proj}(\mathcal{V}) \cong \{\ker \varphi;\ \varphi \in \mathrm{Hom}_{\mathrm{gr}}(R, k[X]),\ \mathrm{im}\,\varphi \neq k\}$$

is computable via non-trivial graded homomorphisms with values in $k[X]$. Thus, the hope is that $\alpha \mapsto \ker \alpha^\bullet$ provides a connection between $P(\mathcal{G})$ and $\mathrm{Proj}(\mathcal{V}_\mathcal{G}(k))$.

The following proposition elicits the cohomological aspects of the equivalence relation on p-points. The key notion is that of the Carlson module associated to a cohomology class $\zeta \in \mathrm{H}^n(\mathcal{G}, k) \setminus \{0\}$. By general theory, we have an isomorphism $\mathrm{H}^n(\mathcal{G}, k) \cong \mathrm{Hom}_\mathcal{G}(\Omega^n_\mathcal{G}(k), k)$, allowing us to view the representing cocycle as a map $\hat{\zeta} : \Omega^n_\mathcal{G}(k) \longrightarrow k$. Then

$$L_\zeta := \ker \hat{\zeta}$$

is called the *Carlson module* of ζ.

Proposition 3.8 ([60]). *Let $\mathcal{G}$ be a finite group scheme over k.*

(1) *If $\alpha : \mathfrak{A}_p \longrightarrow k\mathcal{G}$ is flat, then*

$$\begin{aligned} &\ker \alpha^\bullet \cap (\mathrm{H}^{2n}(\mathcal{G}, k) \setminus \{0\}) \\ &\quad = \{\zeta \in \mathrm{H}^{2n}(\mathcal{G}, k) \setminus \{0\};\ \alpha^*(L_\zeta) \text{ is not projective}\} \end{aligned}$$

for every $n \geq 0$.

(2) *If $\alpha, \beta : \mathfrak{A}_p \longrightarrow k\mathcal{G}$ are flat maps such that $\ker \alpha^\bullet \not\subseteq \ker \beta^\bullet$, then there exist $n \geq 1$ and $\zeta \in \mathrm{H}^{2n}_\mathcal{G}(k), k) \setminus \{0\}$ such that $\alpha^*(L_\zeta)$ is not projective and $\beta^*(L_\zeta)$ is projective.*

Proof. (1) As α is flat, the corresponding exact functor

$$\alpha^* : \mathrm{mod}\,\mathcal{G} \longrightarrow \mathrm{mod}\,\mathfrak{A}_p$$

takes projectives to projectives. Consequently, α^* sends a minimal projective resolution of $M \in \mathrm{mod}\,\mathcal{G}$ to a (not necessarily minimal) projective resolution of $\alpha^*(M)$, so that

$$\alpha^*(\Omega^m_\mathcal{G}(M)) \cong \Omega^m_{\mathfrak{A}_p}(\alpha^*(M)) \oplus (\text{proj.}) \quad \forall\ m \in \mathbb{N}.$$

Let $\zeta \in \mathrm{H}^{2n}(\mathcal{G}, k)\setminus\{0\}$ be an arbitrary element. Upon application of α^* to the exact sequence $(0) \longrightarrow L_\zeta \longrightarrow \Omega^{2n}_\mathcal{G}(k) \stackrel{\hat{\zeta}}{\longrightarrow} k \longrightarrow (0)$ we obtain, observing $\Omega^{2n}_{\mathfrak{A}_p}(k) \cong k$, an exact sequence

$$(*) \qquad (0) \longrightarrow \alpha^*(L_\zeta) \longrightarrow k \oplus (\text{proj.}) \stackrel{\alpha^*(\hat{\zeta})}{\longrightarrow} k \longrightarrow (0).$$

The definition of $\alpha^\bullet$ yields

$$\alpha^\bullet(\zeta) = 0 \Leftrightarrow k \subseteq \ker \alpha^*(\hat{\zeta}).$$

If $\alpha^*(L_\zeta)$ is projective, then it is injective and $(*)$ splits. Consequently, k is not a direct summand of $\alpha^*(L_\zeta)$, so that $k \not\subseteq \ker \alpha^*(\hat{\zeta})$, whence $\zeta \notin \ker \alpha^\bullet$. Conversely, the assumption $\zeta \notin \ker \alpha^\bullet$ yields the splitting of the sequence $(*)$, whence

$$k \oplus (\text{proj.}) \cong \alpha^*(L_\zeta) \oplus k.$$

By the Theorem of Krull-Remak-Schmidt, the $\mathfrak{A}_p$-module $\alpha^*(L_\zeta)$ is projective.

(2) Since $\alpha^\bullet$ and $\beta^\bullet$ are homomorphisms of degree 0, our assumption provides an element $\zeta \in \mathrm{H}^{2n}(\mathcal{G}, k)$ with $\zeta \in \ker \alpha^\bullet \setminus \ker \beta^\bullet$. Thus, $\zeta \neq 0$, and (1) shows that L_ζ has the requisite properties. □

In view of Proposition 3.8, equivalent p-points yield the same kernels in cohomology. This is the first step towards proving the existence of a map

$$\Psi_{\mathcal{G}} : P(\mathcal{G}) \longrightarrow \mathrm{Proj}(\mathcal{V}_{\mathcal{G}}(k)); \quad [\alpha] \mapsto \ker \alpha^\bullet.$$

The missing point is to show that $\operatorname{im} \alpha^\bullet \neq k$ for all p-points α. This is actually a rather deep result.

Lemma 3.9 ([60]). *Let $\mathcal{G}$ be a finite k-group scheme. If $\alpha : \mathfrak{A}_p \longrightarrow k\mathcal{G}$ is a flat homomorphism, then* $\operatorname{im} \alpha^\bullet \neq k$.

Proof. Since α is flat, $\alpha^*(k\mathcal{G})$ is a projective $\mathfrak{A}_p$-module, so that the Eckmann-Shapiro Lemma furnishes an isomorphism

$$\mathrm{H}^\bullet(\mathfrak{A}_p, k) \cong \mathrm{H}^\bullet(\mathcal{G}, \mathrm{Hom}_{\mathfrak{A}_p}(k\mathcal{G}, k))$$

of $\mathrm{H}^\bullet(\mathcal{G}, k)$-modules, with $\mathrm{H}^\bullet(\mathcal{G}, k)$ operating on $\mathrm{H}^\bullet(\mathfrak{A}_p, k)$ via $\alpha^\bullet$. Thanks to the Friedlander-Suslin Theorem, the space $\mathrm{H}^\bullet(\mathfrak{A}_p, k) \cong k[X]$ is a finitely generated $\mathrm{H}^\bullet(\mathcal{G}, k)$-module, so that $\operatorname{im} \alpha^\bullet \neq k$. □

Following Friedlander-Pevtsova [60], we associate to every $\mathcal{G}$-module M the set

$$P(\mathcal{G})_M := \{[\alpha] \in P(\mathcal{G});\ \alpha^*(M) \text{ is not projective}\}.$$

If these sets were a good model for support varieties, then they should interact nicely with tensor products. The subtlety of the following result resides in p-points not being Hof algebra homomorphisms, so that α^* does not necessarily commute with tensor products. Nevertheless, one can show:

Theorem 3.10 ([60]). *Let $\mathcal{G}$ be a finite group scheme. Then we have*

$$P(\mathcal{G})_{M \otimes_k N} = P(\mathcal{G})_M \cap P(\mathcal{G})_N$$

for all $M, N \in \mathrm{mod}\, k\mathcal{G}$.

Proof. Suppose first that $\mathcal{G} = \mathcal{U}$ is abelian and unipotent. Recall that there exist $r_i \geq 1$ such that

$$k\mathcal{U} \cong k[X_1, \dots, X_n]/(X_1^{p^{r_1}}, \dots, X_n^{p^{r_n}}).$$

We denote the canonical generators by the corresponding lower case letters. Thus, by defining a comultiplication $\Delta_1 : k\mathcal{U} \longrightarrow k\mathcal{U} \otimes_k k\mathcal{U}$ via

$$\Delta_1(x_i) = x_i \otimes 1 + 1 \otimes x_i; \quad 1 \leq i \leq n$$

we see that $k\mathcal{U}$ can be considered as the restricted enveloping algebra $U_0(\mathfrak{u})$ of the abelian restricted Lie algebra $\mathfrak{u} := \bigoplus_{i=1}^n \bigoplus_{j=0}^{r_i-1} kx_i^{p^j}$. Let $\Delta_0 : k\mathcal{U} \longrightarrow k\mathcal{U} \otimes_k k\mathcal{U}$ be the originally given comultiplication. Since tensor products of projective modules over Hopf algebras are projective, both comultiplications are flat. The well-known identity $\Delta_0(u) \equiv u \otimes 1 + 1 \otimes u \mod(\mathrm{Rad}(k\mathcal{U}) \otimes_k \mathrm{Rad}(k\mathcal{U}))$, in conjunction with $k\mathcal{U}$ being generated by Δ_1-primitive elements, yields

$$\Delta_0(u) \equiv \Delta_1(u) \mod(\mathrm{Rad}(k\mathcal{U}) \otimes_k \mathrm{Rad}(k\mathcal{U})) \quad \forall\ u \in k\mathcal{U}.$$

We let $\Delta_{i,*} : P(\mathcal{U}) \longrightarrow P(k\mathcal{U} \otimes_k k\mathcal{U})$ be the canonical map induced by Δ_i, sending p-points of $\mathcal{U}$ to flat maps $\mathfrak{A}_p \longrightarrow k\mathcal{U} \otimes_k k\mathcal{U}$. Then Proposition 3.7 yields

$$P(\mathcal{U})_{M \otimes N} = \Delta_{0,*}^{-1}(P(k\mathcal{U} \otimes_k k\mathcal{U})_{M \otimes_k N}) = \Delta_{1,*}^{-1}(P(k\mathcal{U} \otimes_k k\mathcal{U})_{M \otimes_k N}),$$

so that we may consider the tensor product relative to the Hopf structure given by Δ_1. In other words, $k\mathcal{U}$ is a restricted enveloping algebra.

Let $[\alpha]$ be an element of $P(\mathcal{U})$. By our observations concerning p-points of restricted enveloping algebras, we may assume that $\alpha : \mathfrak{A}_p \longrightarrow k\mathcal{U}$ is a homomorphism of Hopf algebras, where $\mathfrak{A}_p = U_0(kt)$ is the restricted enveloping algebra of the one-dimensional nil-cyclic Lie algebra kt. Thus, the functor α^* commutes with tensor products, and our result follows from the representation theory of $U_0(kt)$.

In the general case, a p-point $\alpha : \mathfrak{A}_p \longrightarrow k\mathcal{G}$ factors through an abelian, unipotent subgroup $\mathcal{U} \subseteq \mathcal{G}$. Hence there exists a p-point $\beta : \mathfrak{A}_p \longrightarrow k\mathcal{U}$ such that $\iota_{\mathcal{U}} \circ \beta = \alpha$. Here $\iota_{\mathcal{U}}$ denotes the canonical inclusion, whose associated pull-back functor commutes with tensor products. □

Remark. As is shown in [43, §1], Theorem 3.10 loses its validity, if one considers arbitrary flat maps $\mathfrak{A}_p \longrightarrow k\mathcal{G}$ instead of p-points.

The conceptual importance of Theorem 3.10 manifests itself in the following result that justifies the notion of a support space:

Theorem 3.11 ([60]). *Let $\mathcal{G}$ be a finite group scheme. Then*

$$\{P(\mathcal{G})_M;\ M \in \operatorname{mod} k\mathcal{G}\}$$

is the set of closed subsets of a noetherian topology on $P(\mathcal{G})$. □

In particular, we can speak of the Krull dimension of the closed subspace $P(\mathcal{G})_M$ of $P(\mathcal{G})$.

Theorem 3.12 ([60]). *The map*

$$\Psi_{\mathcal{G}} : \begin{cases} P(\mathcal{G}) \longrightarrow \operatorname{Proj}(\mathcal{V}_{\mathcal{G}}(k)) \\ \quad \alpha \quad \mapsto \quad \ker \alpha^{\bullet} \end{cases}$$

is a homeomorphism such that

$$\Psi_{\mathcal{G}}^{-1}(\operatorname{Proj}(\mathcal{V}_{\mathcal{G}}(M))) = P(\mathcal{G})_M$$

for every $M \in \operatorname{mod} \mathcal{G}$. *Moreover,* $\Psi_{\mathcal{G}}$ *is natural relative to flat homomorphisms* $\mathcal{G} \longrightarrow \mathcal{H}$. □

Remark. By definition, the Carlson module L_ζ belongs to the principal block $\mathcal{B}_0(\mathcal{G})$ of $k\mathcal{G}$. Theorem 3.12 and Proposition 3.8 thus state that the equivalence relation $\sim$ is completely determined on $\operatorname{mod} \mathcal{B}_0(\mathcal{G})$.

Corollary 3.13. *We have* $\mathcal{V}_{\mathcal{G}}(L_\zeta) = Z(\zeta)$ *for any* $\zeta \in \mathrm{H}^{2n}(\mathcal{G}, k) \setminus \{0\}$.

Proof. Let $\alpha : \mathfrak{A}_p \longrightarrow k\mathcal{G}$ be a p-point of $\mathcal{G}$, and consider the induced map $\alpha^{\bullet}$: $\mathrm{H}^{\bullet}(\mathcal{G}, k) \longrightarrow \mathrm{H}^{\bullet}(\mathfrak{A}_p, k)$. In view of Proposition 3.8(1), we have $\zeta \in \ker \alpha^{\bullet}$ if and only if the pull-back $\alpha^*(L_\zeta)$ is not projective. Thanks to Theorem 3.12, the latter condition is equivalent to $\ker \alpha^{\bullet} \supseteq \ker \Phi_{L_\zeta}$, so that

$$(\zeta) \subseteq \ker \alpha^{\bullet} \Leftrightarrow \ker \Phi_{L_\zeta} \subseteq \ker \alpha^{\bullet}.$$

By virtue of Theorem 3.12, this implies $\operatorname{Proj}(Z(\zeta)) = \operatorname{Proj}(\mathcal{V}_{\mathcal{G}}(L_\zeta))$, whence $Z(\zeta) = \mathcal{V}_{\mathcal{G}}(L_\zeta)$. □

Given a restricted Lie algebra $(\mathfrak{g}, [p])$, we recall the bijection

$$\operatorname{Proj}(\widehat{\mathcal{V}}_{\mathfrak{g}}) \longrightarrow P(\mathfrak{g}); \quad ku \mapsto [\alpha_u].$$

If M is a $U_0(\mathfrak{g})$-module, then the pre-image of $P(\mathfrak{g})_M$ under this map is the projective space of the *rank variety*

$$\widehat{\mathcal{V}}_{\mathfrak{g}}(M) := \{u \in \widehat{\mathcal{V}}_{\mathfrak{g}}; \; M|_{U_0(ku)} \text{ is not projective}\} \cup \{0\}$$

of M, which is a closed, conical subset of the nullcone $\widehat{\mathcal{V}}_{\mathfrak{g}}$. The above map is actually a homeomorphism; see also the work by Jantzen [79] and Friedlander-Parshall [57] concerning the relationship between the nullcone $\widehat{\mathcal{V}}_{\mathfrak{g}}$ and the cohomological support variety $\mathcal{V}_{\mathfrak{g}}(k)$. We record the following consequence:

Corollary 3.14. *Let* $(\mathfrak{g}, [p])$ *be a restricted Lie algebra. Then we have*

$$\dim \widehat{\mathcal{V}}_{\mathfrak{g}}(M) = \mathrm{cx}_{U_0(\mathfrak{g})}(M)$$

for every $M \in \operatorname{mod} U_0(\mathfrak{g})$. □

3.3 Tameness and complexity

In this section we shall employ support varieties to verify a useful necessary condition for the tameness of blocks of cocommutative Hopf algebras. This result was first announced by Rickard [103] in 1990, yet his proof is not correct. We therefore follow the approach of [42], which is valid for blocks of Hopf algebras associated to finite group schemes and related classes of algebras. Given such a group scheme $\mathcal{G}$ and a $\mathcal{G}$-module M, we recall that $\mathrm{cx}_{\mathcal{G}}(M) := \mathrm{cx}_{k\mathcal{G}}(M)$ denotes the complexity of M.

Throughout, our base field k is assumed to be algebraically closed.

Theorem 3.15 ([42]). *Let $\mathcal{G}$ be a finite group scheme, $\mathcal{B} \subseteq k\mathcal{G}$ be a tame block. Then $\mathrm{cx}_{\mathcal{G}}(M) \leq 2$ for every $M \in \mathrm{mod}\,\mathcal{B}$.*

Proof. We outline the proof, since it rests on the geometric notions we have studied in the foregoing sections. Let $\{S_1, \ldots, S_n\}$ be a complete set of representatives for the simple $\mathcal{B}$-modules. We set

$$\mathcal{V}_{\mathcal{B}} := \mathcal{V}_{\mathcal{G}}(\bigoplus_{i=1}^{n} S_i).$$

Our Theorem is a consequence of the following assertion, valid for an arbitrary block $\mathcal{B} \subseteq k\mathcal{G}$:

(∗) *If $m := \dim \mathcal{V}_{\mathcal{B}} \geq 3$, then $\mathcal{B}$ is not tame.*

The idea is to produce a family of indecomposable $\mathcal{B}$-modules that violates Theorem 2.10. Using the Noether Normalization Lemma, one first proves the existence of $d > 0$ and a family $(\zeta_\alpha)_{\alpha \in k}$ with $\zeta_\alpha \in \mathrm{H}^{2d}(\mathcal{G}, k) \setminus \{0\}$ for all $\alpha \in k$, such that

(a) $\dim(Z(\zeta_\alpha) \cap \mathcal{V}_{\mathcal{B}}) = m - 1$ for $\alpha \in k$, and

(b) $\dim(Z(\zeta_\alpha) \cap Z(\zeta_\beta) \cap \mathcal{V}_{\mathcal{B}}) = m - 2$ for $\alpha \neq \beta \in k$.

Now consider the $\mathcal{B}$-module $M_{\mathcal{B}} := \bigoplus_{i=1}^{n} S_i$ as well as the Carlson modules $(L_{\zeta_\alpha})_{\alpha \in k}$. These give rise to exact sequences

$$(0) \longrightarrow L_{\zeta_\alpha} \otimes_k M_{\mathcal{B}} \longrightarrow \Omega^{2d}_{\mathcal{G}}(k) \otimes_k M_{\mathcal{B}} \xrightarrow{\hat{\zeta}_\alpha \otimes \mathrm{id}_{M_{\mathcal{B}}}} M_{\mathcal{B}} \longrightarrow (0).$$

Owing to Theorem 3.10, Theorem 3.12 and Corollary 3.13 we have

$$\mathcal{V}_{\mathcal{G}}(L_{\zeta_\alpha} \otimes_k M_{\mathcal{B}}) = \mathcal{V}_{\mathcal{G}}(L_{\zeta_\alpha}) \cap \mathcal{V}_{\mathcal{G}}(M_{\mathcal{B}}) = Z(\zeta_\alpha) \cap \mathcal{V}_{\mathcal{B}}.$$

By considering indecomposable summands of $L_{\zeta_\alpha} \otimes_k M_{\mathcal{B}}$ belonging to $\mathcal{B}$, we finally obtain a family $(N_\alpha)_{\alpha \in k}$ of indecomposable $\mathcal{B}$-modules with

(i) $\dim_k N_\alpha \leq (\dim_k M_{\mathcal{B}})(\dim_k \Omega^{2d}_{\mathcal{G}}(k))$, and

(ii) $\dim \mathcal{V}_{\mathcal{G}}(N_\alpha) = \dim \mathcal{V}_{\mathcal{B}} - 1$, and

(iii) $\dim(\mathcal{V}_{\mathcal{G}}(N_\alpha) \cap \mathcal{V}_{\mathcal{G}}(N_\beta)) = \dim \mathcal{V}_{\mathcal{B}} - 2$ for all $\alpha \neq \beta$.

In view of (iii), these modules are pairwise non-isomorphic, and Lemma 3.2 yields $\mathrm{cx}_{\mathcal{G}}(N_\alpha) = \dim \mathcal{V}_{\mathcal{G}}(N_\alpha) \geq 2$. Owing to (i), Theorem 2.10 shows that $\mathcal{B}$ is not tame.

Now let $\mathcal{B}$ be a tame block. Thanks to (∗), we obtain $\mathrm{cx}_{\mathcal{G}}(S_i) \leq \mathrm{cx}_{\mathcal{G}}(M_{\mathcal{B}}) \leq 2$, so that an application Proposition 2.4 completes the proof. □

By Proposition 2.4(4) we have $\mathrm{cx}_{\mathbb{G}_{a(2)}}(M) \leq \mathrm{cx}_{\mathbb{G}_{a(2)}}(k) = 2$ for every $M \in \operatorname{mod} k\mathbb{G}_{a(2)}$. However, $k\mathbb{G}_{a(2)} \cong k[X,Y]/(X^p, Y^p)$ is tame only if $p = 2$. Consequently, the converse of Theorem 3.15 does not hold in general.

Remark. By combining Theorem 1.6, Proposition 2.4 and Theorem 3.15, we see that Hopf subalgebras of tame or finite representation type "come close to" being tame or representation-finite. In fact, by special properties of extensions of group algebras, the representation type is inherited by subgroups of finite groups (cf. [15, Prop.2]). As we shall see later, such a result does not hold in the context of infinitesimal group schemes.

3.4 Blocks of Frobenius kernels of smooth groups

Throughout this section, we fix a reduced (=smooth) affine group scheme $\mathcal{G}$ and assume that $\mathcal{G}$ is defined over the algebraically closed field k of characteristic $\mathrm{char}(k) = p \geq 3$. We will illustrate how rank varieties and nilpotent orbits can be employed to classify the representation-finite and tame blocks of the Frobenius kernels of such groups. Let us begin by looking at an important example.

Example. Consider the first Frobenius kernel $\mathrm{SL}(2)_1$ of $\mathrm{SL}(2)$. According to Theorem 1.20, the algebra $k\,\mathrm{SL}(2)_1$ is isomorphic to $U_0(\mathfrak{sl}(2))$, the restricted enveloping algebra of the Lie algebra of (2×2)-matrices of trace zero.

The block structure of $U_0(\mathfrak{sl}(2))$ was determined by Pollack [98]. The algebra $U_0(\mathfrak{sl}(2))$ possesses exactly $\frac{p+1}{2}$ blocks. There is one simple block, corresponding to the p-dimensional Steinberg module $\mathrm{St} = L(p-1)$, and $\ell := \frac{p-1}{2}$ blocks $\mathcal{B}_0, \ldots, \mathcal{B}_{\ell-1}$, with $\mathcal{B}_i$ having two simple modules $L(i)$ and $L(p-2-i)$ of dimensions $i+1$ and $p-1-i$, respectively. Pollack also proved that $U_0(\mathfrak{sl}(2))$ has infinite representation type, which of course also follows from Corollary 3.3 and the fact that $\widehat{\mathcal{V}}_{\mathfrak{sl}(2)}$ has dimension 2. The tameness of these blocks was verified independently by Drozd [34], Fischer [55], and Rudakov [109]. The indecomposable $U_0(\mathfrak{sl}(2))$-modules were determined by Premet [99].

We shall return to these blocks once we have discussed path algebras. For the time being, we just record

$$\dim_k \mathrm{Ext}^1_{U_0(\mathfrak{sl}(2))}(L(i), L(j)) = \begin{cases} 2 & i+j = p-2, \\ 0 & \text{otherwise.} \end{cases}$$

For every $r \in \mathbb{N}$, we define

$$\ell_r(\mathcal{G}) := r \dim \mathcal{G}/B(\mathcal{G}),$$

where $B(\mathcal{G})$ is a Borel subgroup of $\mathcal{G}$. Since all Borel subgroups of $\mathcal{G}$ are conjugate, the number $\ell_r(\mathcal{G})$ does not depend on the choice of $B(\mathcal{G})$.

Given $n \geq 0$, we denote by

$$\mathfrak{n}_n := \bigoplus_{n=0}^{n-1} kx^{[p]^i}; \quad x^{[p]^n} = 0 \neq x^{[p]^{n-1}},$$

the n-dimensional *nil-cyclic* Lie algebra. The smooth group giving rise to this Lie algebra is the group $\mathcal{W}_n$ of *Witt vectors* of length n (cf. [31, V,§5]). Thus, we have

$$\mathrm{Lie}(\mathcal{W}_n) = \mathfrak{n}_n \qquad \forall\, n \geq 0.$$

The following result shows in particular that the representation-finite blocks of $k\mathcal{G}_r$ are Nakayama algebras.

Theorem 3.16 ([40]). *Let $\mathcal{G}$ be a smooth group with unipotent radical $\mathcal{U}$ of dimension $n := \dim \mathcal{U}$. Let $\mathcal{B} \subseteq k\mathcal{G}_r$ be a block.*

(1) *Suppose that $\mathcal{B}$ is representation-finite.*

(a) *If $r \geq 2$, then $\mathcal{B}$ is simple, and $\mathcal{G}$ is reductive (i.e., $\mathcal{U} = e_k$),*

(b) *If $r = 1$, then there exists $s \in \{0,1\}$ such that $\mathcal{B} \cong \mathrm{Mat}_{p^{\ell_1(\mathcal{G})}}(k((\mathcal{W}_n)_1 \rtimes \mu_{(p^s)}))$.*

(2) *If $\mathcal{B}$ is tame, then $\mathcal{G}$ is reductive and $\mathcal{B} \cong \mathrm{Mat}_{p^{\ell_r(\mathcal{G})-1}}(\mathcal{C})$ for a (tame) block $\mathcal{C} \subseteq k\,\mathrm{SL}(2)_1$.*

Proof. By way of illustration, we shall only consider (1) for $r = 1$, and in the case where the group $\mathcal{G}$ is reductive. Then $n = 0$ and we have to show that the block $\mathcal{B}$ is simple. Since the Frobenius kernels of $\mathcal{G}$ and its connected component $\mathcal{G}^0$ coincide, we may also assume that $\mathcal{G}$ is connected.

The group $\mathcal{G}(k)$ acts on $\mathfrak{g} := \mathrm{Lie}(\mathcal{G})$ via the adjoint representation, which we may also consider as an action

$$\mathrm{Ad} : \mathcal{G}(k) \longrightarrow \mathrm{Aut}_k(U_0(\mathfrak{g})).$$

Let S be a simple module belonging to the block $\mathcal{B} \subseteq U_0(\mathfrak{g})$. For $g \in \mathcal{G}(k)$ we set

$$S^{(g)} := ((\mathrm{Ad}\, g)^{-1})^*(S).$$

Direct computation shows that

$$\widehat{\mathcal{V}}_{\mathfrak{g}}(S^{(g)}) = (\mathrm{Ad}\, g)(\widehat{\mathcal{V}}_{\mathfrak{g}}(S)) \qquad \forall\, g \in \mathcal{G}(k).$$

Since S is simple and $\mathcal{G}(k)$ is connected, we have $S \cong S^{(g)}$ $\forall\, g \in \mathcal{G}(k)$, so that the variety $\widehat{\mathcal{V}}_{\mathfrak{g}}(S) \subseteq \widehat{\mathcal{V}}_{\mathfrak{g}} \subseteq \mathfrak{g}$ is invariant under the adjoint action.

We first show that $\dim \widehat{\mathcal{V}}_{\mathfrak{g}}(S) \neq 1$. Otherwise, each irreducible component of the conical variety $\widehat{\mathcal{V}}_{\mathfrak{g}}(S)$ is a $\mathcal{G}(k)$-stable line, and $\mathfrak{g}$ contains a line $kx \neq (0)$ with $x^{[p]} = 0$ that is invariant under the adjoint representation. This contradicts the fact that the group $\mathcal{G}$ is reductive (cf. [73]).

Since $\mathcal{B}$ has finite representation type, the rank variety $\widehat{\mathcal{V}}_{\mathfrak{g}}(S)$ has dimension ≤ 1 (cf. Corollary 3.3). By what we have just seen, this implies $\dim \widehat{\mathcal{V}}_{\mathfrak{g}}(S) = 0$. Consequently, S is projective, and the self-injective algebra $\mathcal{B}$ is simple. □

Remark. The scarcity of tame or representation-finite blocks for Frobenius kernels of smooth reductive groups closely parallels the situation for blocks of groups of Lie type in the defining characteristic. The reader is referred to [74, §8] for more details.

3.5 π-points and Jordan type

The concept of a p-point has turned out to be merely the beginning of far-reaching investigations involving π-points and Jordan types. While π-points incorporate the treatment of infinite-dimensional modules for group schemes over arbitrary fields, the notion Jordan type refines the defining property of support spaces by investigating the isomorphism types of the $\mathfrak{A}_p$-modules $\alpha^*(M)$. In this section we sketch a few results of the relevant articles [61, 62, 24].

Throughout, k denotes a field of positive characteristic $p > 0$. If $\mathcal{G} := \operatorname{Spec}_k(k[\mathcal{G}])$ is an affine group scheme over k and $R \in M_k$ is a commutative k-algebra, then

$$\mathcal{G}_R : M_R \longrightarrow \mathrm{Gr}; \quad S \mapsto \mathcal{G}(S)$$

is a group functor, which is the composite of $\mathcal{G}$ with the forgetful functor $M_R \longrightarrow M_k$ induced by the structure map $k \longrightarrow R$. In fact, being represented by the R-algebra $k[\mathcal{G}] \otimes_k R$, the group functor $\mathcal{G}_R$ is affine. If K is an extension field of k, and $\mathcal{G}$ is a finite k-group, then $\mathcal{G}_K$ is a finite K-group with algebra of measures

$$K\mathcal{G} := k\mathcal{G} \otimes_k K.$$

The isomorphism $\mathfrak{A}_p \otimes_k K \cong K[T]/(T^p)$ can be interpreted as a special case. We shall denote this algebra by $\mathfrak{A}_{p,K}$. The main innovation, expounded in [61], is the consideration of p-points over arbitrary field extensions:

Definition. Let $\mathcal{G}$ be a finite group scheme over k. Given a field extension $K:k$, a *π-point*

$$\alpha_K : \mathfrak{A}_{p,K} \longrightarrow K\mathcal{G}$$

is a left flat homomorphism of K-algebras, such that there exists an abelian, unipotent subgroup $\mathcal{U} \subseteq \mathcal{G}_K$ with $\operatorname{im} \alpha_K \subseteq K\mathcal{U}$.

We may now investigate $k\mathcal{G}$-modules M by considering their extensions $M_K := M \otimes_k K$ as $K\mathcal{G}$-modules as well as the structure of their pullbacks $\alpha_K^*(M_K) \in \operatorname{mod} \mathfrak{A}_{p,K}$. In their paper [61], Friedlander and Pevtsova define an equivalence relation via

$$\alpha_K \sim \beta_L \;:\Leftrightarrow\; \alpha_K^*(M_K) \text{ is projective if and only if } \beta_L^*(M_L) \text{ is projective} \\ \forall\, M \in \operatorname{mod} k\mathcal{G}$$

for π-points $\alpha_K : \mathfrak{A}_{p.K} \longrightarrow K\mathcal{G}$ and $\beta_L : \mathfrak{A}_{p,L} \longrightarrow L\mathcal{G}$. We let $\Pi(\mathcal{G})$ be the set of equivalence classes of π-points.

If $k \subseteq K \subseteq L$ are fields, then $\mathfrak{A}_{p,K} \otimes_K L \cong \mathfrak{A}_{p,L}$, and base change associates to a π-point $\alpha_K : \mathfrak{A}_{p,K} \longrightarrow K\mathcal{G}$ a π-point $\alpha_L : \mathfrak{A}_{p,L} \longrightarrow L\mathcal{G}$. Given any $k\mathcal{G}$-module M, we have $M_K \otimes_K L \cong M_L$, whence

$$\alpha_L^*(M_L) \cong \alpha_K^*(M_K) \otimes_K L.$$

This implies:

Proposition 3.17. *Let $\mathcal{G}$ be a finite group scheme over k. Two π-points $\alpha_K : \mathfrak{A}_{p,K} \longrightarrow K\mathcal{G}$ and $\beta_L : \mathfrak{A}_{p,L} \longrightarrow L\mathcal{G}$ are equivalent if and only if $\alpha_F \sim \beta_F$ for some field F containing K and L.* □

Let me comment on the technical aspect of the definition involving the existence of an abelian, unipotent subgroup $\mathcal{U} \subseteq \mathcal{G}_K$. By not insisting on $\mathcal{U}$ to be defined over k, we introduce a degree of freedom that enables us to reduce many questions to the consideration of p-points over algebraically closed fields. Here is the relevant result:

Lemma 3.18. *Let $\mathcal{G}$ be a finite k-group scheme, K be an extension field of k.*

(1) *A π-point $\alpha_L : \mathfrak{A}_{p,L} \longrightarrow L\mathcal{G}_K$ of $\mathcal{G}_K$ can be considered as a π-point $\widehat{\alpha}_L : \mathfrak{A}_{p,L} \longrightarrow L\mathcal{G}$ of $\mathcal{G}$.*

(2) *The map $\Pi(\mathcal{G}_K) \longrightarrow \Pi(\mathcal{G})$; $[\alpha_L] \mapsto [\widehat{\alpha}_L]$ is well-defined.*

Proof. Since $L\mathcal{G}_K \cong L\mathcal{G}$, it is clear that $\widehat{\alpha}_L$ is a flat homomorphism. By assumption there exists an abelian, unipotent subgroup $\mathcal{U} \subseteq (\mathcal{G}_K)_L$ such that $\operatorname{im}\alpha_L \subseteq L\mathcal{U}$. Consequently, $\operatorname{im}\widehat{\alpha}_L \subseteq L\mathcal{U}$, so that $\widehat{\alpha}_L$ is a π-point for $\mathcal{G}$. □

Given a finite k-group scheme $\mathcal{G}$, the Universal Coefficient Theorem (cf. [108, (8.22)]) provides an isomorphism

$$\mathrm{H}^*(\mathcal{G}_K, K) \cong \mathrm{H}^*(\mathcal{G}, k) \otimes_k K$$

for any field $K \supseteq k$. We may thus consider $\mathrm{H}^\bullet(\mathcal{G}, k)$ as a k-subalgebra of the K-algebra $\mathrm{H}^\bullet(\mathcal{G}_K, K)$. Using the arguments of Section 3.2, we see that every π-point $\alpha_K : \mathfrak{A}_{p,K} \longrightarrow K\mathcal{G}$ defines a non-trivial homomorphism

$$\alpha_K^\bullet : \mathrm{H}^\bullet(\mathcal{G}_K, K) \longrightarrow \mathrm{H}^\bullet(\mathfrak{A}_{p,K}, K)$$

of graded k-algebras. In particular, $\ker\alpha_K^\bullet \cap \mathrm{H}^\bullet(\mathcal{G}, k)$ is a homogeneous prime ideal of $\mathrm{H}^\bullet(\mathcal{G}, k)$ which is properly contained in the augmentation ideal $\bigoplus_{n\geq 1} \mathrm{H}^{2n}(\mathcal{G}, k)$ of $\mathrm{H}^\bullet(\mathcal{G}, k)$.

Let $K : k$ be a field extension. Given an element $\zeta \in \mathrm{H}^{2n}(\mathcal{G}, k) \setminus \{0\}$, we denote by $L_{\zeta,K}$ the kernel of the map $\hat{\zeta}_K := \hat{\zeta} \hat{\otimes} \mathrm{id}_K : \Omega_{\mathcal{G}}^{2n}(k) \otimes_k K \longrightarrow K$. The corresponding cohomology class ζ_K belongs to $\mathrm{H}^\bullet(\mathcal{G}, k) \subseteq \mathrm{H}^\bullet(\mathcal{G}_K, K)$. The following subsidiary result indicates the significance of the ideals $\ker\alpha_K^\bullet \cap \mathrm{H}^\bullet(\mathcal{G}, k)$:

Proposition 3.19 ([61]). *Let $\mathcal{G}$ be a finite group scheme over k, $\alpha_K : \mathfrak{A}_{p.K} \longrightarrow K\mathcal{G}$ be a π-point. Given an element $\zeta \in \mathrm{H}^{2n}(\mathcal{G}, k) \setminus \{0\}$, the following statements are equivalent:*

(1) $\zeta_K \in \ker\alpha_K^\bullet \cap \mathrm{H}^\bullet(\mathcal{G}, k)$.

(2) *$\alpha_K^*(L_{\zeta,K})$ is not projective.*

Proof. By applying α_K^* to the map $\hat{\zeta}_K$, we obtain a short exact sequence

$$(0) \longrightarrow \alpha_K^*(L_{\zeta,K}) \longrightarrow \alpha_K^*(\Omega_{\mathcal{G}}^{2n}(k) \otimes_k K) \xrightarrow{\alpha_K^*(\hat{\zeta}_K)} K \longrightarrow (0).$$

Since α_K^* sends projectives to projectives, the isomorphisms $\Omega^{2n}_{\mathcal{G}}(k) \otimes_k K \cong \Omega^{2n}_{\mathcal{G}_K}(K) \oplus (\text{proj.})$ and $\Omega^{2n}_{\mathfrak{A}_{p,K}}(K) \cong K$ give rise to a short exact sequence

$$(0) \longrightarrow \alpha_K^*(L_{\zeta,K}) \longrightarrow K \oplus (\text{proj.}) \stackrel{(f,g)}{\longrightarrow} K \longrightarrow (0),$$

where the cohomology class of f corresponds to $\alpha_K^\bullet(\zeta_K)$. We may now argue as in the proof of Proposition 3.8. □

In view of Proposition 3.17, the equivalence relation of π-points interacts nicely with field extensions. As this also holds for the associated prime ideals, the results of Section 3.2 are amenable to applications. Using Theorem 3.12, one obtains for π-points α_k, β_L:

$$\alpha_K \sim \beta_L \;\Leftrightarrow\; \ker \alpha_K^\bullet \cap \mathrm{H}^\bullet(\mathcal{G}, k) = \ker \beta_L^\bullet \cap \mathrm{H}^\bullet(\mathcal{G}, k).$$

Consequently, $[\alpha_K] \mapsto \ker \alpha_K^\bullet \cap \mathrm{H}^\bullet(\mathcal{G}, k)$ is a well-defined, injective map. Friedlander and Pevtsova also show that one obtains all relevant homogeneous prime ideals of $\mathrm{H}^\bullet(\mathcal{G}, k)$ in this fashion.

Let $R = \bigoplus_{n\geq 0} R_n$ be a commutative, graded k-algebra such that $R_0 = k$. Recall that

$$\mathrm{Proj}(R) := \{\mathfrak{p} \trianglelefteq R;\ \mathfrak{p} \text{ homogeneous prime ideal, } \mathfrak{p} \subsetneq \bigoplus_{n\geq 1} R_n\}$$

is the projective space of R. If k is algebraically closed and $R = k[\mathcal{V}]$ is the coordinate ring of a conical variety $\mathcal{V}$, then $\mathrm{Proj}(\mathcal{V})$ is given by the maximal elements of $\mathrm{Proj}(k[\mathcal{V}])$.

Lemma 3.20. *Suppose $R = \bigoplus_{n\geq 0} R_n$ be a finitely generated k-algebra with $R_0 = k$. Given $\mathfrak{p} \in \mathrm{Proj}(R)$, let K be an algebraically closed extension field of the field of fractions $Q(R/\mathfrak{p})$. Then there exists a maximal element $\mathfrak{P} \in \mathrm{Proj}(R\otimes_k K)$ such that $\mathfrak{P} \cap R = \mathfrak{p}$.*

Proof. We first consider the case where $\mathfrak{p} = (0)$, so that R is an integral domain and $K \supseteq Q(R)$ contains its field of fractions. The canonical inclusion $R \hookrightarrow K$ induces an injective homomorphism

$$\gamma : R \longrightarrow K[T];\ \sum_{n\geq 0} r_n \mapsto \sum_{n\geq 0} r_n T^n$$

of graded k-algebras. We consider the extended graded K-homomorphism

$$\hat{\gamma} : R \otimes_k K \longrightarrow K[T];\ \ x \otimes \alpha \mapsto \alpha\gamma(x).$$

Since K is algebraically closed, the observations of Section 3.2 show that the ideal $\mathfrak{P} := \ker \hat{\gamma}$ is a maximal element of $\mathrm{Proj}(R \otimes_k K)$. By construction, we have $\mathfrak{P} \cap R = (0)$.

In the general case, we consider the finitely generated, graded integral domain $R' := R/\mathfrak{p}$. By the above, there exists a maximal element $\mathfrak{P}' \in \operatorname{Proj}(R' \otimes_k K)$ with $\mathfrak{P}' \cap R' = (0)$. We let $\pi : R \otimes_k K \longrightarrow R' \otimes_k K$ be the canonical graded homomorphism with $\ker \pi = \mathfrak{p} \otimes_k K$ and consider the maximal element $\mathfrak{P} := \pi^{-1}(\mathfrak{P}') \in \operatorname{Proj}(R \otimes_k K)$. Directly from the definitions we obtain

$$\mathfrak{P} \cap R = \pi^{-1}(\mathfrak{P}') \cap \pi^{-1}(R') \cap R = \pi^{-1}(\mathfrak{P}' \cap R') \cap R = \ker \pi \cap R = \mathfrak{p},$$

as desired. □

Theorem 3.21 ([61]). *Let $\mathcal{G}$ be a finite group scheme over k. Then*

$$\Psi_{\mathcal{G}} : \Pi(\mathcal{G}) \longrightarrow \operatorname{Proj}(\mathrm{H}^\bullet(\mathcal{G}, k)); \quad [\alpha_K] \mapsto \ker \alpha_K^\bullet \cap \mathrm{H}^\bullet(\mathcal{G}, k)$$

is a bijective map.

Proof. It remains to verify the surjectivity of the map $\Psi_{\mathcal{G}}$. Given $\mathfrak{p} \in \operatorname{Proj}(\mathrm{H}^\bullet(\mathcal{G}, k))$, we let K be the algebraic closure of the field of fractions of $R/\mathfrak{p}$. We have a commutative diagram

$$\begin{array}{ccc} \Pi(\mathcal{G}_K) & \xrightarrow{\Psi_{\mathcal{G}_K}} & \operatorname{Proj}(\mathrm{H}^\bullet(\mathcal{G}_K, K)) \\ \big\downarrow{\scriptstyle \mathrm{res}_1} & & \big\downarrow{\scriptstyle \mathrm{res}_2} \\ \Pi(\mathcal{G}) & \xrightarrow{\Psi_{\mathcal{G}}} & \operatorname{Proj}(\mathrm{H}^\bullet(\mathcal{G}, k)), \end{array}$$

where res_2 is induced by base change and res_1 is defined in Lemma 3.18. Lemma 3.20 provides a maximal element $\mathfrak{P} \in \operatorname{Proj}(\mathrm{H}^\bullet(\mathcal{G}_K, K))$ with $\mathfrak{p} = \mathrm{res}_2(\mathfrak{P})$. Thanks to Theorem 3.12, we can find $x \in \Pi(\mathcal{G}_K)$ with $\Psi_{\mathcal{G}_K}(x) = \mathfrak{P}$. Consequently, $\mathrm{res}_1(x) \in \Pi(\mathcal{G})$ is the desired pre-image of $\mathfrak{p}$ under $\Psi_{\mathcal{G}}$. □

Let M be an arbitrary $\mathcal{G}$-module (not necessarily finite-dimensional). In analogy with Section 3.2 one defines the Π-*support of* M via

$$\Pi(\mathcal{G})_M := \{[\alpha_K] \in \Pi(\mathcal{G});\ \alpha_K^*(M_K) \text{ is not projective}\}.$$

The sets $\Pi(\mathcal{G})_M$ enjoy the usual properties with respect to tensor products and short exact sequences. We summarize the relevant results of [61] in the following:

Theorem 3.22. *Let $\mathcal{G}$ be a finite group scheme over a field k.*

(1) *The set $\{\Pi(\mathcal{G})_M;\ \dim_k M < \infty\}$ is the set of closed sets of a noetherian topology on $\Pi(\mathcal{G})$.*

(2) *For every subset $\mathcal{X} \subseteq \Pi(\mathcal{G})$ there exists a $\mathcal{G}$-module M such that $\mathcal{X} = \Pi(\mathcal{G})_M$.*

(3) *The map $\Psi_{\mathcal{G}} : \Pi(\mathcal{G}) \longrightarrow \operatorname{Proj}(\mathrm{H}^\bullet(\mathcal{G}, k))$ is a homeomorphism such that*

$$\Psi_{\mathcal{G}}^{-1}(\operatorname{Proj}(\mathrm{H}^\bullet(\mathcal{G}, k)/\ker \Phi_M) = \Pi(\mathcal{G})_M$$

for every finite-dimensional $\mathcal{G}$-module M.

(4) *If M is a finite-dimensional $\mathcal{G}$-module, then $\dim \Pi(\mathcal{G})_M = \mathrm{cx}_{\mathcal{G}}(M) - 1$.* □

Instead of asking whether the pull-back $\alpha_K^*(M_K)$ of a finite-dimensional $\mathcal{G}_K$-module is projective, one can equally well investigate the isomorphism type of the $\mathfrak{A}_{p,K}$-module $\alpha_K^*(M_K)$. This point of view is adopted in the articles [24, 62], which study generic properties as well as modules of constant Jordan type.

Let $\alpha_K : \mathfrak{A}_{p,K} \longrightarrow K\mathcal{G}$ be a π-point. If $M \in \operatorname{mod} \mathcal{G}$, then $\alpha_K^*(M_K) \in \operatorname{mod} \mathfrak{A}_{p,K}$ uniquely decomposes as

$$\alpha_K^*(M_K) \cong \bigoplus_{i=1}^{p} \alpha_{K,i}(M)[i],$$

where $[i]$ represents the (up to isomorphism) unique indecomposable $\mathfrak{A}_{p,K}$-module of dimension i. We will interpret the right-hand side as a base change of an $\mathfrak{A}_{p,k}$-module N, that is,

$$\alpha_K^*(M_K) \cong N_K,$$

with $N = \bigoplus_{i=1}^{p} \alpha_{K,i}(M)[i] \in \operatorname{mod} \mathfrak{A}_{p,k}$. The isomorphism class of N is the *Jordan type of M with respect to α_K*, denoted $\operatorname{Jt}(M, \alpha_K)$.

Let α_K and β_L be π-points, $M \in \operatorname{mod} \mathcal{G}$ be a $\mathcal{G}$-module. In [62], the authors introduce a relation by setting

$$\alpha_K \trianglelefteq_M \beta_L \;:\Leftrightarrow \dim_K \operatorname{im} t^m_{\alpha_K^*(M_K)} \leq \dim_K \operatorname{im} t^m_{\beta_L^*(M_K)} \quad \forall\, m \in \{1, \ldots, p\}.$$

This condition is readily seen to be equivalent to

$$\alpha_K \trianglelefteq_M \beta_L \;:\Leftrightarrow \sum_{i=j}^{p} (i-j)\alpha_{K,i}(M_K) \leq \sum_{i=j}^{p} (i-j)\beta_{L,i}(M_L) \quad 1 \leq j \leq p.$$

In view of [27, (6.2.2)], the latter relation corresponds to the usual dominance ordering on the partitions of $\dim_k M$, associated to $\alpha_K^*(M_K)$ and $\beta_L^*(M_L)$, respectively. More precisely,

$$\alpha_K \sim_M \beta_L :\Leftrightarrow \operatorname{Jt}(M, \alpha_K) = \operatorname{Jt}(M, \beta_L)$$

defines an equivalence relation on the set $\operatorname{Pt}(\mathcal{G})$ of π-points and $\trianglelefteq_M$ is a partial ordering on the set of its equivalence classes.

This point of view provides new invariants of $\mathcal{G}$-modules and leads to the definition of classes of $\mathcal{G}$-modules that are only beginning to be understood. In [62] the authors show:

Theorem 3.23. *Let $\mathcal{G}$ be a finite k-group scheme and M be a finite-dimensional $\mathcal{G}$-module. Then*

$$\tilde{\Pi}(\mathcal{G})_M := \{x \in \Pi(\mathcal{G}); \exists\, \alpha_K \in x \textit{ such that } \operatorname{Jt}(M, \alpha_K) \textit{ is not of maximal for } \trianglelefteq_M\}$$

is a closed subspace of $\Pi(\mathcal{G})$. □

The non-maximal support space $\tilde{\Pi}(\mathcal{G})_M$ coincides with $\Pi(\mathcal{G})_M$ if and only if $\Pi(\mathcal{G})_M \neq \Pi(\mathcal{G})$.

Recall that a finite-dimensional $\mathcal{G}$-module M is projective if and only if $\Pi(\mathcal{G})_M = \emptyset$. If $\tilde{\Pi}(\mathcal{G})_M = \emptyset$, then M is of *constant Jordan type*, that is, $\mathrm{Jt}(M, \alpha_K) = \mathrm{Jt}(M, \beta_L)$ for any two π-points α_K and β_L of $\mathcal{G}$. This interesting class of modules was investigated in recent papers by Carlson-Friedlander-Pevtsova [24] and Carlson-Friedlander [23]. The full subcategory of $\mathrm{mod}\,\mathcal{G}$, whose objects are of constant Jordan type, is closed under direct sums and direct summands.

4 Varieties of tori

As we have seen in Sections 3.2 and 3.3, the determination of the representation type of finite algebraic groups leads to conditions on rank varieties. This leaves us with the problem of interpreting the ramifications of these conditions for the structure of the underlying groups. For Frobenius kernels of smooth groups, our knowledge of nilpotent orbits suffices to reduce the problem to the study of SL(2), a group whose representation theory is well enough understood to provide us with complete answers. The failure of the classical Lie-Kolchin Theorem for arbitrary infinitesimal groups already indicates the problems one encounters when leaving the classical environment of Frobenius kernels associated to smooth group schemes. Schemes of tori and their associated algebraic families of Lie algebras help us deal with this problem for arbitrary infinitesimal groups.

The material of this section is taken from [48, 49].

4.1 Algebraic families of vector spaces

Throughout, we let A be a finitely generated integral domain over the algebraically closed field k with associated scheme $\mathcal{X} := \mathrm{Spec}_k(A)$. Given a finite-dimensional k-vector space V, we consider the free A-module $V \otimes_k A$. For an A-submodule $P \subseteq V \otimes_k A$, and $x \in \mathcal{X}(k)$, we denote by $P(x) := (\mathrm{id}_V \hat{\otimes} x)(P) \subseteq V$ the subspace of V obtained by specialization along x. If P is a direct summand of the A-module $V \otimes_k A$, then $P(x) \cong P \otimes_A k_x$, where k_x denotes the one-dimensional A-module afforded by x. We will be studying the algebraic family $(P(x))_{x \in \mathcal{X}(k)}$ of subspaces of V.

Let X be a variety, $x \in X$ be a point. Then

$$\dim_x X := \max\{\dim Y;\ Y \subseteq X \text{ is an irreducible component of } X \text{ containing } x\}$$

is called the *local dimension* of X at x. A function $f : X \longrightarrow \mathbb{N}_0$, defined on a variety X, is called *upper semicontinuous* if for every $n \in \mathbb{N}_0$ the subset $f^{-1}(\{m \in \mathbb{N}_0;\ m \geq n\}) \subseteq X$ is closed. Given a morphism $\varphi : X \longrightarrow Y$ of varieties, we denote by $\varepsilon_\varphi(x) := \dim_x \varphi^{-1}(\varphi(x))$ the local dimension of $\varphi^{-1}(\varphi(x))$ at the point $x \in X$. The following basic result is referred to as semicontinuity of fiber dimension, see [35, §14].

Proposition 4.1. *Let $\varphi : X \longrightarrow Y$ be a morphism of varieties. Then the map*

$$X \longrightarrow \mathbb{N}_0;\quad x \mapsto \varepsilon_\varphi(x)$$

is upper semicontinuous. □

Proposition 4.2. *Let $Y \subseteq V$ be a conical closed subset, and suppose that $P \subseteq V \otimes_k A$ is an A-direct summand. Then the following statements hold:*

(1) *The function $\delta_{P\cap Y} : \mathcal{X}(k) \longrightarrow \mathbb{N}_0;\ x \mapsto \dim P(x) \cap Y$ is upper semicontinuous.*

(2) *If $d := \min\{\dim P(x) \cap Y;\ x \in \mathcal{X}(k)\}$, then $U := \delta^{-1}_{P\cap Y}(d)$ is a dense, open subset of $\mathcal{X}(k)$.*

Proof. (1) Consider the variety $\mathcal{X}(k) \times Y$ as well as the subset

$$Z := \{(x, y) \in \mathcal{X}(k) \times Y;\ \exists\ p \in P \text{ such that } (\mathrm{id}_V \hat{\otimes} x)(p) = y\}.$$

By assumption, there exists a submodule $Q \subseteq V \otimes_k A$ such that $V \otimes_k A = P \oplus Q$. Let $\{v_1, \ldots, v_n\}$ be a basis of V. We thus have projection maps

$$\mathrm{pr}_j : V \longrightarrow k;\ \sum_{i=1}^{n} \alpha_i v_i \mapsto \alpha_j \qquad 1 \le j \le n$$

as well as

$$\mathrm{pr}_Q : V \otimes_k A \longrightarrow Q;\ \ p + q \mapsto q.$$

Let $(a_{ij}) \in \mathrm{Mat}_n(A)$ be the $(n \times n)$-matrix, given by

$$\mathrm{pr}_Q(v_j \otimes 1) = \sum_{i=1}^{n} v_i \otimes a_{ij} \qquad 1 \le i \le n.$$

Let $(x, y) \in \mathcal{X}(k) \times Y$. Direct computation now shows that

$$(x, y) \in Z \ \text{ if and only if } \sum_{j=1}^{n} x(a_{ij})\mathrm{pr}_j(y) = 0 \qquad 1 \le i \le m.$$

Since $\mathrm{pr}_j|_Y \in k[Y]$ for $1 \le j \le n$ and $\tilde{a} : \mathcal{X}(k) \longrightarrow k; x \mapsto x(a)$ belongs to $k[\mathcal{X}(k)]$ for every $a \in A$, it follows that

$$Z = \{(x, y) \in \mathcal{X}(k) \times Y;\ [\sum_{j=1}^{n} \tilde{a}_{ij} \otimes \mathrm{pr}_j](x, y) = 0 \quad 1 \le i \le n\}$$

is closed.

We consider the restriction $\pi : Z \longrightarrow \mathcal{X}(k)$ of the projection onto the first factor. Given $x \in \mathcal{X}(k)$, the canonical map

$$\iota_x : Y \longrightarrow \mathcal{X}(k) \times Y;\ \ y \mapsto (x, y)$$

induces an isomorphism $\pi^{-1}(x) \cong P(x) \cap Y$, so that $\varepsilon_\pi(x, y) = \dim_y P(x) \cap Y$ for every point $(x, y) \in Z$. Since Y is a conical, closed subset of V, the neutral element 0 belongs to Y and $\dim_0 Y = \dim Y$. This also applies to the closed, conical subset $P(x) \cap Y$. Consider the morphism $\gamma : \mathcal{X}(k) \longrightarrow Z;\ x \mapsto (x, 0)$. The above observations yield

$$\varepsilon_\pi \circ \gamma = \delta_{P\cap Y},$$

and Proposition 4.1 ensures that this map is upper semicontinuous.

(2) By definition of d, the set $U = \mathfrak{X}(k) \setminus \{x \in \mathfrak{X}(k); \delta_{P\cap Y}(x) \geq d+1\}$ is a non-empty open subset of $\mathfrak{X}(k)$. Since A is an integral domain, the variety $\mathfrak{X}(k)$ is irreducible, so that U lies dense in $\mathfrak{X}(k)$. □

Corollary 4.3. *Suppose that $P \subseteq V \otimes_k A$ is an A-direct summand.*

(1) *The function $\delta_P : \mathfrak{X}(k) \longrightarrow \mathbb{N}_0;\ x \mapsto \dim_k P(x)$ is constant.*

(2) *If $P \neq (0)$, then $P(x) \neq (0)$ for every $x \in \mathfrak{X}(k)$.*

Proof. (1) By assumption, there exists an A-submodule $Q \subseteq V \otimes_k A$ such that $V \otimes_k A = P \oplus Q$. Consequently, $V = P(x) \oplus Q(x)$ for every $x \in \mathfrak{X}(k)$. Let $d_P := \min\{\dim_k P(x);\ x \in \mathfrak{X}(k)\}$ and $d_Q := \min\{\dim_k Q(x);\ x \in \mathfrak{X}(k)\}$. Since $\mathfrak{X}(k)$ is irreducible, Proposition 4.2 yields $d_P + d_Q = \dim_k V$. Consequently, $d_P \leq \dim_k P(x) = \dim_k V - \dim_k Q(x) \leq \dim_k V - d_Q = d_P$ for an arbitrary $x \in \mathfrak{X}(k)$.

(2) Suppose there is $x_0 \in \mathfrak{X}(k)$ such that $P(x_0) = (0)$. According to (1) we then have $P(x) = (0)$ for every $x \in \mathfrak{X}(k)$. Let $\{v_1, \dots, v_n\}$ be a basis of V. Given $p \in P$, we write $p = \sum_{i=1}^n v_i \otimes a_i$. Since

$$0 = (\mathrm{id}_V \hat{\otimes} x)(p) = \sum_{i=1}^n x(a_i)v_i \qquad \forall\ x \in \mathfrak{X}(k)$$

we see that the zero locus $Z(I)$ of the ideal $I := (\{a_1, \dots, a_n\}) \subseteq A$ is all of $\mathfrak{X}(k)$. Hilbert's Nullstellensatz now yields $I = (0)$, whence $p = 0$. Consequently, $P = (0)$, a contradiction. □

Given an A-direct summand $P \subseteq V \otimes_k A$, and a subset $U \subseteq \mathfrak{X}(k)$, we define

$$I_U(P) := \bigcap_{x \in U} P(x).$$

In the sequel, we are going to provide conditions ensuring that $I_U(P)$ is a non-zero subspace of V which is stable under certain group actions.

Definition. Let V be a finite-dimensional vector space, $\mathfrak{F}$ be a finite set of closed subsets of V. A family $(W_i)_{i\in I}$ of subspaces of V is said to be *$\mathfrak{F}$-regular*, provided every subspace W_i contains an element of $\mathfrak{F}$.

Lemma 4.4. *Let $\mathfrak{F}$ be a finite set of closed subsets of the finite-dimensional k-vector space V. If $P \subseteq V \otimes_k A$ is an A-submodule such that*

(a) *P is an A-direct summand of $V \otimes_k A$, and*

(b) *there exists a dense subset $U \subseteq \mathfrak{X}(k)$ such that $(P(x))_{x\in U}$ is $\mathfrak{F}$-regular,*

then the subspace $I_{\mathfrak{X}(k)}(P) \subseteq V$ is $\mathfrak{F}$-regular.

Proof. We write $\mathfrak{F} = \{Y_1, \dots, Y_n\}$. In view of (a), there exists an A-submodule $Q \subseteq V \otimes_k A$ with $V \otimes_k A = P \oplus Q$, so that $V = P(x) \oplus Q(x)$ for every $x \in \mathfrak{X}(k)$. Given $y \in V$, we write $y \otimes 1 = p + q$ with $p \in P$ and $q \in Q$. Then $y \in P(x)$ if and only if $(\mathrm{id}_V \hat{\otimes} x)(q) = 0$. By choosing a basis $\{v_1, \dots, v_m\}$ of V and writing

$q = \sum_{i=1}^{m} v_i \otimes a_i$, we see that the latter condition is equivalent to $x(a_i) = 0$ for $1 \leq i \leq m$. Consequently, the set $Z[y] := \{x \in \mathcal{X}(k);\ y \in P(x)\}$ is closed, and $Z_j := \{x \in \mathcal{X}(k);\ Y_j \subseteq P(x)\} = \bigcap_{y \in Y_j} Z[y]$ has the same property. Condition (b) implies $\mathcal{X}(k) = \bar{U} \subseteq \bigcup_{j=1}^{n} Z_j$. Since A is an integral domain, the variety $\mathcal{X}(k)$ is irreducible, and there exists $j_0 \in \{1, \ldots, n\}$ such that $Z_{j_0} = \mathcal{X}(k)$. It follows that $Y_{j_0} \subseteq I_{\mathcal{X}(k)}(P)$. □

Let $\mathcal{G}$ be an affine algebraic k-group, V be a $\mathcal{G}$-module. We assume that $\mathcal{G}$ acts on A via algebra homomorphisms. Consider the $\mathcal{G}$-module $V \otimes_k A$ with the diagonal operation

$$g.[(v \otimes_k r) \otimes_R (a \otimes_k s)] := g(v \otimes_k r) \otimes_R g(a \otimes_k s)$$

for $g \in \mathcal{G}(R)$, $r, s \in R$, $v \in V$, $a \in A$.

Lemma 4.5. *Let $P \subseteq V \otimes_k A$ be a $\mathcal{G}$-stable A-submodule that is also an A-direct summand of $V \otimes_k A$. If $U \subseteq \mathcal{X}(k)$ is dense, then $I_U(P) \subseteq V$ is a $\mathcal{G}$-submodule.*

Proof. One verifies the identity

$$V \cap P = I_U(P)$$

and observes that the left-hand side is a $\mathcal{G}$-submodule of V. □

4.2 Schemes of tori

Recall that k is an algebraically closed field of characteristic $p > 0$. Throughout this section we will only consider finite-dimensional restricted Lie algebras. Such a Lie algebra $(\mathfrak{t}, [p])$ is called a *torus* if and only if $\widehat{\mathcal{V}}_{\mathfrak{t}} = \{0\}$. Tori are necessarily abelian and linearly reductive: Every $U_0(\mathfrak{t})$-module is completely reducible.

If $\mathfrak{g} = \mathrm{Lie}(\mathcal{G})$ is the Lie algebra of a smooth group, then much of the structure of $\mathfrak{g}$ can be detected via the so-called *root space decomposition.* One picks a maximal torus $\mathfrak{t} \subseteq \mathfrak{g}$ and decomposes $\mathfrak{g}$ into its eigenspaces relative to $\mathfrak{t}$:

$$\mathfrak{g} = C_{\mathfrak{g}}(\mathfrak{t}) \oplus \bigoplus_{\alpha \in R} \mathfrak{g}_\alpha; \qquad R \subseteq \mathfrak{g}^* \setminus \{0\}.$$

Here $C_{\mathfrak{g}}(\mathfrak{t}) := \{x \in \mathfrak{g};\ [t, x] = 0 \quad \forall\ t \in \mathfrak{t}\}$ is the *centralizer of* $\mathfrak{t}$ *in* $\mathfrak{g}$, and $\mathfrak{g}_\alpha := \{x \in \mathfrak{g};\ [t, x] = \alpha(t)x \quad \forall\ t \in \mathfrak{t}\} \neq (0)$ is the *root space* for the *root* $\alpha \in R$. Since $\mathcal{G}$ is smooth, any two maximal tori are conjugate under the adjoint representation, so it doesn't really matter which maximal torus we take. The following example shows that this is no longer true for arbitrary restricted Lie algebras.

Example. For $p \geq 5$ we consider the *Witt algebra* $W(1) := \mathrm{Der}_k(k[X]/(X^p))$ of the derivations of the truncated polynomial ring $k[X]/(X^p)$. Since the p-th power of a derivation is again a derivation, $(W(1), p)$ is a restricted Lie algebra. Let ∂

be the derivation induced by $\frac{d}{dX}$ and set $x := X + (X^p)$ as well as $e_i := x^{i+1}\partial$ for $-1 \le i \le p-2$. Then $\{e_{-1}, \dots, e_{p-2}\}$ is a basis of $W(1)$, and we have

$$[e_i, e_j] = (j-i)e_{i+j}; \qquad e_i^p = \delta_{i,0}e_0,$$

where the product is understood to be zero whenever $i+j$ does not lie within $\{-1, \dots, p-2\}$. It follows that the p-subalgebra $\mathfrak{t} := ke_0$ is self-centralizing and hence is a maximal torus of $W(1)$.

We define another basis via $f_i := (x+1)^{i+1}\partial$ for $-1 \le i \le p-2$. Then $f_0^p = f_0$ and $[f_i, f_j] = (j-i)f_{i+j}$. In particular, $\mathfrak{t}' := kf_0$ is another maximal torus of $W(1)$. However, now the subscripts have to be interpreted mod(p), e.g., $[f_1, f_{p-2}] = -3f_{-1}$. Thus, while the root space decomposition relative to $\mathfrak{t}$ induces a $\mathbb{Z}$-grading, we have a grading with respect to the group $\mathbb{Z}/(p)$ in the latter case. The $\mathbb{Z}$-grading is better to work with because we can for instance read off that $\operatorname{ad} e_1$ is a nilpotent transformation.

These observations already indicate that $\mathfrak{t}$ and $\mathfrak{t}'$ are really different. In fact, they cannot be mapped onto each other by any automorphism of $W(1)$. Direct computation shows that $W(1)_{(0)} := \sum_{i=0}^{p-2} k\, e_i$ is the unique p-subalgebra of codimension 1 (here we need $p \ge 5$). Hence it is fixed by any automorphism $\varphi \in \operatorname{Aut}_p(W(1))$, and $\varphi(\mathfrak{t}) \subseteq W(1)_{(0)}$. In particular, $\varphi(\mathfrak{t}) \ne \mathfrak{t}'$.

The foregoing example illustrates our predicament. We have to choose a maximal torus without knowing which choice is good for our purposes. Schemes of tori obviate this difficulty by simultaneously studying all tori of a certain isomorphism type.

Let $(\mathfrak{g}, [p])$ be a restricted Lie algebra over k, R be a commutative k-algebra. Recall that $\mathfrak{g} \otimes_k R$ obtains the structure of a restricted Lie algebra over R via

$$[x \otimes r, y \otimes s] := [x, y] \otimes rs;$$
$$(x \otimes r)^{[p]} = x^{[p]} \otimes r^p \qquad \forall\ x, y \in \mathfrak{g},\ r, s \in R.$$

Now let $(\mathfrak{t}, \mathfrak{g})$ be a pair of restricted Lie algebras over k. We consider the k-functor $\mathcal{T}_{\mathfrak{g}} : M_k \longrightarrow \text{Ens}$ that associates to each commutative k-algebra R the set $\mathcal{T}_{\mathfrak{g}}(R)$ of those homomorphisms $\varphi : \mathfrak{t} \otimes_k R \longrightarrow \mathfrak{g} \otimes_k R$ of restricted Lie algebras over R that are split injective R-linear maps. Observe that the set $\mathcal{T}_{\mathfrak{g}}(k)$ of k-rational points is just the set of embeddings $\mathfrak{t} \hookrightarrow \mathfrak{g}$.

Theorem 4.6 ([48]). *Let $\mathfrak{t}$ be a torus. Given a restricted Lie algebra $\mathfrak{g}$, the following statements hold:*

(1) *$\mathcal{T}_{\mathfrak{g}}$ is a smooth, affine, algebraic scheme.*

(2) *If $\mathcal{X} \subseteq \mathcal{T}_{\mathfrak{g}}$ is an irreducible component, then*

$$\dim \mathcal{X} = \dim_k \mathfrak{g} - \dim_k C_{\mathfrak{g}}(\varphi(\mathfrak{t})) \qquad \forall\ \varphi \in \mathcal{X}(k). \qquad \square$$

One main point of (1) is that the connected components of $\mathcal{T}_{\mathfrak{g}}$ coincide with its irreducible components. Thus, if $\mathfrak{t} \subseteq \mathfrak{g}$ is a torus with the embedding $\mathfrak{t} \hookrightarrow$

$\mathfrak{g}$ corresponding to a rational point $x_0 \in \mathcal{T}_{\mathfrak{g}}(k)$, then there exists exactly one irreducible component $\mathcal{X}_{\mathfrak{t}} \subseteq \mathcal{T}_{\mathfrak{g}}$ such that $x_0 \in \mathcal{X}_{\mathfrak{t}}(k)$.

We illustrate Theorem 4.6 by considering Lie algebras of smooth groups. Let $\mathcal{G}$ be an affine algebraic group, $\mathrm{Ad} : \mathcal{G}(k) \longrightarrow \mathrm{Aut}_k(\mathrm{Lie}(\mathcal{G}))$ be its adjoint representation. Then $\mathcal{G}(k)$ operates on the affine variety $\mathcal{T}_{\mathfrak{g}}(k)$ via

$$g.\varphi := \mathrm{Ad}(g) \circ \varphi \qquad \forall\, g \in \mathcal{G}(k),\ \varphi \in \mathcal{T}_{\mathfrak{g}}(k).$$

Proposition 4.7. *Let* $\mathfrak{g} = \mathrm{Lie}(\mathcal{G})$ *be the Lie algebra of a smooth, connected, affine algebraic group,* $\mathfrak{t}$ *be a torus. Then the connected components of* $\mathcal{T}_{\mathfrak{g}}(k)$ *are the* $\mathcal{G}(k)$*-orbits of* $\mathcal{T}_{\mathfrak{g}}(k)$. □

Remarks. (1) Let $\mathfrak{t} \overset{x_0}{\hookrightarrow} \mathfrak{g}$ be the canonical embedding. Under the assumptions of Proposition 4.7 the morphism $g \mapsto \mathrm{Ad}(g) \circ x_0$ induces a bijective $\mathcal{G}(k)$-equivariant map $\mathcal{G}(k)/\mathrm{Stab}_{\mathcal{G}(k)}(x_0) \longrightarrow \mathcal{X}_{\mathfrak{t}}(k)$ of homogeneous spaces. Owing to [48, (1.4)], its differential $T_{\bar{1}}(\mathcal{G}(k)/\mathrm{Stab}_{\mathcal{G}(k)}(x_0)) \longrightarrow \mathrm{T}_{x_0}(\mathcal{X}_{\mathfrak{t}})$ is given by the bijection

$$\mathfrak{g}/C_{\mathfrak{g}}(\mathfrak{t}) \longrightarrow \mathrm{Der}_k(\mathfrak{t}, \mathfrak{g});\quad [x] \mapsto \mathrm{ad}\, x|_{\mathfrak{t}}.$$

Consequently, we have an isomorphism $\mathcal{G}(k)/C_{\mathcal{G}(k)}(\mathfrak{t}) \cong \mathcal{X}_{\mathfrak{t}}(k)$. Here $C_{\mathcal{G}(k)}(\mathfrak{t}) = \{g \in \mathcal{G}(k);\ \mathrm{Ad}(g)(t) = t\ \ \forall\, t \in \mathfrak{t}\}$ is the *centralizer of* $\mathfrak{t}$ *in* $\mathcal{G}(k)$.

(2) One can use schemes of tori to recover the classical conjugacy theorem for maximal tori of Lie algebras of smooth groups, see [39] for more details.

Let $(\mathfrak{g}, [p])$ be a restricted Lie algebra, $\mathfrak{t} \subseteq \mathfrak{g}$ be a torus with embedding $\mathfrak{t} \hookrightarrow \mathfrak{g}$ corresponding to a rational point $x_0 \in \mathcal{T}_{\mathfrak{g}}(k)$. Thanks to Theorem 4.6, the irreducible component $\mathcal{X}_{\mathfrak{t}}$ is representable: There exists a finitely generated integral domain A such that $\mathcal{X}_{\mathfrak{t}} \cong \mathrm{Spec}_k(A)$. We consider the restricted Lie algebra $\tilde{\mathfrak{g}} := \mathfrak{g} \otimes_k A$. Under the above identification $\mathrm{id}_A \in \mathrm{Spec}_k(A)(A)$ corresponds to an embedding $j : \mathfrak{t} \longrightarrow \tilde{\mathfrak{g}}$ of restricted k-Lie algebras such that the A-submodule $Aj(\mathfrak{t}) \subseteq \tilde{\mathfrak{g}}$ is a direct summand of $\tilde{\mathfrak{g}}$. Since $\mathcal{X}_{\mathfrak{t}} \cong \mathrm{Spec}_k(A)$, any element $\varphi : \mathfrak{t} \longrightarrow \mathfrak{g} \otimes_k R$ of $\mathcal{X}_{\mathfrak{t}}(R)$ is obtained via specialization: If $\varphi \in \mathcal{X}_{\mathfrak{t}}(R)$ corresponds to $x \in \mathrm{Spec}_k(A)(R)$, then we have

$$\varphi = (\mathrm{id}_{\mathfrak{g}} \otimes x) \circ j.$$

For that reason, we call $j : \mathfrak{t} \hookrightarrow \tilde{\mathfrak{g}}$ the *universal embedding*.

Note that j endows the extended Lie algebra $\tilde{\mathfrak{g}}$ with the structure of an infinite-dimensional $U_0(\mathfrak{t})$-module. Since $U_0(\mathfrak{t})$ is commutative and semi-simple, there results a weight space decomposition

$$\tilde{\mathfrak{g}} = \tilde{\mathfrak{g}}_0 \oplus \bigoplus_{\alpha \in \Phi} \tilde{\mathfrak{g}}_\alpha$$

of $\tilde{\mathfrak{g}}$ relative to $\mathfrak{t}$. Here $\Phi \subseteq \mathfrak{t}^* \setminus \{0\}$ is the set of weights, and for $\alpha \in \Phi \cup \{0\}$ the weight space $\tilde{\mathfrak{g}}_\alpha = \{v \in \tilde{\mathfrak{g}};\ [j(t), v] = \alpha(t)v\ \ \forall\, t \in \mathfrak{t}\} \neq (0)$ is an A-direct summand of $\tilde{\mathfrak{g}}$. Given an arbitrary element $x \in \mathcal{X}_{\mathfrak{t}}(k)$, we have

$$\mathfrak{g} = \tilde{\mathfrak{g}}_0(x) \oplus \bigoplus_{\alpha \in \Phi} \tilde{\mathfrak{g}}_\alpha(x),$$

where $\tilde{\mathfrak{g}}_\alpha(x) = \{v \in \mathfrak{g};\ [(\mathrm{id}_\mathfrak{g} \hat{\otimes} x)(j(t)), v] = \alpha(t)v \ \ \forall\, t \in \mathfrak{t}\}$ for $\alpha \in \Phi \cup \{0\}$. In other words, if $\varphi := (\mathrm{id}_\mathfrak{g} \hat{\otimes} x) \circ j$ is the embedding corresponding to $x \in \mathfrak{X}_\mathfrak{t}(k)$, then $\tilde{\mathfrak{g}}_\alpha(x)$ is the root space with root $\alpha \circ \varphi^{-1}$ relative to the torus $\varphi(\mathfrak{t}) \subseteq \mathfrak{g}$. In particular, specialization along x_0 yields the root space decomposition

$$\mathfrak{g} = C_\mathfrak{g}(\mathfrak{t}) \oplus \bigoplus_{\alpha \in \Phi} \mathfrak{g}_\alpha$$

of $\mathfrak{g}$ relative to $\mathfrak{t}$. Thanks to Corollary 4.3, this also shows that Φ is a finite set.

Recall that the rank variety of the trivial module of a restricted Lie algebra $(\mathfrak{g}, [p])$ is given by

$$\widehat{\mathcal{V}}_\mathfrak{g} = \{x \in \mathfrak{g};\ x^{[p]} = 0\}.$$

We will occasionally refer to $\dim \widehat{\mathcal{V}}_\mathfrak{g}$ as the *null-rank* of the restricted Lie algebra $\mathfrak{g}$.

Returning to our general set-up, we let $P \subseteq \tilde{\mathfrak{g}}$ be an A-direct summand of $\tilde{\mathfrak{g}}$. By virtue of Proposition 4.2, there exists a non-empty open subset $U_P \subseteq \mathfrak{X}_\mathfrak{t}(k)$, and a natural number $c_P(\mathfrak{g}, \mathfrak{t}) \in \mathbb{N}_0$ such that

$$\dim P(x) \cap \widehat{\mathcal{V}}_\mathfrak{g} = c_P(\mathfrak{g}, \mathfrak{t}) \qquad \forall\, x \in U_P.$$

The number $c_P(\mathfrak{g}, \mathfrak{t})$ is the *generic null-rank* of the algebraic family $(P(x))_{x \in \mathfrak{X}_\mathfrak{t}(k)}$ of subspaces of $\mathfrak{g}$. Given a subset $\Psi \subseteq \mathfrak{t}^*$, the A-submodule $\tilde{\mathfrak{g}}^\Psi := \bigoplus_{\alpha \in \Psi} \tilde{\mathfrak{g}}_\alpha$ is a direct summand of $\tilde{\mathfrak{g}}$, and we write $c_\Psi(\mathfrak{g}, \mathfrak{t}) := c_{\tilde{\mathfrak{g}}^\Psi}(\mathfrak{g}, \mathfrak{t})$.

A nilpotent, self-normalizing p-subalgebra of $\mathfrak{g}$ is called a *Cartan subalgebra*. According to general theory, the centralizer $C_\mathfrak{g}(\mathfrak{t})$ of the torus $\mathfrak{t} \subseteq \mathfrak{g}$ is a Cartan subalgebra of $\mathfrak{g}$ if and only if $\mathfrak{t}$ is a maximal torus (cf. [115, Chapter II]).

We illustrate the utility of the generic null-rank by verifying the following subsidiary result:

Lemma 4.8. *Let $\mathfrak{t} \subseteq \mathfrak{g}$ be a torus. Then $c_0(\mathfrak{g}, \mathfrak{t}) = 0$ if and only if $C_\mathfrak{g}(\mathfrak{t}) = \mathfrak{t}$.*

Proof. Let $A := k[\mathfrak{X}_\mathfrak{t}]$ be the coordinate ring of the component $\mathfrak{X}_\mathfrak{t}$. We consider the A-direct summands $Aj(\mathfrak{t})$ and $\tilde{\mathfrak{g}}_0$ of $\tilde{\mathfrak{g}}$, and observe that for $x \in \mathfrak{X}(k)$ and its associated embedding $\varphi := (\mathrm{id}_\mathfrak{g} \hat{\otimes} x) \circ j$, the specializations along x are given by

$$(Aj(\mathfrak{t}))(x) = \varphi(\mathfrak{t}) \ \text{ and } \ \tilde{\mathfrak{g}}_0(x) = C_\mathfrak{g}(\varphi(\mathfrak{t})),$$

respectively. By definition, there exists a dense open subset $U \subseteq \mathfrak{X}_\mathfrak{t}(k)$ such that

$$\dim \widehat{\mathcal{V}}_{\tilde{\mathfrak{g}}_0(x)} = c_0(\mathfrak{g}, \mathfrak{t}) \quad \forall\, x \in U.$$

Suppose that $c_0(\mathfrak{g}, \mathfrak{t}) = 0$ and let $x \in U$. Then $\widehat{\mathcal{V}}_{\tilde{\mathfrak{g}}_0(x)} = \{0\}$ and $C_\mathfrak{g}(\varphi(\mathfrak{t}))$ is a torus. Since $C_\mathfrak{g}(\varphi(\mathfrak{t}))$ is self-normalizing, it is a Cartan subalgebra of $\mathfrak{g}$. Thus, $(Aj(\mathfrak{t}))(x) = \varphi(\mathfrak{t})$ is a maximal torus of $\mathfrak{g}$, and we conclude that $(Aj(\mathfrak{t}))(x) = \tilde{\mathfrak{g}}_0(x)$. As $Aj(\mathfrak{t})$ and $\tilde{\mathfrak{g}}_0$ are A-direct summands of $\tilde{\mathfrak{g}}$, Corollary 4.3 yields

$$\begin{aligned}\dim_k C_\mathfrak{g}(\mathfrak{t}) &= \dim_k \tilde{\mathfrak{g}}_0(x_0) = \dim_k \tilde{\mathfrak{g}}_0(x) = \dim_k (Aj(\mathfrak{t}))(x) \\ &= \dim_k (Aj(\mathfrak{t}))(x_0) = \dim_k \mathfrak{t},\end{aligned}$$

so that $\mathfrak{t} = C_{\mathfrak{g}}(\mathfrak{t})$.

Conversely, assume that $\mathfrak{t} = C_{\mathfrak{g}}(\mathfrak{t})$. Then we have $\dim_k(Aj(\mathfrak{t}))(x_0) = \dim_k \tilde{\mathfrak{g}}_0(x_0)$. As $Aj(\mathfrak{t})$ and $\tilde{\mathfrak{g}}_0$ are A-direct summands of $\tilde{\mathfrak{g}}$, Corollary 4.3 implies $(Aj(\mathfrak{t}))(x) = \tilde{\mathfrak{g}}_0(x)$ for every $x \in \mathfrak{X}_{\mathfrak{t}}(k)$. This shows that $c_0(\mathfrak{g}, \mathfrak{t}) = 0$. □

Let $(\mathfrak{g}, [p])$ be a restricted Lie algebra. We denote by $\mathcal{AUT}(\mathfrak{g})$ the automorphism scheme of $\mathfrak{g}$. For every commutative k-algebra R, $\mathcal{AUT}(\mathfrak{g})(R)$ is the set of automorphisms of the restricted R-Lie algebra $\mathfrak{g}\otimes_k R$. The connected component of $\mathcal{AUT}(\mathfrak{g})$ will be denoted $\mathcal{G}_{\mathfrak{g}}$. A subspace $\mathfrak{n} \subseteq \mathfrak{g}$ is $\mathcal{G}_{\mathfrak{g}}$-*invariant* if $g(\mathfrak{n}\otimes_k R) = \mathfrak{n}\otimes_k R$ for every $R \in M_k$ and $g \in \mathcal{G}_{\mathfrak{g}}(R)$.

The natural operation of $\mathcal{G}_{\mathfrak{g}}$ on $\mathfrak{g}$ induces an action of $\mathcal{G}_{\mathfrak{g}}$ on $\mathcal{T}_{\mathfrak{g}}$: For $g \in \mathcal{G}_{\mathfrak{g}}(R)$ and $\varphi \in \mathcal{T}_{\mathfrak{g}}(R)$ we have $g.\varphi := g \circ \varphi$.

Lemma 4.9. *The following statements hold:*

(1) *Every irreducible component* $\mathfrak{X} \subseteq \mathcal{T}_{\mathfrak{g}}$ *is* $\mathcal{G}_{\mathfrak{g}}$*-invariant.*

(2) *Let* $\Psi \subseteq \mathfrak{t}^*$. *Then* $I_{\mathfrak{X}_{\mathfrak{t}}(k)}(\tilde{\mathfrak{g}}^{\Psi})$ *is* $\mathcal{G}_{\mathfrak{g}}$*-invariant.*

(3) *Let* $\Psi \subseteq \mathfrak{t}^*$. *Then* $I_{\mathfrak{X}_{\mathfrak{t}}(k)}(\tilde{\mathfrak{g}}^{\Psi})$ *is an ideal of* $\mathfrak{g}$. □

We illustrate the use of these techniques by considering representation-finite restricted Lie algebras. Recall that $\mathfrak{g}$ is referred to as *solvable* if its *derived series* $(\mathfrak{g}^{(n)})_{n\geq 0}$, which is inductively defined via $\mathfrak{g}^{(0)} := \mathfrak{g}$ and $\mathfrak{g}^{(n+1)} = [\mathfrak{g}^{(n)}, \mathfrak{g}^{(n)}]$, contains the ideal (0)of $\mathfrak{g}$. If this holds for the *descending central series* $(\mathfrak{g}^n)_{n\geq 1}$ with $\mathfrak{g}^1 := \mathfrak{g}$ and $\mathfrak{g}^{n+1} := [\mathfrak{g}, \mathfrak{g}^n]$, then $\mathfrak{g}$ is called *nilpotent.*

Proposition 4.10. *Let* $(\mathfrak{g}, [p])$ *be a restricted Lie algebra such that the principal block* $\mathcal{B}_0(\mathfrak{g}) \subseteq U_0(\mathfrak{g})$ *has finite representation type. Then* $\mathfrak{g}$ *is solvable.*

Proof. According to Theorem 2.7 and Corollary 3.14 the null-rank of $\mathfrak{g}$ is bounded by 1, that is, $\dim \widehat{\mathcal{V}}_{\mathfrak{g}} \leq 1$. If $\mathfrak{g}$ has null-rank 0, then $\mathfrak{g}$ is a torus, and thus is in particular abelian. We therefore assume $\dim \widehat{\mathcal{V}}_{\mathfrak{g}} = 1$ and proceed by induction on $\dim_k \mathfrak{g}$, the case $\dim_k \mathfrak{g} = 1$ being trivial.

Let $\mathfrak{t} \subseteq \mathfrak{g}$ be a maximal torus, $\mathfrak{X}_{\mathfrak{t}} \subseteq \mathcal{T}_{\mathfrak{g}}$ be the irreducible component containing the embedding $\mathfrak{t} \overset{x_0}{\hookrightarrow} \mathfrak{g}$. We have the weight space decomposition

$$\tilde{\mathfrak{g}} = \tilde{\mathfrak{g}}_0 \oplus \bigoplus_{\alpha\in\Phi} \tilde{\mathfrak{g}}_\alpha$$

of $\tilde{\mathfrak{g}}$ relative to $\mathfrak{t}$. In our current situation, the generic null-rank $c_0(\mathfrak{g}, \mathfrak{t})$ satisfies

$$0 \leq c_0(\mathfrak{g}, \mathfrak{t}) \leq 1.$$

If $c_0(\mathfrak{g}, \mathfrak{t}) = 1$, then there exists a dense open subset $U \subseteq \mathfrak{X}_{\mathfrak{t}}(k)$ such that the family $(\tilde{\mathfrak{g}}_0(x))_{x\in U}$ is regular with respect to the one-dimensional irreducible components of the variety $\widehat{\mathcal{V}}_{\mathfrak{g}}$. A consecutive application of Lemma 4.4 and Lemma 4.9 now implies that

$$\mathfrak{n} := I_{\mathfrak{X}_{\mathfrak{t}}(k)}(\tilde{\mathfrak{g}}_0) \subseteq C_{\mathfrak{g}}(\mathfrak{t})$$

is a non-zero p-ideal of $\mathfrak{g}$ that is contained in the Cartan subalgebra $C_{\mathfrak{g}}(\mathfrak{t})$. Thus, $\mathfrak{n}$ is nilpotent and hence solvable. Moreover, the principal block $\mathcal{B}_0(\mathfrak{g}/\mathfrak{n})$ is representation-finite, so that the inductive hypothesis implies the solvability of $\mathfrak{g}/\mathfrak{n}$. Consequently, the Lie algebra $\mathfrak{g}$ is solvable.

It remains to consider the case $c_0(\mathfrak{g}, \mathfrak{t}) = 0$. Let $\alpha \in \Phi$ and consider $\Psi := \mathbb{F}_p\alpha \cap \Phi$. Then $c_\Psi(\mathfrak{g}, \mathfrak{t}) = 1$, and the foregoing arguments in conjunction with Lemma 4.8 provide a non-zero p-ideal

$$\mathfrak{n} \subseteq \mathfrak{t} \oplus \bigoplus_{i=1}^{p-1} \mathfrak{g}_{i\alpha} =: \mathfrak{h}$$

of $\mathfrak{g}$. We shall show that the p-subalgebra $\mathfrak{h}$ is solvable. Note that $\ker \alpha$ is contained in the center $C(\mathfrak{h})$ of $\mathfrak{h}$. Direct computation shows that $\mathfrak{h}' := \mathfrak{h}/\ker\alpha$ is a restricted Lie algebra with a one-dimensional maximal torus and a one-dimensional nullcone. Hence we may assume that the torus $\mathfrak{t}$ of $\mathfrak{h}$ has dimension $\dim_k \mathfrak{t} = 1$, so that $\mathfrak{t} = kt_0$. Let $\beta \in \mathbb{F}_p\alpha$ be a root. Given $x \in \mathfrak{g}_\beta \setminus \{0\}$, there exists $\omega(x) \in k$ with $x^{[p]} = \omega(x)t_0$. Since $\beta(t_0) \neq 0$, the identities

$$0 = [x^{[p]}, x] = \omega(x)[t_0, x] = \omega(x)\beta(t_0)x$$

imply $\omega(x) = 0$, so that $\mathfrak{g}_\beta \subseteq \widehat{\mathcal{V}}_{\mathfrak{g}}$. As a result, $\dim_k \mathfrak{g}_\beta = 1$ and $\mathfrak{h}_{(\beta)} := \mathfrak{g}_{-\beta} \oplus \mathfrak{t} \oplus \mathfrak{g}_\beta$ is a p-subalgebra of $\mathfrak{h}$. The assumption $[\mathfrak{g}_\beta, \mathfrak{g}_{-\beta}] \neq (0)$ implies $\mathfrak{h}_{(\beta)} \cong \mathfrak{sl}(2)$, which contradicts $\dim \widehat{\mathcal{V}}_{\mathfrak{h}} = 1$.

As a result, the derived algebra $[\mathfrak{h}, \mathfrak{h}] = \bigoplus_{i=1}^{p-1} \mathfrak{g}_{i\alpha}$ is a p-subalgebra of $\mathfrak{h}$ that is generated by the nil Lie set $\bigcup_{i=1}^{p-1} \mathfrak{g}_{i\alpha}$. The Engel-Jacobson Theorem now implies that $[\mathfrak{h}, \mathfrak{h}]$ is nilpotent. Consequently, $\mathfrak{h}$ is solvable.

Hence $\mathfrak{n}$ is solvable and we may argue as before. □

Refinements of the above arguments yield stronger results. A non-zero Lie algebra $\mathfrak{g}$ which does not possess any non-zero abelian ideals is called *semi-simple.* For future reference, we record the following result:

Theorem 4.11 ([48]). *Let $(\mathfrak{g}, [p])$ be a semi-simple restricted Lie algebra of characteristic $p \geq 3$. If $\dim \widehat{\mathcal{V}}_{\mathfrak{g}} = 2$, then $\mathfrak{g} \cong \mathfrak{sl}(2)$.* □

5 Quivers and path algebras

In this section we turn to the combinatorial aspects of the representation theory of group schemes. The methods originate in the abstract representation theory of algebras, where one is interested in understanding module categories up to Morita equivalence (or other equivalences, such as stable equivalence or derived equivalence). Over algebraically closed fields, this amounts to studying the so-called *basic algebras*, whose simple modules are one-dimensional. Such an algebra is given by a quiver and certain relations, giving rise to a presentation that encapsulates a lot of information concerning the structure of the module category. We shall only give a rather short introduction to this wide subject. A detailed account may be found in [8, 4].

A *quiver* $Q = (Q_0, Q_1, s, t)$ is an oriented graph on a set of vertices Q_0 and with a set Q_1 of arrows, whose starting points and terminal points are given by the functions $s, t : Q_1 \longrightarrow Q_0$. The *path algebra* $k[Q]$ of the quiver has an underlying vector space, whose basis is the set of all paths of Q. The product of two paths is given by concatenation if this is possible, and is defined to be zero otherwise. We compose arrows like maps. Thus, if p is a path starting in $u \in Q_0$ and terminating $v \in Q_0$ and q is a path starting in v and terminating in w, then qp starts in u and ends in w. We shall denote by $k[Q]^+$ the two-sided ideal of $k[Q]$ that is generated by all arrows. Then $(k[Q]^+)^n$ is the ideal of $k[Q]$ consisting of all linear combinations of all paths of length $\geq n$. For each vertex $v \in Q_0$ there is a path e_v of length zero starting and ending in v. The corresponding element of $k[Q]$ is an idempotent with $e_v p = p$ ($qe_v = q$) for all paths p ending in v (for all paths q starting in v).

Let me illustrate the advantages of path algebras by considering a simple example.

Example. Let $\text{char}(k) = p > 0$, and consider the two-dimensional, non-abelian restricted Lie algebra $\mathfrak{b} := kt \oplus kx$, whose multiplication and p-map are given by

$$[t, x] = x; \quad t^{[p]} = t, \ x^{[p]} = 0.$$

The corresponding presentation of its restricted enveloping algebra is

$$U_0(\mathfrak{b}) \cong k\langle t, x\rangle/(\{tx - xt - x, t^p - t, x^p\}).$$

The path algebra presentation, which yields an isomorphic algebra in this case, is

$$U_0(\mathfrak{b}) \cong k[\tilde{A}_{p-1}]/(k[\tilde{A}_{p-1}]^+)^p; \quad \tilde{A}_{p-1} \text{ clockwise oriented.}$$

Here $\tilde{A}_{p-1}$ is a circle with p vertices. This more complicated presentation immediately tells us that the algebra $U_0(\mathfrak{b})$ is a Nakayama algebra and hence has finite representation type (cf. Proposition 2.6).

5.1 Gabriel's Theorem

Let k be an algebraically closed field. It turns out that every associative algebra is Morita equivalent to an algebra that possesses a presentation like the one given above. The first question is how to associate a quiver Q_Λ to an algebra Λ.

Definition. A k-algebra Λ is called *basic* if $\Lambda = \bigoplus_{i=1}^{n} P_i$ with the P_i being indecomposable and pairwise non-isomorphic.

Since k is algebraically closed, an algebra is basic if and only if all of its simple modules are one-dimensional.

Let us look again at the above example. The Hopf algebra $U_0(\mathfrak{b})$ is basic, and its simple modules are of the form k_λ, where the algebra homomorphism

$\lambda : U_0(\mathfrak{b}) \longrightarrow k$ is uniquely determined by $\lambda(x) = 0$ and $\lambda(t) \in \mathbb{F}_p$. Hence there are exactly p simple $U_0(\mathfrak{b})$-modules, corresponding to the vertices of $\tilde{A}_{p-1}$. Note that the convolution product on the character group $X(U_0(\mathfrak{b}))$ corresponds to the addition of linear forms on $\mathfrak{b}$. We thus obtain

$$\operatorname{Ext}^1_{U_0(\mathfrak{g})}(k_\lambda, k_\gamma) \cong \mathrm{H}^1(U_0(\mathfrak{b}), k_{\gamma-\lambda}).$$

According to early results by Hochschild [70], the restricted cohomology groups on the right-hand side can be computed via Lie cocycles, whence

$$\dim_k \operatorname{Ext}^1_{U_0(\mathfrak{g})}(k_\lambda, k_\gamma) = \delta_{\gamma-\lambda,\alpha},$$

where $\alpha : U_0(\mathfrak{b}) \longrightarrow k$ is given by $\alpha(x) = 0$ and $\alpha(t) = 1$. By identifying the simple module k_λ with the value $\lambda(t) \in \mathbb{F}_p$ and taking $\dim_k \operatorname{Ext}_{U_0(\mathfrak{b})}(k_\lambda, k_\gamma)$ arrows from $\lambda(t)$ to $\gamma(t)$, we obtain the quiver $\tilde{A}_{p-1}$.

Definition. Let Λ be a finite-dimensional associative k-algebra, $\mathcal{S}$ a complete set of representatives of the simple Λ-modules. The quiver Q_Λ with set of vertices $\mathcal{S}$ and $\dim_k \operatorname{Ext}^1_\Lambda(S,T)$ arrows from S to T is called the *Ext-quiver* of Λ.

Suppose that $\Lambda = k[Q]/I$ is a finite-dimensional algebra such that $I \subseteq (k[Q]^+)^2$. Basic homological algebra then shows that $Q = Q_\Lambda$. Gabriel's Theorem asserts that any k-algebra is essentially of this type:

Theorem 5.1 ([64]). *Let Λ be a finite-dimensional k-algebra. Then there exist $n \geq 2$ and an ideal $(k[Q_\Lambda]^+)^n \subseteq I \subseteq (k[Q_\Lambda]^+)^2$ such that Λ is Morita equivalent to $k[Q_\Lambda]/I$.*

Proof. To ease notation, we put $k[Q_\Lambda]_{\geq n} := (k[Q_\Lambda]^+)^n$. By general theory, the algebra Λ is Morita equivalent to a basic algebra, so that we may assume that all simple Λ-modules are one-dimensional.

Let $\{\varepsilon_1, \ldots, \varepsilon_r\} \subseteq k[Q_\Lambda]$ be the set of paths of length zero. By assumption, there exist orthogonal primitive idempotents $e_1, \ldots, e_r \in \Lambda$ with $1 = \sum_{i=1}^r e_i$. Let J be the Jacobson radical of Λ. Given $i, j \in \{1, \ldots, r\}$, we pick elements $a_{ij\ell} \in e_i J e_j$ such that the residue classes $\bar{a}_{ij\ell}$ form a basis of $e_i J e_j / e_i J^2 e_j$.

Proceeding in several steps, we define a k-linear map $f : k[Q_\Lambda] \longrightarrow \Lambda$. By general theory, we have

$$\operatorname{Ext}^1_\Lambda(S_i, S_j) \cong e_j J e_i / e_j J^2 e_i.$$

Letting $n_{ij} := \dim_k \operatorname{Ext}^1_\Lambda(S_i, S_j)$, we put

- $f(\varepsilon_i) := e_i$ for $1 \leq i \leq r$.
- $f(\alpha_{ij\ell}) := a_{ji\ell}$ for every arrow $\alpha_{ij\ell} : i \to j$ with $1 \leq \ell \leq n_{ij}$.
- $f(p) := \prod_{j=1}^m f(\alpha_j)$ for every path $p = \alpha_1 \cdots \alpha_m$.

In this fashion, we obtain a linear map $f : k[Q_\Lambda] \longrightarrow \Lambda$.

Let $\alpha : i \to j$ and $\beta : \ell \longrightarrow m$ be arrows such that $\ell \neq j$. Then $\beta\alpha = 0$ and $f(\beta)f(\alpha) \in e_m J e_\ell e_j J e_j = (0)$. This shows that f is a homomorphism of k-algebras.

Since every simple Λ-module is one-dimensional, Wedderburns Theorem provides an isomorphism $\Lambda/J \cong \bigoplus_{i=1}^{r} ke_i$. Moreover, $J/J^2 \cong \bigoplus_{i,j=1}^{r} e_iJe_j/e_iJ^2e_j$. As a result, the map $\bar{f} : k[Q_\Lambda]/k[Q_\Lambda]_{\geq 2} \longrightarrow \Lambda/J^2$, induced by f, is an isomorphism of k-vector spaces. Consequently, $f(k[Q_\Lambda]) + J^2 = \Lambda$, so that Proposition 1.19 ensures the surjectivity of f. By the same token, the ideal $I := \ker f$ is contained in $k[Q_\Lambda]_{\geq 2}$. By definition of f, we have $f(k[Q_\Lambda]_{\geq m}) \subseteq J^m$ for all $m \geq 1$. Thus, if $J^n = (0)$, then $k[Q_\Lambda]_{\geq n} \subseteq I$. □

Remarks. (1) As every simple module of the *bound quiver algebra* $k[Q_\Lambda]/I$ is one-dimensional, we cannot expect to obtain isomorphisms $k\mathcal{G} \cong k[Q_\Lambda]/I$. In fact, this is only the case if the group scheme $\mathcal{G}$ is *trigonalizable*, that is, if $\mathcal{G}$ can be embedded into a group of upper triangular matrices. Such groups are necessarily solvable, and the failure of Lie's Theorem in positive characteristic means that there also are solvable groups that are not trigonalizable.

(2) Passage to Morita equivalent algebras not only trivializes the structure of the simple modules, but it also does not interact well with the comultiplication. Thus, we usually do not have a tensor product of modules available when replacing $k\mathcal{G}$ by its bound quiver algebra.

5.2 Hereditary algebras

In this section we briefly review basic results concerning representations of hereditary algebras. Aside from group algebras of finite groups, the study of these algebras has been the other major focal point of much of the early work in representation theory. *Throughout, k is assumed to be an algebraically closed field.*

Definition. A finite-dimensional k-algebra Λ is *hereditary* if submodules of projective Λ-modules are projective.

The following result partly explains why hereditary algebras are of interest. In terms of their presentation by quivers and relations, they are the most tractable algebras.

Theorem 5.2. *Let Λ be a finite-dimensional k-algebra. Then the following statements hold:*

(1) *If Λ is hereditary and $I \subseteq \mathrm{Rad}(\Lambda)^2$ is an ideal such that Λ/I is hereditary, then $I = (0)$.*

(2) *Λ is hereditary if and only if Λ is Morita equivalent to $k[Q_\Lambda]$.*

Proof. (1) Let $I \trianglelefteq \Lambda$ be as stated in (1) and put $J := \mathrm{Rad}(\Lambda)$. Using $I \subseteq J$, we obtain an exact sequence

$$(0) \longrightarrow I/IJ \longrightarrow J/IJ \longrightarrow J/I \longrightarrow (0)$$

of (Λ/I)-modules. As $J/I \subseteq \Lambda/I$ is a projective (Λ/I)-module, the sequence splits. Our condition $I \subseteq J^2$ implies that the submodule $I/IJ \subseteq \mathrm{Rad}_{\Lambda/I}(J/IJ)$ is superfluous, so that $I/IJ = (0)$. Consequently, $I = IJ$, whence $I = (0)$.

(2) Direct computation shows that $k[Q]$ is hereditary for any quiver Q without oriented cycles (that is, for a quiver giving rise to a finite-dimensional path algebra).

Suppose that Λ is hereditary with simple modules $S_1, \dots, S_n$ and corresponding projective covers $P_1, \dots, P_n$. In view of

$$\mathrm{Ext}^1_\Lambda(S_i, S_j) \cong \mathrm{Hom}_\Lambda(\mathrm{Rad}(P_i), S_j)$$

the condition $\mathrm{Ext}^1_\Lambda(S_i, S_j) \neq (0)$ implies the existence of a surjective homomorphism $\mathrm{Rad}(P_i) \longrightarrow S_j$. As $\mathrm{Rad}(P_i)$ is projective, this map lifts to a surjection $\mathrm{Rad}(P_i) \longrightarrow P_j$ onto the projective cover of S_j. Consequently, $\ell(P_j) < \ell(P_i)$, so that the quiver Q_Λ affords no oriented cycles. We conclude that $k[Q_\Lambda]$ is finite-dimensional with $k[Q_\Lambda]^+ = \mathrm{Rad}(k[Q_\Lambda])$. Our assertion now follows from (1) and Theorem 5.1. □

The representation theory of hereditary algebras is very well understood and turns out to be closely related to Lie theory. Given a quiver Q, we denote by $\bar{Q}$ the *underlying graph*, in which every arrow of Q is replaced by a bond.

A finite-dimensional k-algebra Λ is *connected* provided Λ affords only one block. This condition is equivalent to Q_Λ being connected.

Theorem 5.3 ([64]). *A finite-dimensional connected path algebra $k[Q]$ is of finite representation type if and only if the underlying graph $\bar{Q}$ of Q is isomorphic to a Dynkin diagram A_n, D_n, or $E_{6,7,8}$. In that case, the dimension vectors of the indecomposable $k[Q]$-modules correspond to the positive roots of the root system of $\bar{Q}$.* □

Remarks. (1) Recall that the *dimension vector* of a Λ-module M is the coordinate vector of its class $[M]$ in the Grothendieck group $K_0(\Lambda)$ relative to the standard basis given by the simple Λ-modules. Thus, if Λ is hereditary of finite representation type, then the lengths of the indecomposable Λ-modules are known. For instance, if Q_Λ is of type A_n, then there exist $n-j+1$ indecomposable Λ-modules of length j.

(2) Owing to Theorem 5.3, the representation type of $k[Q]$ does not depend on the orientation of the quiver Q, even though the algebras involved may be rather different. For instance, the module categories of the algebras

$$\Lambda_1 := k[1 \to 2 \to 3] \quad \text{and} \quad \Lambda_2 := k[1 \leftarrow 2 \to 3]$$

have the same distribution of indecomposable modules, yet they are not Morita equivalent: Λ_1 is a Nakayama algebra, while Λ_2 is not. The relationship of these algebras is investigated in tilting theory, whose starting point is the seminal article [14] (see also [4, (VII.5)]).

Shortly after Gabriel's determination of the representation-finite hereditary algebras, Donovan-Freislich and Nazarova independently classified the tame hereditary algebras:

Theorem 5.4 ([32], [90]). *A finite-dimensional connected path algebra $k[Q]$ is tame if and only if the underlying graph $\bar{Q}$ of Q is isomorphic to a Euclidean diagram $\tilde{A}_n$, $\tilde{D}_n$, or $\tilde{E}_{6,7,8}$.* □

The foregoing result can be used to obtain information on the Ext-quivers of arbitrary algebras of finite or tame representation type.

Recall that the *separated quiver* Q_s of a quiver Q with vertex set $\{1,\ldots,n\}$ has $2n$ vertices $\{1,\ldots,n,1',\ldots,n'\}$ and an arrow $\ell \to m'$ for every arrow $\ell \to m$ of Q.

Corollary 5.5. *Let Λ be a finite-dimensional k-algebra. If Λ is tame, then each connected component of the separated quiver $(Q_\Lambda)_s$ of the Ext-quiver Q_Λ is a Dynkin diagram of type A, D, E or a Euclidean diagram of type $\tilde{A}$, $\tilde{D}$, $\tilde{E}$.*

Proof. Let J be the Jacobson radical of Λ. Then the algebra $\Lambda' := \Lambda/J^2$ is representation-finite or tame, has Jacobson radical $J' = J/J^2$ and Ext-quiver $Q_{\Lambda'} = Q_\Lambda$. The triangular matrix algebra

$$\Sigma := \begin{pmatrix} \Lambda'/J' & 0 \\ J' & \Lambda'/J' \end{pmatrix}$$

is hereditary with Ext-quiver $Q_\Sigma \cong (Q_{\Lambda'})_s$, and the functor

$$F : \operatorname{mod} \Lambda' \longrightarrow \operatorname{mod} \Sigma; \quad M \mapsto \begin{pmatrix} M/J'M \\ J'M \end{pmatrix}$$

reflects isomorphisms and indecomposability. Moreover, F reaches all but finitely many indecomposable Σ-modules, so that Σ is tame or representation-finite. Our assertion now follows from Theorem 5.3 and Theorem 5.4. □

5.3 A criterion for wildness

As before, we assume k to be algebraically closed. The results of Section 5.2 yield simple criteria implying the wildness of an algebra Λ. By way of illustration, we consider the following:

Lemma 5.6. *Let Λ be a k-algebra such that $\dim_k \operatorname{Ext}^1_\Lambda(S,T) \geq 3$ for two simple Λ-modules S and T. Then Λ is wild.*

Proof. By assumption, the graph $(\bar{Q}_\Lambda)_s$ underlying the separated quiver has two vertices that are joined by at least 3 bonds. Thus, the connected component of $(\bar{Q}_\Lambda)_s$ containing these vertices is neither a Dynkin diagram nor a Euclidean diagram of the types given in Corollary 5.5. Consequently, Λ is neither representation-finite nor tame, and Theorem 2.9 yields the result. □

Unfortunately, there are many algebras, whose wildness cannot be recognized via the foregoing criterion. More sophisticated results can be established by studying Galois coverings of quivers, a subject that lies beyond the scope of these notes. We shall be content with one basic fact, involving the quiver and the relations, which works in situations of interest where Lemma 5.6 does not apply. The interested reader may consult [36, Chap. I] for more details.

Lemma 5.7. *Let Λ be a finite-dimensional k-algebra such that each vertex of Q_Λ is the starting point and end point of at least two arrows. If there exist three arrows α, β, γ such that neither $\beta\alpha$ nor $\gamma\alpha$ are summands of a relation of Λ, then the algebra Λ is wild.* □

5.4 Trivial extensions

From a homological perspective, hereditary algebras and algebras of measures are far apart. While the former have global dimension ≤ 1, the global dimension of self-injective algebras, such as $k\mathcal{G}$, is finite if and only if $k\mathcal{G}$ is semi-simple. The concept of a trivial extension allows us to pass from any algebra to a symmetric algebra. Moreover, there is a close connection between the representation theory of an algebra Λ and that of its associated trivial extension.

Definition. Let Λ be a finite-dimensional k-algebra. The algebra $T(\Lambda) := \Lambda \ltimes \Lambda^*$, whose multiplication is given by

$$(a, \varphi).(b, \psi) := (ab, a.\psi + \varphi.b) \quad \forall\, a, b \in \Lambda,\ \varphi,\ \psi \in \Lambda^*$$

is called the *trivial extension* of Λ.

In the above definition, we have endowed the dual space Λ^* with the usual structure of a (Λ, Λ)-bimodule by setting

$$(a.\varphi)(x) = \varphi(xa) \ \text{ and } \ (\varphi.a)(x) = \varphi(ax) \qquad \forall\, a, x \in \Lambda,\ \varphi \in \Lambda^*.$$

We record the symmetry of $T(\Lambda)$:

Lemma 5.8. *The linear map*

$$\pi : T(\Lambda) \longrightarrow k; \quad (a, \varphi) \mapsto \varphi(1)$$

endows $T(\Lambda)$ with the structure of a symmetric algebra. □

Under certain conditions, the quiver and the relations describing $T(\Lambda)$ can be deduced from the corresponding data for Λ. This applies in particular for the so-called *radical square zero hereditary algebras*. By definition, the Ext-quiver of such an algebra is oriented in such a way that there are no paths of length 2. Such an orientation always exists if Q_Λ is a tree, but for a quiver of type $\tilde{A}_n$ this is only possible if n is odd (that is, when the number $n+1$ of vertices is even).

Given a quiver Q, we let $\widehat{Q}$ be the *double of* Q. The quiver $\widehat{Q}$ has the same set of vertices as Q, and for every arrow $\alpha : i \to j$, we add an arrow $\alpha^* : j \to i$. Here is the relevant basic result:

Proposition 5.9. *Let $Q = (Q_0, Q_1, s, t)$ be a connected quiver with $|Q_0|+|Q_1| \geq 4$ and without paths of length ≥ 2. Then the algebra $T(Q) := T(k[Q])$ is isomorphic to $k[\widehat{Q}]/I$, where I is generated by the set $\{\alpha.\alpha^* - \beta.\beta^*, (t(\alpha) = t(\beta));\ \alpha^*.\alpha - \beta^*.\beta, (s(\alpha) = s(\beta));\ \alpha^*.\beta,\ \alpha.\beta^*(\alpha \neq \beta)\}$.* □

Example. Let $Q = \tilde{A}_{2n-1}$ be the circle with $2n$ vertices and such that there are no paths of length 2. Then $T(\tilde{A}_{2n-1}) \cong k[\widehat{Q}]/I$, where $\widehat{Q}$ is the double

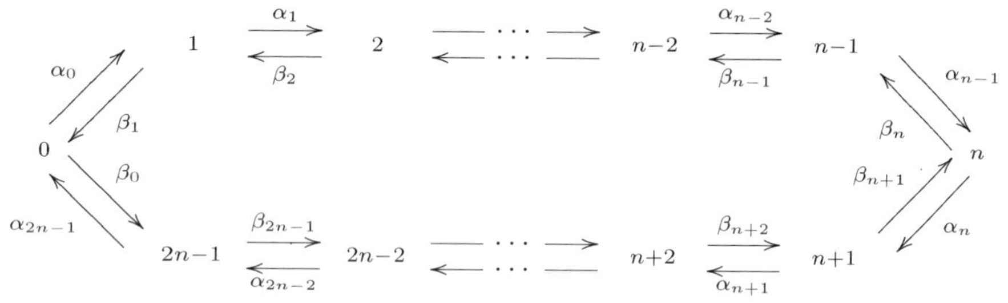

and $I \subseteq k[\widehat{Q}]$ is the ideal generated by

$$\{\beta_{i+1}\alpha_i - \alpha_{i-1}\beta_i,\quad \alpha_{i+1}\alpha_i,\quad \beta_i\beta_{i+1};\quad i \in \mathbb{Z}/(2n)\}.$$

The algebra $T(\tilde{A}_{2n-1})$ belongs to a particularly tractable class of tame algebras, the so-called *special biserial* algebras that were introduced by Skowroński and Waschbüsch in [111]. They are bound quiver algebras $k[Q]/I$, whose presentation satisfies the following conditions:

- Each vertex of Q is the starting point and end point of at most two arrows.
- For any arrow $\alpha \in Q_1$, there exist at most one arrow β and one arrow γ with $\alpha\beta,\ \gamma\alpha \notin I$.

Special biserial algebras are biserial; they appear in the classification of blocks of group algebras with cyclic or dihedral defect groups (cf. [36, 105]) as well as in connection with the Gel'fand-Ponomarev classification of the singular Harish-Chandra modules over the Lorentz group [65].

The representation theory of special biserial algebras is completely understood. In view of work by Tachikawa [119] and Ringel [107] (cf. also [66]), the algebra $T(Q)$ affords "twice as many" indecomposable modules as $k[Q]$, and thus has the same representation type.

5.5 McKay quivers

Up to now, the presence of tensor products in the module category of a finite group scheme has only entered via the definition of support varieties. We shall see in this section how knowledge of tensor products of simple modules helps us understand the block structure of a given Hopf algebra. Our approach is based on McKay's seminal papers [87, 88], in which he systematically studied tensor products within the context of complex representations of finite groups. Subsequently, McKay quivers appeared in other areas of representation theory, see for instance [5, 6].

Throughout this section, k is an arbitrary field. Given a finite-dimensional Hopf algebra H, we let $K_0(H)$ denote the *Grothendieck group* of $\operatorname{mod} H$. By definition, the group $K_0(H)$ is generated by the set of isomorphism classes of finite-dimensional H-modules with relations $[Y]-[X]-[Z]$, corresponding to short exact sequences $(0) \longrightarrow X \longrightarrow Y \longrightarrow Z \longrightarrow (0)$ of H-modules. Let $\{S_1, \dots, S_n\}$ be a complete set of representatives for the isomorphism classes of the simple H-modules. Recall that

(a) $K_0(H)$ is a free abelian group with basis $\{[S_1], \dots, [S_n]\}$, and

(b) every exact functor $F : \operatorname{mod} H \longrightarrow \operatorname{mod} H$ determines an endomorphism F of $K_0(H)$.

If V is a finite-dimensional H-module, such that

$$[V] = \sum_{j=1}^{n} m_j [S_j],$$

then the coefficient m_j is the Jordan-Hölder multiplicity $[V\!:\!S_j]$.

In view of (b), the tensor product endows $K_0(H)$ with the structure of a ring. Given a finite-dimensional H-module V, we consider the integral $(n \times n)$-matrix $(m_{ij})_{1\le i,j\le n}$ representing the endomorphism $V \otimes_k -$ of $K_0(H)$ relative to the above basis:

$$[V \otimes_k S_j] = \sum_{i=1}^{n} m_{ij} [S_i].$$

We are interested in the quiver associated to this matrix:

Definition. The quiver $\Theta_V(H)$ with set of vertices $\{1, \dots, n\}$ and m_{ij} arrows $i \to j$ is called the *McKay quiver of H relative to V*.

Remark. Our definition differs from McKay's original one in that he considered the opposite quiver. However, the above orientation can also be found in the literature, see [5, 6, 67].

Examples. (1) Suppose that every simple H-module is one-dimensional and let $G := X(H)$ be the character group of H. Then the Grothendieck ring $K_0(H)$ is the integral group ring $\mathbb{Z}G$. In that case, the McKay quiver of H relative to the simple H-module k_λ, defined by $\lambda \in G$, is just a union of $\frac{\operatorname{ord}(G)}{\operatorname{ord}(\lambda)}$ oriented cycles $\tilde{A}_{\operatorname{ord}(\lambda)-1}$.

(2) Assuming $p \ge 3$, we consider the restricted enveloping algebra $H = U_0(\mathfrak{sl}(2))$ as well as the standard $\mathfrak{sl}(2)$-module $V = L(1)$. By the modular Clebsch-Gordan formula [11], we have

$$[L(1) \otimes_k L(0)] = [L(1)]; \quad [L(1) \otimes_k L(j)] = [L(j-1)] + [L(j+1)] \qquad 1 \le j \le p-2$$

as well as

$$[L(1) \otimes_k L(p-1)] = [P(p-2)] = 2[L(p-2)] + 2[L(0)].$$

In particular, the McKay quiver $\Theta_V(U_0(\mathfrak{sl}(2)))$ is connected. The following results show that this is not an accident. We begin with the generalization of Burnside's classical result [19] (see also [16, 114, 104, 94]):

Theorem 5.10 (Burnside's Theorem for Hopf algebras). *Let V be a finite-dimensional module for a finite-dimensional Hopf algebra H. If the annihilator $\mathrm{ann}_H(V)$ of V in H does not contain any non-zero Hopf ideals, then every simple H-module S is a submodule of a tensor power $V^{\otimes m}$ for some $m \geq 1$.*

Proof. Setting $M := \bigoplus_{m\geq 1} V^{\otimes m}$, we obtain

$$M \otimes_k M \cong \bigoplus_{m,n\geq 1} V^{\otimes m} \otimes_k V^{\otimes n} \cong \bigoplus_{m\geq 2} (m-1)V^{\otimes m},$$

so that the inclusion

$$\mathrm{ann}_H(M) = \bigcap_{m\geq 1} \mathrm{ann}_H(V^{\otimes m}) \subseteq \mathrm{ann}_H(M \otimes_k M)$$

holds. Consequently, $\Delta(\mathrm{ann}_H(M)) \subseteq \mathrm{ann}_H(M) \otimes_k H + H \otimes_k \mathrm{ann}_H(M)$. Since H is finite-dimensional, Lemma 1.2 implies that $\mathrm{ann}_H(M) \subseteq \mathrm{ann}_H(V)$ is a Hopf ideal. In view of our current assumption, the H-module M is faithful.

Let $I \subseteq H$ be a minimal left ideal. Then $I.M \neq (0)$, and there exist $m \in \mathbb{N}$ and $v \in V^{\otimes m}$ such that $I.v \neq (0)$. Since I is a simple H-module, the map $x \mapsto x.v$ thus defines an embedding $I \hookrightarrow V^{\otimes m}$.

As a result, every simple module belonging to $\mathrm{Soc}(H)$ is a submodule of a suitable tensor power of V. By the Theorem of Larson-Sweedler (Theorem 1.3), the algebra H is self-injective. Consequently, every simple H-module occurs in $\mathrm{Soc}(H)$, so that our assertion follows. □

Given any quiver Q and vertices $i, j \in Q_0$, we let $Q(i, j; m)$ be the set of paths of length m starting at the vertex i and terminating at the vertex j.

Corollary 5.11. *Let H be a finite-dimensional Hopf algebra, V be a finite-dimensional H-module.*

(1) *We have $[V^{\otimes m} \otimes_k S_j : S_i] = |\Theta_V(H)(i, j; m)|$.*

(2) *If $\mathrm{ann}_H(V)$ contains no non-zero Hopf ideals, then $\Theta_V(H)$ is connected.*

Proof. (1) Using induction on m, we assume that $m \geq 2$. Note that

$$\Theta_V(H)(i, j; m) \cong \bigsqcup_{\ell=1}^{n} \Theta_V(H)(\ell, j; m-1) \times \Theta_V(H)(i, \ell; 1). \qquad (*)$$

The inductive hypothesis provides a decomposition

$$V^{\otimes(m-1)} \otimes_k S_j \cong \bigoplus_{t=1}^{n} b_t S_t,$$

where $b_t = |\Theta_V(H)(t,j;m-1)|$. Consequently,

$$V^{\otimes m} \otimes_k S_j \cong \bigoplus_{t=1}^{n} b_t(V \otimes_k S_t) \cong \bigoplus_{r=1}^{n}(\sum_{t=1}^{n} m_{rt}b_t)S_r,$$

and $(*)$ implies

$$[V^{\otimes m} \otimes_k S_j : S_i] = \sum_{t=1}^{n} m_{it}b_t = |\Theta_V(H)(i,j;m)|,$$

as desired.

(2) Let $S_1 = k$ be the trivial H-module. Then we have $V^{\otimes m} \otimes_k S_1 \cong V^{\otimes m} \ \forall\ m \geq 1$. Given a vertex $i \in \{1,\dots,n\}$, Burnside's Theorem provides $m \in \mathbb{N}$ with

$$0 \neq [V^{\otimes m} : S_i] = [V^{\otimes m} \otimes_k S_1 : S_i] = |\Theta_V(H)(i,1;m)|.$$

Hence there is a path from i to 1. □

Suppose that $\mathrm{char}(k) \geq 3$. By general theory, the Hopf ideals of $U_0(\mathfrak{sl}(2))$ correspond to the p-ideals of the restricted Lie algebra $\mathfrak{sl}(2)$. As $\mathfrak{sl}(2)$ is simple, every $L(i) \neq L(0)$ satisfies the hypothesis of Corollary 5.11(2). The example of a p-group shows that the converse of Corollary 5.11(2) does not hold in general.

Definition. Let $C = (c_{ij}) \in \mathrm{Mat}_n(\mathbb{Z})$ be an integral $(n \times n)$-matrix. We say that C is a *generalized Cartan matrix* if

(1) $c_{ii} \leq 2$ for $1 \leq i \leq n$, and

(2) $c_{ij} \leq 0$ for $1 \leq i \neq j \leq n$, and

(3) $c_{ij} = 0$ if and only if $c_{ji} = 0$.

A function $d : \{1,\dots,n\} \longrightarrow \mathbb{N};\ \ i \mapsto d_i$ is an *additive function for* C if

$$\sum_{i=1}^{n} d_i c_{ij} = 0; \qquad 1 \leq j \leq n.$$

To each generalized Cartan matrix $C \in \mathrm{Mat}_n(\mathbb{Z})$ we associate a valued graph Γ_C, whose vertices are the elements $\{1,\dots,n\}$. There is a valued edge

$$i \overset{(|c_{ij}|,|c_{ji}|)}{\text{———}} j$$

between i and j whenever $c_{ij} \neq 0$ and $i \neq j$. In addition, there are $2 - c_{ii}$ loops at the vertex i. By property (3), the graph of a generalized Cartan matrix is well-defined.

In the following, we let $I_n \in \mathrm{Mat}_n(\mathbb{Z})$ be the identity matrix.

Lemma 5.12. *Let k be algebraically closed. Suppose that H is a semi-simple, cocommutative Hopf algebra with simple modules $S_1, \ldots, S_n$. If V is a two-dimensional self-dual H-module, and (m_{ij}) is the matrix representing $V \otimes_k -$, then $C := 2I_n - (m_{ij})$ is a symmetric generalized Cartan matrix, and $i \mapsto \dim_k S_i$ defines an additive function for C.*

Proof. The defining conditions (1) and (2) follow immediately. Since H is semi-simple, we have

$$V \otimes_k S_j \cong \bigoplus_{i=1}^{n} m_{ij} S_i,$$

so that commutativity of the tensor product, adjointness, self-duality and Schur's Lemma give

$$\begin{aligned} m_{ij} &= \dim_k \mathrm{Hom}_H(V \otimes_k S_j, S_i) = \dim_k \mathrm{Hom}_H(S_j, V^* \otimes_k S_i) \\ &= \dim_k \mathrm{Hom}_H(S_j, V \otimes_k S_i) = m_{ji}. \end{aligned}$$

Accordingly, C is symmetric and (3) holds. Since $\dim_k V = 2$, the map $d : \{1, \ldots, n\} \longrightarrow \mathbb{N};\ i \mapsto \dim_k S_i$ is an additive function for C. □

If $H = U_0(\mathfrak{sl}(2))$ and $V = L(1)$, then $C := 2I_p - (m_{ij})$ is not a generalized Cartan matrix, as $c_{p-1,0} = 0$ and $c_{0,p-1} = -2$.

The connected generalized Cartan matrices affording an additive function were classified by Happel, Preiser, and Ringel (see [67]). We will only need the following special case of their result:

Theorem 5.13. *Let C be a connected, symmetric generalized Cartan matrix having an additive function. Then the following statements hold:*

(1) *The graph Γ_C belongs to the following list: $\tilde{A}_n\ (n \geq 1)$, $\tilde{D}_n\ (n \geq 4)$, $\tilde{E}_{6,7,8}$, $\tilde{L}_n$, $\widetilde{DL}_n\ (n \geq 2)$.*

(2) *There exists an additive function d_C of C such that any additive function for C is an integral multiple of d_C.* □

Example. If $C = \tilde{A}_n$, then $d_C(i) = 1$ for $1 \leq i \leq n+1$.

6 Representation-finite and tame group schemes

This section employs the techniques discussed so far to obtain information on those group schemes $\mathcal{G}$, whose Hopf algebras $k\mathcal{G}$ are of finite or tame representation type. Let me begin by recalling what is known for group algebras of finite groups. In view of Maschke's Theorem [86], group algebras are semi-simple in case $\mathrm{char}(k) = 0$. We thus assume $\mathrm{char}(k) = p > 0$. To a block $\mathcal{B} \subseteq kG$ one associates a p-group $D_{\mathcal{B}} \subseteq G$ to $\mathcal{B}$, the so-called *defect group*. It turns out that the representation type of $\mathcal{B}$ is reflected by structural properties of $D_{\mathcal{B}}$:

- The block $\mathcal{B}$ is simple if and only if $D_{\mathcal{B}} = \{1\}$ is trivial. This is the generalization of Maschke's classical result.

- Group algebras of finite representation type were first investigated by Higman [69], who showed that kG is representation-finite if and only if the Sylow-p-subgroups of G are cyclic. More generally, the block $\mathcal{B} \subseteq kG$ is representation-finite if and only if its defect group $D_{\mathcal{B}}$ is cyclic.
- Thanks to work by Bondarenko and Drozd [15], a block $\mathcal{B} \subseteq kG$ is tame if and only if $p = 2$, and if $D_{\mathcal{B}}$ is dihedral, semidihedral, or generalized quaternion.

Defect theory and other special features of group algebras also yield the following facts that proved to be of importance:

- If $H \subseteq G$ is a subgroup and kG is representation-finite (tame), then kH is representation-finite (tame or representation-finite).
- If the principal block $\mathcal{B}_0(G)$ is representation-finite (tame), then any other block of kG is representation-finite (tame or representation-finite).

The latter fact is related to the eponymous property of $\mathcal{B}_0(G)$: The defect group of $\mathcal{B}_0(G)$ is a Sylow-p-subgroup of G, and thus the largest among all defect groups. Consequently, $\mathcal{B}_0(G)$ is commonly thought of as the most complicated block of kG.

Let H be a Hopf algebra. Recall from Lemma 1.5 that the tensor product $P \otimes_k M$ of a projective H-module P with any H-module M is again projective. Thus, if the principal block $\mathcal{B}_0(H)$ is simple, then k is a projective H-module and hence $M \cong k \otimes_k M$ is projective for every $M \in \operatorname{mod} H$. Consequently, H is semi-simple, so that the representation theory of $\mathcal{B}_0(H)$ governs that of H. We shall see later, however, that the paradigm of $\mathcal{B}_0(H)$ being the most complicated block fails for infinitesimal group schemes of tame representation type.

6.1 Nagata's Theorem

Let H be a finite-dimensional cocommutative Hopf algebra. We know from Corollary 1.12 that H is semi-simple whenever $\operatorname{char}(k) = 0$. For fields of positive characteristic, Nagata's Theorem determines those group schemes $\mathcal{G}$, whose algebras of measures are semi-simple. Such group schemes are also referred to as *linearly reductive*.

Throughout this section, k denotes an algebraically closed field of characteristic $p > 0$.

Definition. A finite group scheme $\mathcal{G}$ is called *diagonalizable* if its coordinate ring $k[\mathcal{G}]$ is the group algebra of its character group $X(\mathcal{G})$.

If $\mathcal{G}$ is diagonalizable, then we have $\dim_k k\mathcal{G} = \operatorname{ord}(X(\mathcal{G}))$. On the other hand, $\bigoplus_{\lambda \in X(\mathcal{G})} k_\lambda$ is a submodule of $k\mathcal{G}/\operatorname{Rad}(k\mathcal{G})$, so that $\operatorname{Rad}(k\mathcal{G}) = (0)$ and $k\mathcal{G}$ is semi-simple. The following result, which was first established by Hochschild [71] for enveloping algebras of restricted Lie algebras, determines all linearly reductive infinitesimal group schemes.

Proposition 6.1. *Let $\mathcal{G}$ be an infinitesimal group. If $k\mathcal{G}$ is semi-simple, then $\mathcal{G}$ is diagonalizable.*

Proof. We proceed by induction on the order of $\mathcal{G}$, noting that $\mathfrak{g} := \mathrm{Lie}(\mathcal{G}) \neq (0)$ whenever $\mathcal{G} \neq e_k$. By assumption, the trivial $\mathcal{G}$-module k is projective. Since $k\mathcal{G}$ is a free left $k\mathcal{G}_1$-module (cf. Theorem 1.6), the trivial $\mathcal{G}_1$-module $k|_{\mathcal{G}_1}$ is also projective, so that $k\mathcal{G}_1$ is semi-simple. According to Theorem 1.20, this implies the semi-simplicity of the restricted enveloping algebra $U_0(\mathfrak{g})$.

Let $x \in \mathfrak{g}$ with $x^{[p]} = 0$. Owing to Corollary 1.18 we have

$$U_0(kx) \cong \begin{cases} k[X]/(X^p) & \text{for } x \neq 0, \\ k & \text{otherwise.} \end{cases}$$

By the above arguments, k is a projective $U_0(kx)$-module, so that $x = 0$.

Given an arbitrary element $x \in \mathfrak{g}$, we consider the abelian p-subalgebra $\mathfrak{h} := \sum_{i \geq 0} kx^{[p]^i}$. Thus, the p-map $[p]$ is semilinear on $\mathfrak{h}$, and the above shows that its restriction $[p]|_{\mathfrak{h}}$ is injective and hence bijective. Consequently, there exist $\alpha_1, \dots, \alpha_n \in k$ with $x = \sum_{i=1}^n \alpha_i x^{[p]^i}$. As a result, the left multiplication

$$\mathrm{ad}\, x : \mathfrak{g} \longrightarrow \mathfrak{g}; \quad y \mapsto [x, y]$$

satisfies the polynomial $\sum_{i=1}^n \alpha_i X^{p^i} - X$ and is therefore diagonalizable. Let $\alpha \in k$ be an eigenvalue for $\mathrm{ad}\, x$. Then there exists $y \in \mathfrak{g} \setminus \{0\}$ such that $[x, y] = \alpha y$. Accordingly, the subspace $V := kx + ky$ is invariant under $\mathrm{ad}\, y$, and $(\mathrm{ad}\, y)|_V$ is nilpotent and diagonalizable. Hence $(\mathrm{ad}\, y)|_V = 0$, so that $\alpha = 0$. Consequently, the Lie algebra $\mathfrak{g}$ is abelian, and, being a commutative semi-simple algebra, $U_0(\mathfrak{g}) \cong k^{\dim_k U_0(\mathfrak{g})}$ is a product of fields. In particular, $U_0(\mathfrak{g})^* \cong kX(U_0(\mathfrak{g}))$ is the group algebra of the character group of $U_0(\mathfrak{g})$, so that $\mathcal{G}_1$ is diagonalizable.

Suppose that $\mathcal{G} \neq \mathcal{G}_1$. By inductive hypothesis, the group $\mathcal{G}_1$ and its factor group $\mathcal{G}/\mathcal{G}_1$ are diagonalizable. By rigidity of tori (cf. [123, (7.7)]), the algebra $k\mathcal{G}_1$ is contained in the center of $k\mathcal{G}$.

Let S be a simple $k\mathcal{G}$-module. According to Schur's Lemma, the central subalgebra $k\mathcal{G}_1$ of $k\mathcal{G}$ operates on S via a character. Consequently, $k\mathcal{G}_1$ acts trivially on $\mathrm{End}_k(S)$, and the latter space has the structure of a $\mathcal{G}/\mathcal{G}_1$-module. Since $\mathcal{G}/\mathcal{G}_1$ is linearly reductive, $\mathrm{End}_k(S)$ is a semi-simple $k\mathcal{G}$-module, and there results a decomposition

$$\mathrm{End}_k(S) = \bigoplus_{\lambda \in C} \mathrm{End}_k(S)_\lambda,$$

where $C \subseteq X(\mathcal{G})$, and $\mathrm{End}_k(S)_\lambda = \{\varphi \in \mathrm{End}_k(S);\ h.\varphi = \lambda(h)\varphi \ \ \forall\, h \in k\mathcal{G}\}$. One readily verifies that $\varphi \circ \psi \in \mathrm{End}_k(S)_{\lambda * \gamma}$ for $\varphi \in \mathrm{End}_k(S)_\lambda$ and $\psi \in \mathrm{End}_k(S)_\gamma$.

Let φ be an element of $\mathrm{End}_k(S)_\lambda$. Since

$$\varphi(hs) = \sum_{(h)} h_{(1)}(\eta(h_{(2)}).\varphi)(s) = (\mathrm{id}_{k\mathcal{G}} * (\lambda \circ \eta))(h)\varphi(s)$$

for $h \in k\mathcal{G}$ and $s \in S$, we see that $\ker \varphi$ is a $\mathcal{G}$-submodule of S. Accordingly, every non-zero element of $\mathrm{End}_k(S)_\lambda$ is invertible.

Direct computation shows that $\operatorname{tr}(h.\varphi) = \varepsilon(h)\operatorname{tr}(\varphi)$ for $\varphi \in \operatorname{End}_k(S)$ and $h \in k\mathcal{G}$. This implies that $\operatorname{tr}(\operatorname{End}_k(S)_\lambda) = (0)$ whenever $\lambda \neq \varepsilon$.

Let $\varphi \in \operatorname{End}_k(S)_\lambda \setminus \{0\}$ for some $\lambda \neq \varepsilon$. Owing to Theorem 1.6 and Proposition 1.21, the character λ has order p^n for some $n \geq 1$, whence $\varphi^{p^n} \in \operatorname{End}_k(S)_\varepsilon$. By Schur's Lemma, the latter space coincides with $k\,\mathrm{id}_S$, so that there exists $\alpha \in k$ such that $\varphi^{p^n} = \alpha\,\mathrm{id}_S$. Since φ is invertible, we have $\alpha \neq 0$. From the identity

$$\operatorname{tr}(\alpha\,\mathrm{id}_S) = \operatorname{tr}(\varphi^{p^n}) = \operatorname{tr}(\varphi)^{p^n} = 0,$$

we conclude that $\operatorname{tr}(\operatorname{End}_k(S)) = \operatorname{tr}(\operatorname{End}_k(S)_\varepsilon) = (0)$, a contradiction. As a result, $\operatorname{End}_k(S) = \operatorname{End}_k(S)_\varepsilon$ is one-dimensional, so that $\dim_k S = 1$.

In view of the above, the semi-simple $k\mathcal{G}$-module $k[\mathcal{G}] = k\mathcal{G}^*$ decomposes into one-dimensional constituents and we obtain

$$k[\mathcal{G}] = \bigoplus_{\lambda \in D} k[\mathcal{G}]_\lambda,$$

where $D \subseteq X(\mathcal{G})$ and $k[\mathcal{G}]_\lambda = \{x \in k[\mathcal{G}];\ h.x = \lambda(h)x \ \ \forall\, h \in k\mathcal{G}\}$ is the λ-weight space of $k[\mathcal{G}]$. Direct computation shows that $k[\mathcal{G}]_\lambda = k\,(\lambda \circ \eta)$. Hence the Hopf algebra $k[\mathcal{G}]$ is generated by group-like elements, and $\mathcal{G}$ is diagonalizable. □

Theorem 6.2 (Nagata). *Let $\mathcal{G}$ be a finite group scheme. Then the following statements are equivalent:*

(1) *The algebra $k\mathcal{G}$ is semi-simple.*

(2) *The group $\mathcal{G}^0$ is diagonalizable and $p \nmid \operatorname{ord}(\mathcal{G}(k))$.*

Proof. Let $\Lambda := k\mathcal{G}^0$ be the distribution algebra of the connected component $\mathcal{G}^0$ of $\mathcal{G}$. By general theory (cf. Theorem 1.13),

$$k\mathcal{G} \cong \Lambda\mathcal{G}(k)$$

is a skew group algebra.

If $k\mathcal{G}$ is semi-simple, then the factor algebra $k\mathcal{G}(k)$ is semi-simple, so that $p \nmid \operatorname{ord}(\mathcal{G}(k))$. Moreover, k is a projective $k\mathcal{G}$-module. Since $k\mathcal{G}$ is a free left Λ-module, it follows that k is also projective over $k\mathcal{G}^0$. Hence $k\mathcal{G}^0$ is semi-simple and Proposition 6.1 implies that $\mathcal{G}^0$ is diagonalizable.

Conversely, assuming (2), we note that $P := \Lambda\mathcal{G}(k) \otimes_\Lambda k$ is a projective $\mathcal{G}$-module, which contains the trivial $\Lambda\mathcal{G}(k)$-module $k(\sum_{g\in\mathcal{G}(k)} g \otimes 1)$ as a direct summand. Consequently, k is a projective $\mathcal{G}$-module and $k\mathcal{G}$ is semi-simple (see Lemma 1.5). □

6.2 Tensor products of simple and principal indecomposable modules

Throughout, we will be working over an algebraically closed field k. Let $\mathcal{G}$ be a finite algebraic k-group, $\mathcal{N} \trianglelefteq \mathcal{G}$ be a closed, normal subgroup. Under favorable circumstances, which will be seen to hold in all cases of interest, the simple and principal indecomposable $k\mathcal{G}$-modules are given by tensor products of the corresponding modules for $k\mathcal{N}$ and $k(\mathcal{G}/\mathcal{N})$. In particular, the Ext-quiver of $k\mathcal{G}$ will be computable from the structure of tensor products of simple $k(\mathcal{G}/\mathcal{N})$-modules.

In the sequel, we shall often view $(\mathcal{G}/\mathcal{N})$-modules as $\mathcal{G}$-modules via pull-back along the canonical quotient map $\mathcal{G} \longrightarrow \mathcal{G}/\mathcal{N}$. We begin by studying the underlying set of vertices of the Ext-quiver of $k\mathcal{G}$ in a setting that will turn out to be appropriate for our purposes.

Lemma 6.3. *Let $\mathcal{N} \trianglelefteq \mathcal{G}$ be a normal subgroup of the finite algebraic k-group $\mathcal{G}$. If $L_1, \ldots, L_n$ are simple $\mathcal{G}$-modules such that $\{L_n|_{\mathcal{N}}, \ldots, L_n|_{\mathcal{N}}\}$ is a complete set of representatives for the isoclasses of the simple $\mathcal{N}$-modules, then every simple $\mathcal{G}$-module S is of the form*

$$S \cong L_i \otimes_k M$$

for a unique $i \in \{1, \ldots, n\}$ and a unique (up to isomorphism) simple $(\mathcal{G}/\mathcal{N})$-module M.

Proof. Let $L_1, \ldots, L_n$ and $M_1, \ldots, M_m$ be complete sets of representatives for the isoclasses of the simple $\mathcal{N}$-modules and simple $(\mathcal{G}/\mathcal{N})$-modules, respectively. By our current assumption, each $\mathcal{N}$-module L_i is the restriction of a simple $\mathcal{G}$-module, which we will also denote by L_i. Given a simple $\mathcal{G}$-module S, there exists $i \in \{1, \ldots, n\}$ such that $L_i \hookrightarrow S|_{\mathcal{N}}$. Consequently, the $\mathcal{G}$-linear map

$$\varphi : \mathrm{Hom}_{\mathcal{N}}(L_i, S) \otimes_k L_i \longrightarrow S; \quad f \otimes x \mapsto f(x)$$

is surjective. The image $\mathrm{im}\,\varphi$ is the L_i-isotypic component S_{L_i} of the module $S|_{\mathcal{N}}$. By Schur's Lemma, the $\mathcal{N}$-module S_{L_i} has dimension $\dim_k \mathrm{Hom}_{\mathcal{N}}(L_i, S) \otimes_k L_i$. As a result, φ is also injective, and

$$S \cong \mathrm{Hom}_{\mathcal{N}}(L_i, S) \otimes_k L_i \cong L_i \otimes_k \mathrm{Hom}_{\mathcal{N}}(L_i, S).$$

Note that $\mathrm{Hom}_{\mathcal{N}}(L_i, S)$ has the structure of a $(\mathcal{G}/\mathcal{N})$-module, which is necessarily simple.

Next, we consider the $\mathcal{G}$-module $S := L_i \otimes_k M_r$. By the above, S contains a simple submodule $T \cong L_j \otimes_k M_s$. Upon restriction to $\mathcal{N}$ we obtain

$$(\dim_k M_s)L_j \cong T|_{\mathcal{N}} \hookrightarrow S|_{\mathcal{N}} \cong (\dim_k M_r)L_i,$$

so that $i = j$. Moreover, we have homomorphisms

$$M_s \cong \mathrm{Hom}_{\mathcal{N}}(L_i, L_i \otimes_k M_s) \cong \mathrm{Hom}_{\mathcal{N}}(L_i, T) \hookrightarrow \mathrm{Hom}_{\mathcal{N}}(L_i, S) \cong M_r$$

of $(\mathcal{G}/\mathcal{N})$-modules, whence $r = s$. As a result, the module $S = L_i \otimes_k M_r$ is simple and the pair (i, r) is uniquely determined by S. □

Remark. If $\mathcal{N} = \mathcal{U}$ is a unipotent normal subgroup of $\mathcal{G}$, then, setting $L_1 = k$, our result specializes to the well-known bijection between the simple $\mathcal{G}/\mathcal{U}$-modules and the simple $\mathcal{G}$-modules.

The technical condition of Lemma 6.3 is known to hold for Frobenius kernels of semi-simple, simply connected groups (cf. [80, (II.3.15)]). In our projected applications, the normal subgroup $\mathcal{N}$ will be the first Frobenius kernel of the special linear group SL(2). The following subsidiary result shows how this fact can be exploited in the computation of the Ext-quiver of $k\mathcal{G}$.

Lemma 6.4. *Let $\mathcal{N} \trianglelefteq \mathcal{G}$ be a normal subgroup of a finite algebraic k-group $\mathcal{G}$. Suppose that L_1, L_2 and M_1, M_2 are simple $\mathcal{G}$-modules and simple $(\mathcal{G}/\mathcal{N})$-modules, respectively, such that $L_i|_{\mathcal{N}}$ is simple for $i \in \{1, 2\}$.*

(1) *If $\mathrm{Ext}^1_{\mathcal{N}}(S, S) = (0)$ for every simple $\mathcal{N}$-module S, then*

$$\mathrm{Ext}^1_{\mathcal{G}}(L_1 \otimes_k M_1, L_2 \otimes_k M_2) \cong \begin{cases} \mathrm{Ext}^1_{\mathcal{G}/\mathcal{N}}(M_1, M_2) & \text{if } L_1 \cong L_2, \\ \mathrm{Hom}_{\mathcal{G}/\mathcal{N}}(M_1, \mathrm{Ext}^1_{\mathcal{N}}(L_1, L_2) \otimes_k M_2) & \text{if } L_1 \not\cong L_2. \end{cases}$$

(2) *If $\mathcal{G}/\mathcal{N}$ is linearly reductive, then*

$$\mathrm{Ext}^1_{\mathcal{G}}(L_1 \otimes_k M_1, L_2 \otimes_k M_2) \cong \mathrm{Hom}_{\mathcal{G}/\mathcal{N}}(M_1, \mathrm{Ext}^1_{\mathcal{N}}(L_1, L_2) \otimes_k M_2). \qquad \square$$

The following consequence gives a first illustration of the relationship between the Ext-quivers and the McKay quivers of certain groups.

Corollary 6.5. *Let $\mathcal{G}$ be a finite algebraic group, $\mathcal{U} \trianglelefteq \mathcal{G}$ be a unipotent, normal subgroup such that $\mathcal{G}/\mathcal{U}$ is linearly reductive. Then the following statements hold:*

(1) *The Ext-quiver $Q_{k\mathcal{G}}$ is isomorphic to the McKay quiver $\Theta_{\mathrm{H}^1(\mathcal{U},k)}(\mathcal{G}/\mathcal{U})$ of $\mathcal{G}/\mathcal{U}$ relative to the $(\mathcal{G}/\mathcal{U})$-module $\mathrm{H}^1(\mathcal{U}, k)$.*

(2) *If $\mathrm{H}^1(\mathcal{U}, k)$ is a faithful $(\mathcal{G}/\mathcal{U})$-module, then $k\mathcal{G} = \mathcal{B}_0(\mathcal{G})$.*

Proof. Let $S_1, \ldots, S_n$ be a complete set of representatives for the simple $\mathcal{G}$-modules. The unipotent normal subgroup $\mathcal{U}$ acts trivially on the simple modules S_i, so that each S_i is a simple $(\mathcal{G}/\mathcal{U})$-module. Thanks to Lemma 6.4(2), we have

$$\mathrm{Ext}^1_{\mathcal{G}}(S_i, S_j) \cong \mathrm{Hom}_{\mathcal{G}/\mathcal{U}}(S_i, \mathrm{H}^1(\mathcal{U}, k) \otimes_k S_j).$$

Thus, if $\mathrm{H}^1(\mathcal{U}, k) \otimes_k S_j \cong \bigoplus_{i=1}^n m_{ij} S_i$, then Schur's Lemma yields

$$m_{ij} = \dim_k \mathrm{Hom}_{\mathcal{G}/\mathcal{U}}(S_i, \mathrm{H}^1(\mathcal{U}, k) \otimes_k S_j) = \dim_k \mathrm{Ext}^1_{\mathcal{G}}(S_i, S_j),$$

so that (1) follows. Assertion (2) is now a consequence of Corollary 5.11. $\square$

Our final result of this section provides a criterion for the construction of principal indecomposable modules. For large p, condition (b) of the following proposition is known to hold in the classical context of Frobenius kernels [10, 78].

Proposition 6.6. *Let $\mathcal{G}$ be a finite algebraic group, $\mathcal{N} \trianglelefteq \mathcal{G}$ be a normal subgroup. Suppose that*

(a) *every simple $\mathcal{N}$-module is the restriction of a $\mathcal{G}$-module, and*
(b) *every principal indecomposable $\mathcal{N}$-module is the restriction of a $\mathcal{G}$-module.*

If $P_1, \dots, P_n$ are $\mathcal{G}$-modules such that $\{P_1|_{\mathcal{N}}, \dots, P_n|_{\mathcal{N}}\}$ is a complete set of representatives for the isoclasses of the principal indecomposable $\mathcal{N}$-modules, and $\{Q_1, \dots, Q_m\}$ is a complete set of representatives for the isoclasses of the principal indecomposable $(\mathcal{G}/\mathcal{N})$-modules, then the $\mathcal{G}$-modules $(P_i \otimes_k Q_r)_{1 \leq i \leq n,\ 1 \leq r \leq m}$ form a complete set of principal indecomposable $\mathcal{G}$-modules. □

Remark. If $\mathcal{G} = \mathcal{U} \rtimes \mathcal{H}$ is a semidirect product of a unipotent group $\mathcal{U}$ and a linearly reductive group $\mathcal{H}$, then $P_1 := k\mathcal{G} \otimes_{k\mathcal{H}} k$ is a projective $k\mathcal{G}$-module such that $P_1|_{\mathcal{U}} \cong k\mathcal{U}$. By Proposition 6.6, the principal indecomposable $\mathcal{G}$-modules are of the form $P_1 \otimes_k S \cong k\mathcal{G} \otimes_{k\mathcal{H}} S$, where S is a simple $\mathcal{H}$-module. If $\mathcal{H}$ is diagonalizable, then the simple $\mathcal{H}$-modules are of the form k_λ, with $\lambda \in X(\mathcal{G}) \cong X(\mathcal{H})$ (see also [47, (2.4)]).

6.3 Groups of finite representation type

Let k be an algebraically closed field of characteristic $p > 0$. We now consider those finite algebraic k-groups, whose principal blocks have finite representation type. Recall that Nagata's Theorem gives a complete answer for groups, whose principal blocks are simple.

By Higman's Theorem [69], the finite groups with representation-finite principal blocks are precisely those having a cyclic Sylow-p-subgroup. The structure of these groups was investigated by Brauer [17, 18]. For instance, the factor group $G/O_{p'}(G)$ of G by its largest normal subgroup of order prime to p, is either trivial, or an automorphism group of a simple group, or its center is cyclic of order p. General theory shows that every simple group possesses some cyclic Sylow subgroup.

Turning to infinitesimal groups, we have already seen in Proposition 4.10 that schemes of tori do provide information on the structure of representation-finite Lie algebras.

Definition. An infinitesimal group $\mathcal{G}$ is called *supersolvable* if there exists a chain $e_k = \mathcal{G}_{[0]} \subseteq \mathcal{G}_{[1]} \subseteq \cdots \subseteq \mathcal{G}_{[n]} = \mathcal{G}$ of normal subgroups of $\mathcal{G}$ such that $\mathcal{G}_{[i]}/\mathcal{G}_{[i-1]} \cong \mathbb{G}_{a(1)}, \mu_{(p)}$ for every $i \in \{1, \dots, n\}$.

Given an infinitesimal group $\mathcal{G}$ with Lie algebra $\mathfrak{g} := \operatorname{Lie}(\mathcal{G})$, we recall from Corollary 3.3, Lemma 3.2, Theorem 1.20 and Corollary 3.14 that $\dim \widehat{\mathcal{V}}_{\mathfrak{g}} \leq 1$ whenever $\mathcal{B}_0(\mathcal{G})$ has finite representation type. The following result (cf. [47, (2.1)]), exploits this property:

Proposition 6.7. *Let $\mathcal{G}$ be an infinitesimal group such that $\dim \widehat{\mathcal{V}}_{\mathfrak{g}} \leq 1$. Then $\mathcal{G}$ is supersolvable.*

Proof. We shall only sketch a proof of the weaker statement: *If $\mathcal{B}_0(\mathcal{G})$ is representation-finite, then $\mathcal{G}$ is supersolvable.* Proceeding by induction on the order of $\mathcal{G}$, we let $\mathfrak{n} \trianglelefteq \mathfrak{g}$ be a minimal $\mathcal{G}$-invariant p-ideal of $\mathfrak{g}$, relative to the adjoint representation $\mathcal{G} \longrightarrow \operatorname{Aut}_p(\mathfrak{g})$. As in Proposition 4.10 one shows that $\mathfrak{g}$ is solvable. Consequently, $\mathfrak{n}$ is abelian, and $\mathfrak{n}^{[p]} \subseteq \mathfrak{n} \cap C(\mathfrak{g})$ is a $\mathcal{G}$-invariant p-ideal of $\mathfrak{g}$. We thus have $\mathfrak{n}^{[p]} = \mathfrak{n}$ or $\mathfrak{n}^{[p]} = (0)$. In the former case, $\mathfrak{n}$ is a torus, so that the connected group $\mathcal{G}$ acts trivially on $\mathfrak{n}$. Alternatively, $\mathfrak{n} \subseteq \widehat{V}_{\mathfrak{g}}$ has dimension 1. Thus, the normal subgroup $\mathcal{N} \trianglelefteq \mathcal{G}$ corresponding to $\mathfrak{n}$ is isomorphic to $\mu_{(p)}$ or $\mathbb{G}_{a(1)}$. Since the principal block of $\mathcal{G}/\mathcal{N}$ is representation-finite, an application of the inductive hypothesis completes the proof. □

Proposition 6.7 allows us to bring our knowledge of supersolvable groups to bear: The factor group $\mathcal{G}/\mathcal{M}(\mathcal{G})$ of such a group $\mathcal{G}$ by its unique maximal normal diagonalizable subgroup $\mathcal{M}(\mathcal{G})$ is trigonalizable, and thus decomposes into a semidirect product

$$\mathcal{G}/\mathcal{M}(\mathcal{G}) \cong \mathcal{U} \rtimes \mathcal{D},$$

with a normal unipotent subgroup $\mathcal{U}$ and a diagonalizable subgroup $\mathcal{D}$. One can show that $\mathcal{B}_0(\mathcal{G}) \cong \mathcal{B}_0(\mathcal{G}/\mathcal{M}(\mathcal{G}))$. Moreover, the algebra $\mathcal{B}_0(\mathcal{U}\rtimes\mathcal{D})$ is representation-finite if and only if $k\mathcal{U}$ enjoys this property. However, $k\mathcal{U}$ is local of dimension a p-power. Thus, $k\mathcal{U}$ is of finite representation type if and only if it is isomorphic to a truncated polynomial ring $k[X]/(X^{p^n})$. In particular, $k\mathcal{U}$ is a Nakayama algebra, and the group $\mathcal{U}$ is commutative.

The commutative unipotent groups that give rise to Nakayama algebras are the so-called *V-uniserial* groups. Here the prefix "V" refers to the *Verschiebung* $V_{\mathcal{U}} : \mathcal{U}^{(1)} \longrightarrow \mathcal{U}$, a homomorphism that is the dual of the Frobenius homomorphism of the *Cartier dual* $\mathcal{D}(\mathcal{U})$ of $\mathcal{U}$. The V-uniserial groups were classified in [44], using Dieudonné modules (see also [83]).

In the above, we have sketched the implication (1) ⇒ (2) of the following result:

Theorem 6.8 ([47, 50]). *Let $\mathcal{G}$ be an infinitesimal k-group. Then the following statements are equivalent:*

(1) *$\mathcal{B}_0(\mathcal{G})$ has finite representation type.*

(2) *$\mathcal{G}/\mathcal{M}(\mathcal{G}) \cong \mathcal{U} \rtimes \mu_{(p^n)}$ is a semidirect product with a V-uniserial normal subgroup $\mathcal{U}$.*

(3) *$k\mathcal{G}$ is a Nakayama algebra.*

(4) $\dim P(\mathcal{G}_2) \leq 0$.

(5) *$k\mathcal{G}_2$ is a Nakayama algebra.* □

In particular, finite representation type may be detected on the second Frobenius kernel of an infinitesimal group.

Remarks. (i) Special cases of Theorem 6.8 as well as related results can be found in [97, 54, 38].

(ii) The equivalence (1) ⇔ (4) readily shows that subgroups of representation-finite groups are representation-finite. For finite groups, this fact follows directly from the Mackey Decomposition Theorem.

(iii) Let G be a finite group. The example of the quaternion group shows that the analogue of (1) ⇔ (4) fails in this context. Moreover, representation-finite group algebras are not necessarily Nakayama algebras. Since the defect group of the principal block of kG is a Sylow-p-subgroup, kG is representation-finite whenever $\mathcal{B}_0(G)$ has this property.

We now turn to the block structure of $k\mathcal{G}$. According to Theorem 6.8, each block $\mathcal{B}$ of a representation-finite infinitesimal group $\mathcal{G}$ is a self-injective Nakayama algebra. In general, indecomposable Nakayama algebras are determined by their *Kupisch series*, that is, by the number of simple modules, and the lengths of their projective covers (cf. [84], [4, (V.3)]). In particular, every self-injective indecomposable Nakayama algebra is Morita equivalent to a bound quiver algebra

$$k[\tilde{A}_{n-1}]/(k[\tilde{A}_{n-1}]^+)^m,$$

whose quiver is a clockwise or counter-clockwise oriented circle with n vertices.

For infinitesimal groups of finite representation type, the parameters n and m of the blocks of $k\mathcal{G}$ were determined in [46]. They turn out to be p-powers, whose exponents can be described as orders of certain subgroups of $\mathcal{G}$.

Let me briefly indicate the connection with McKay quivers, as these will appear again for groups of tame representation type. If $\mathcal{B}_0(\mathcal{G})$ is representation-finite, then Theorem 6.8 tells us that, for the purpose of describing $\mathcal{B}_0(\mathcal{G})$, we may assume that $\mathcal{G} = \mathcal{U} \rtimes \mu_{(p^n)}$. We may apply Corollary 6.5 to see that the Ext-quiver of $k\mathcal{G}$ coincides with the McKay quiver $\Theta_{\mathrm{H}^1(\mathcal{U},k)}(\mu_{(p^n)})$ of $k\mu_{(p^n)}$ relative to the one-dimensional $\mathcal{G}$-module $\mathrm{H}^1(\mathcal{U}, k)$:

$$\mathrm{Ext}^1_{\mathcal{G}}(k_\lambda, k_\gamma) \cong \mathrm{Hom}_{\mu_{(p^n)}}(k_\lambda, \mathrm{H}^1(\mathcal{U}, k) \otimes_k k_\gamma).$$

We have seen in Corollary 5.5 that the connected components of such quivers are of the form $\tilde{A}_{q-1}$, with q being a divisor of $p^n = \mathrm{ord}(X(\mu_{(p^n)}))$.

Theorem 6.9 ([47]). *Let $\mathcal{G}$ be a finite group scheme. Then the following statements are equivalent:*

(1) *The principal block $\mathcal{B}_0(\mathcal{G})$ has finite representation type.*

(2) *The algebras $k\mathcal{G}^0$ and $k\mathcal{G}(k)$ have finite representation type, with at least one of them being semi-simple.*

(3) *$k\mathcal{G}$ has finite representation type.* □

Further information concerning blocks of $k\mathcal{G}$ can be found in [46, 43].

6.4 Binary polyhedral groups

We now turn to binary polyhedral groups and describe the linearly reductive finite algebraic subgroups of the smooth group scheme $\mathrm{SL}(2)$. As before, $T \subseteq \mathrm{SL}(2)$ denotes the standard maximal torus of diagonal matrices. For $m \geq 1$, we let

$$Q_m := \langle x, y \mid x^m = y^2\,;\, yxy^{-1} = x^{-1} \rangle$$

be the *generalized quaternion group* of order $4m$. In our context, Q_m will occur as the subgroup of $\mathrm{SL}(2)(k)$ generated by the matrices

$$\omega = \begin{pmatrix} 0 & 1 \\ -1 & 0 \end{pmatrix} \text{ and } x(\zeta_{2m}) := \begin{pmatrix} \zeta_{2m} & 0 \\ 0 & \zeta_{2m}^{-1} \end{pmatrix},$$

where $\zeta_{2m} \in k$ is a primitive $2m$-th root of unity and $(p, 2m) = 1$. We let $\hat{T}$, $\hat{O}$ and $\hat{I}$ be the *binary tetrahedral group*, the *binary octahedral group*, and the *binary icosahedral group* of orders 24, 48 and 120, respectively (cf. [112, (4.4)]). Realizations of these groups can be found in [41]. When considering binary polyhedral groups we tacitly assume $p \geq 5$ for $\hat{T}$, $\hat{O}$ as well as $p \geq 7$ for $\hat{I}$.

We shall also consider some non-reduced subgroups of SL(2). For $m \in \mathbb{N}$, we let $T_{(m)} \subseteq T$ be the closed subgroup, given by

$$T_{(m)}(R) := \left\{ \begin{pmatrix} x & 0 \\ 0 & x^{-1} \end{pmatrix} ; \ x \in \mu_{(m)}(R) \right\}$$

for every commutative k-algebra R. Note that $T_{(m)}$ is the unique closed subgroup of T of order m.

Let $h := \begin{pmatrix} \zeta_4 & 1 \\ 1 & \zeta_4 \end{pmatrix} \in \mathrm{GL}(2)(k)$. Then $\mathcal{H}_4 := h^{-1} T_{(4)} h$ is a reduced, closed subgroup of SL(2) such that $\mathcal{H}_4(k) = \langle \omega \rangle \subseteq \mathrm{Nor}_{\mathrm{SL}(2)(k)}(T(k))$. Since T is reduced, an application of [80, (I.2.6(11))] shows that the latter group coincides with $\mathrm{Nor}_{\mathrm{SL}(2)}(T)(k)$. As $\mathcal{H}_4$ is reduced, this implies $\mathcal{H}_4 \subseteq \mathrm{Nor}_{\mathrm{SL}(2)}(T)$. Thus, $\mathcal{H}_4$ normalizes every closed subgroup of T, and we define

$$N_{(m)} := T_{(m)} \mathcal{H}_4$$

for $m \geq 2$. The construction of the other subgroups of SL(2) requires a few preparatory remarks.

When dealing with reduced algebraic group schemes we shall often make use of the fact that such a group $\mathcal{G}$ is uniquely determined by its group of k-rational points. In fact,

$$\mathcal{G} \mapsto \mathcal{G}(k)$$

defines an equivalence between the category of reduced algebraic group schemes and the category of algebraic groups in the sense of [113]. Upon restriction to reduced finite algebraic groups, the above functor provides an equivalence with the category of finite groups (cf. [122, (0.16),(0.17)] for more details). Given a finite group G, we let G_k be the reduced finite algebraic group such that $G_k(k) = G$.

Lemma 6.10. *Let $\mathcal{G}$ and $\mathcal{H}$ be algebraic groups. Suppose that $\mathcal{G}$ is reduced and that there exists a closed embedding $\varphi : \mathcal{G}(k) \hookrightarrow \mathcal{H}(k)$. Then there exists a closed embedding $\psi : \mathcal{G} \hookrightarrow \mathcal{H}$ of group schemes such that $\psi_k = \varphi$.* □

According to Lemma 6.10, there exist uniquely determined reduced subgroups $(\hat{T})_k$, $(\hat{O})_k$ and $(\hat{I})_k$ of SL(2) satisfying $(\hat{T})_k(k) = \hat{T}$, $(\hat{O})_k(k) = \hat{O}$ and $(\hat{I})_k(k) = \hat{I}$, respectively. The proof of Theorem 6.11 below, which extends [67, Thm.1] to our context, shows in particular that $N_{(m)} \cong (Q_m)_k$ whenever $(p, 2m) = 1$.

Recall that the McKay graph $\bar{\Theta}_L(\mathcal{G})$ relative to a two-dimensional self-dual module L is the graph associated to the defining generalized Cartan matrix, cf. Corollary 5.5.

Theorem 6.11. *Let $\mathcal{G} \subseteq \mathrm{SL}(2)$ be a finite, linearly reductive subgroup scheme of characteristic $p \geq 3$. Then there exists $g \in \mathrm{SL}(2)(k)$ such that $g\mathcal{G}g^{-1}$ and the McKay graph $\bar{\Theta}_L(\mathcal{G})$ of $\mathcal{G}$ relative to its standard module L belong to the following list:*

$g\mathcal{G}g^{-1}$	$\bar{\Theta}_L(\mathcal{G})$
e_k	$\tilde{A}_0$
$T_{(np^r)}$	$\tilde{A}_{np^r-1}$
$N_{(np^r)}$	$\tilde{D}_{np^r+2}$
$(\hat{T})_k$	$\tilde{E}_6$
$(\hat{O})_k$	$\tilde{E}_7$
$(\hat{I})_k$	$\tilde{E}_8$

where $(n,p) = 1$, $r := \mathrm{ht}(\mathcal{G}^0)$, and $n + r \neq 1$.

Proof. According to [80, (II.2.5)], the two-dimensional, simple SL(2)-module $L = L(1)$ is self-dual. Consequently, L is also a faithful, self-dual $\mathcal{G}$-module.

Recall from Theorem 1.13 that $\mathcal{G} = \mathcal{G}^0 \rtimes \mathcal{G}_{\mathrm{red}}$ is a semidirect product of an infinitesimal group $\mathcal{G}^0$ and a reduced group $\mathcal{G}_{\mathrm{red}}$. Thanks to Nagata's Theorem (Theorem 6.2), the connected component $\mathcal{G}^0$ is diagonalizable and p does not divide the order of $\mathcal{G}(k) = \mathcal{G}_{\mathrm{red}}(k)$. Since $\mathcal{G}^0$ is diagonalizable, there exists an element $g \in \mathrm{SL}(2)(k)$ such that $g\mathcal{G}^0g^{-1} \subseteq T$. If $\mathrm{ht}(\mathcal{G}^0) = r$, then we have $g\mathcal{G}^0g^{-1} \subseteq T_r \cong \mu_{(p^r)}$. On the other hand, $\mathrm{ord}(\mathcal{G}^0) \geq p^r$ (cf. [123, (2.2)]), so that $g\mathcal{G}^0g^{-1} = T_r$. Thus, replacing $\mathcal{G}$ by a suitable conjugate group, we may assume that $\mathcal{G}^0 = T_r = T_{(p^r)}$.

If $r = 0$, then the group $\mathcal{G}$ is reduced, and $\mathcal{G}$ is completely determined by the binary polyhedral group $\mathcal{G}(k) \subseteq \mathrm{SL}(2)(k)$. Directly from [112, (4.4)] we infer that $\mathcal{G}(k)$ is conjugate to e_k, $T_{(n)}(k)$ for $n \geq 2$, $N_{(n)}(k)$ for $n \geq 2$, $\hat{T}$, $\hat{O}$, or $\hat{I}$. As all groups involved are associated to reduced group schemes, we obtain the left-hand column of our list. Thanks to Theorem 5.13 (see also [67, Thm.1]), the McKay graphs of these groups have the asserted structure.

If $r \geq 1$, then the identity $\mathcal{G}^0 = T_r$ in conjunction with $p \neq 2$ and [80, (I.2.6(11))] implies that

$$\mathcal{G}(k) \subseteq \mathrm{Nor}_{\mathrm{SL}(2)}(T_r)(k) = T(k)\langle\omega\rangle = \mathrm{Nor}_{\mathrm{SL}(2)(k)}(T(k)) = \mathrm{Nor}_{\mathrm{SL}(2)}(T)(k).$$

Consequently, $\mathcal{G}_{\mathrm{red}} \subseteq \mathrm{Nor}_{\mathrm{SL}(2)}(T)$, so that $\mathcal{G} \subseteq \mathrm{Nor}_{\mathrm{SL}(2)}(T)$. There results an embedding $\mathcal{G}/(\mathcal{G} \cap T) \hookrightarrow \mathrm{Nor}_{\mathrm{SL}(2)}(T)/T$, with the latter group being the *Weyl group* of SL(2). In particular, $\mathcal{G}/(\mathcal{G}\cap T)$ has order ≤ 2. The finite algebraic group $\mathcal{G}\cap T$ coincides with $T_{(np^r)}$ for some n not divisible by p. By our observations above, we have $\mathcal{G}^0 \subseteq \mathcal{G}\cap T$. If $\mathcal{G}(k) \subseteq T(k)$, then $\mathcal{G} = \mathcal{G}\cap T = T_{(np^r)}$. Alternatively, some basic considerations allow us to assume that $\mathcal{H}_4(k) \subseteq \mathcal{G}(k)$ (cf. [41, §3]). As $\mathcal{H}_4$ is reduced, we conclude that $\mathcal{G}\cap T \subseteq N_{(np^r)} \subseteq \mathcal{G}$, so that $\mathcal{G} = N_{(np^r)}$.

It remains to identify the McKay graphs of $T_{(np^r)}$ and $N_{(np^r)}$. This can be done using the results of [67]. □

6.5 Groups of tame representation type

Throughout, k is assumed to be an algebraically closed field of characteritic $p \geq 3$. In this section, we discuss the determination of the finite group schemes of tame representation type. The theory proceeds progressively by first studying infinitesimal groups of height 1. In view of Theorem 1.20, this amounts to analyzing restricted enveloping algebras $U_0(\mathfrak{g})$ of finite-dimensional restricted Lie algebras $(\mathfrak{g}, [p])$.

The first example people looked at was the special linear Lie algebra $\mathfrak{g} = \mathfrak{sl}(2)$. In early work, Pollack [98] showed that $U_0(\mathfrak{sl}(2))$ has infinite representation type. As mentioned in Section 3, the basic algebra of $U_0(\mathfrak{sl}(2))$ was determined some 15 years later independently by Drozd [34], Fischer [55] and Rudakov [109]. Their work implies that every non-simple block of $U_0(\mathfrak{sl}(2))$ is Morita equivalent to the trivial extension $T(\tilde{A}_1)$ of the Kronecker algebra $k[\bullet \rightrightarrows \bullet]$. In particular, each of these blocks is special biserial and hence tame. At the same time, Pfautsch [95] showed that the tame blocks of the distribution algebras $k\,\mathrm{SL}(2)_r$ are also of this type.

For quite some time, the blocks of the first Frobenius kernel $\mathrm{SL}(2)_1$ remained essentially the only examples of a tame blocks of infinitesimal group. (In characteristic 2, the enveloping algebra of the unipotent Heisenberg algebra is isomorphic to the group algebra of the dihedral group of order 8 and hence tame.) In retrospect, this is not a surprise: We have seen in Theorem 3.16 that tame blocks of Frobenius kernels of smooth groups are Morita equivalent to the tame blocks of $U_0(\mathfrak{sl}(2))$, so finding new examples amounts to working with non-classical restricted Lie algebras.

The following result, which does not hold for fields of characteristic 2, is fundamental for the following.

Theorem 6.12 ([48, 49]). *Let $\mathcal{G}$ be an infinitesimal group with tame principal block. Then the following statements hold:*

(1) *$\mathcal{G}$ is not solvable.*

(2) *If $\mathcal{G}$ is semi-simple, then* $\mathrm{Lie}(\mathcal{G}) \cong \mathfrak{sl}(2)$.

Proof. (1) One picks a counter-example of minimal order and uses Theorem 6.8 to collect enough structural information to compute the quiver and the relations of $\mathcal{B}_0(\mathcal{G})$. These occur in the lists of [120] and are thus wild.

(2) The solvable radical $\mathcal{R}(\mathcal{G}_1)$ of $\mathcal{G}_1$ is a solvable normal subgroup of $\mathcal{G}$ and therefore trivial. As a result, the Lie algebra $\mathfrak{g} := \mathrm{Lie}(\mathcal{G})$ is also semi-simple. Since $\mathcal{B}_0(\mathcal{G})$ is tame, a consecutive application of Theorem 3.15, Theorem 1.6, Proposition 2.4 and Corollary 3.14 gives $\dim \widehat{\mathcal{V}}_{\mathfrak{g}} = \mathrm{cx}_{U_0(\mathfrak{g})}(k) \leq 2$. In view of (1) and Proposition 6.7, we actually have $\dim \widehat{\mathcal{V}}_{\mathfrak{g}} = 2$, and the assertion now follows from Theorem 4.11. □

6.5.1 Restricted Lie algebras

We begin by studying the easiest case concerning restricted Lie algebras. A priori, it is not clear why this strategy should be fruitful. Contrary to the modular representation theory of finite groups, we don't automatically have a good descent theory at our disposal. As we shall see, subgroups of tame infinitesimal groups may be wild! Fortunately, passage to Frobenius kernels turns out to be much better behaved, and the first Frobenius kernel will be seen to encapsulate most structural features of the group $\mathcal{G}$.

Let $(\mathfrak{g}, [p])$ be a finite-dimensional restricted Lie algebra, and denote by $\mathcal{B}_0(\mathfrak{g})$ the principal block of its restricted enveloping algebra $U_0(\mathfrak{g})$.

Problem. We shall be concerned with the following two interrelated tasks:

- Determine those Lie algebras $\mathfrak{g}$ for which $\mathcal{B}_0(\mathfrak{g})$ is tame.
- Find the presentation of $\mathcal{B}_0(\mathfrak{g})$ by its Ext-quiver and relations in case $\mathcal{B}_0(\mathfrak{g})$ is tame.

A few comments are in order. By focussing on the principal block, we make the determination of $\mathfrak{g}$ more tractable. If $T(\mathfrak{g}) \subseteq \mathfrak{g}$ is the maximal toral p-ideal of $\mathfrak{g}$, then $\mathcal{B}_0(\mathfrak{g}) \cong \mathcal{B}_0(\mathfrak{g}/T(\mathfrak{g}))$, so that we work up to extensions by toral ideals. As we shall see below, the presence of toral ideals may complicate the representation theory of $U_0(\mathfrak{g})$ significantly.

Recall that $C(\mathfrak{g}) := \{x \in \mathfrak{g};\ [x, y] = 0 \ \ \forall\, y \in \mathfrak{g}\}$ is the center of $\mathfrak{g}$. Given $n \in \mathbb{N}_0$, we let

$$\mathfrak{n}_n := \bigoplus_{i=0}^{n-1} k x^{[p]^i}; \quad x^{[p]^n} = 0 \neq x^{[p]^{n-1}}$$

be the n-dimensional, nil-cyclic Lie algebra.

Let $\mathfrak{r}$ be the solvable radical of $\mathfrak{g}$. If $\mathcal{B}_0(\mathfrak{g})$ is tame, then $\mathcal{B}_0(\mathfrak{g}/\mathfrak{r})$ is tame or representation-finite. Owing to Theorem 6.12 the algebra $\mathfrak{g}$ is not solvable, so that $\mathfrak{g}/\mathfrak{r}$ is not trivial. Thus, by Proposition 6.7, the block $\mathcal{B}_0(\mathfrak{g}/\mathfrak{r})$ is tame and Theorem 6.12 implies $\mathfrak{g}/\mathfrak{r} \cong \mathfrak{sl}(2)$. A detailed analysis of the radical then yields:

Proposition 6.13. *Suppose that $\mathcal{B}_0(\mathfrak{g})$ has tame representation type. Then $\mathfrak{g}/C(\mathfrak{g}) \cong \mathfrak{sl}(2)$ and $C(\mathfrak{g}) \cong \mathfrak{n}_n \oplus T(\mathfrak{g})$ for some $n \in \mathbb{N}_0$.* □

Let us record one important consequence: Since $\mathfrak{g}/C(\mathfrak{g}) \cong \mathfrak{sl}(2)$, we have $\mathfrak{g} = [\mathfrak{g}, \mathfrak{g}] + C(\mathfrak{g})$, so that $\operatorname{tr}(\operatorname{ad} x) = 0$ for all $x \in \mathfrak{g}$. By Schue's result (cf. [110] and Proposition 1.4), the restricted enveloping algebra $U_0(\mathfrak{g})$ is symmetric.

The foregoing result marks the extent to which our geometric techniques will take us. We are left with the problem of determining those central extensions of $\mathfrak{sl}(2)$, whose principal blocks are tame.

To obtain the structure of $\mathfrak{g}$, one first observes that the exact sequence

$$(0) \longrightarrow C(\mathfrak{g}) \longrightarrow \mathfrak{g} \longrightarrow \mathfrak{sl}(2) \longrightarrow (0)$$

splits when considered a sequence of ordinary Lie algebras. Thus, $\mathfrak{g} = \mathfrak{sl}(2) \oplus C(\mathfrak{g})$, and there exists a p-semilinear map $\psi : \mathfrak{sl}(2) \longrightarrow C(\mathfrak{g})$ such that

$$(x, c)^{[p]} = (x^{[p]}, \psi(x) + c^{[p]}) \qquad \text{for } (x, c) \in \mathfrak{sl}(2) \oplus C(\mathfrak{g}).$$

Let us consider the case where $C(\mathfrak{g}) = kv_0 \neq (0)$ and $v_0^{[p]} = 0$. The group $\mathrm{SL}(2)(k) \times k^\times$ operates on the space of p-semilinear forms such that the orbits of this action correspond to isomorphism classes of restricted Lie algebras. There are exactly three orbits, whose representatives are given in terms of the standard basis $\{e, h, f\} \subseteq \mathfrak{sl}(2)$ by

$$\psi_0 = 0; \quad \psi_n(e) = 0 = \psi_n(h), \ \psi_n(f) = v_0; \quad \psi_s(e) = 0 = \psi_s(f), \ \psi_s(h) = v_0.$$

Thus, the corresponding central extensions are the algebras $\mathfrak{sl}(2)_0$, $\mathfrak{sl}(2)_n$, and $\mathfrak{sl}(2)_s$, which were introduced in Proposition 1.4. Since $\dim \widehat{V}_{\mathfrak{sl}(2)_0} = 3$, Theorem 3.15 tells us that we only have to consider the latter two types.

In our situation the restricted enveloping algebras $U_0(\mathfrak{g})$ and $U_0(\mathfrak{sl}(2))$ have the same simple modules and almost the same Ext-quiver. In particular, there is a correspondence between the blocks, so that each block has either one or two simple modules. It turns out that the representation type depends on the structure of the hearts $\mathrm{Ht}(P) = \mathrm{Rad}(P)/\mathrm{Soc}(P)$ of the principal indecomposable $U_0(\mathfrak{g})$-modules. Their structure can be analyzed by means of filtrations by Verma modules. We denote the (Jacobson) radical of $U_0(\mathfrak{g})$ by J.

The following result uses techniques from the abstract representation theory of quivers.

Proposition 6.14 ([51]). *Let P be a principal indecomposable $U_0(\mathfrak{g})$-module belonging to a block $\mathcal{B} \subseteq U_0(\mathfrak{g})$ with two simple modules.*

(1) *If $\mathfrak{g} = \mathfrak{sl}(2)_n$, then JP/J^3P is indecomposable of length 4, and $\mathcal{B}$ is wild.*

(2) *If $\mathfrak{g} = \mathfrak{sl}(2)_s$, then JP/J^3P is a sum of two uniserial modules, and $\mathcal{B}$ is special biserial.*

Proof. The first part follows from the fact that a Galois covering of $\mathcal{B}/J^3\mathcal{B}$ contains as a subcategory the module category of a one-point extension of the path algebra of the Kronecker quiver given by a regular module of quasi-length 2 in a homogeneous tube.

The second part involves a delicate analysis by which one extends the information on $\mathcal{B}/J^3\mathcal{B}$ to $\mathcal{B}$. Here the fact that $\mathcal{B}$ is symmetric plays an important role. □

The result has an interesting consequence: contrary to the modular representation theory of finite groups, subalgebras of tame restricted Lie algebras may be wild!

Example. Let $\mathfrak{h} := ke \oplus kv_0 \subseteq \mathfrak{sl}(2)_s$. Then $U_0(\mathfrak{h}) \cong k[X,Y]/(X^p, Y^p)$ is wild even though $U_0(\mathfrak{sl}(2)_s)$ is tame.

Let $\mathrm{Rad}_p(\mathfrak{g})$ be the *p-unipotent radical* of $(\mathfrak{g}, [p])$, that is, the largest p-ideal $\mathfrak{n} \trianglelefteq \mathfrak{g}$ such that $x^{[p]^{\dim_k \mathfrak{g}}} = 0$ for every $x \in \mathfrak{n}$. We put $n(\mathfrak{g}) := \dim_k \mathrm{Rad}_p(\mathfrak{g})$.

For $n \in \mathbb{N}_0$, let $\mathcal{K}(n)$ be the bound quiver algebra of the quiver with underlying vertex set $\mathbb{Z}/(2)$ and arrows $\alpha_i, \beta_i : i \to i+1$ $(i = 0, 1)$ subject to the relations

$$\alpha_{i+1}\alpha_i = 0 = \beta_{i+1}\beta_i; \quad (\beta_{i+1}\alpha_i - \alpha_{i+1}\beta_i)^{p^n} = 0.$$

Note that $\mathcal{K}(0) \cong T(\tilde{A}_1)$ is the trivial extension of the Kronecker algebra.

Theorem 6.15 ([51]). *Let $(\mathfrak{g}, [p])$ be a restricted Lie algebra. Then the following statements are equivalent:*

(1) *The principal block $\mathcal{B}_0(\mathfrak{g})$ is tame.*

(2) *Each block of the algebra $U_0(\mathfrak{g}/T(\mathfrak{g}))$ is Morita equivalent to $k[X]/(X^{p^{n(\mathfrak{g})}})$ or to $\mathcal{K}(n(\mathfrak{g}))$, with $\mathcal{B}_0(\mathfrak{g}) \sim_M \mathcal{K}(n(\mathfrak{g}))$.* □

The question of the tameness of $U_0(\mathfrak{g})$ involves the study of representation types of the family $(U_\chi(\mathfrak{g}))_{\chi\in\mathfrak{g}^*}$ of the so-called *reduced enveloping algebras* of $\mathfrak{g}$. In [51, §8], precise criteria for $U_0(\mathfrak{g})$ being tame were given. In particular, one encounters yet another phenomenon, which has no analogue in the modular representation theory of finite groups:

Example. Consider the five-dimensional restricted Lie algebra $\mathfrak{g} := \mathfrak{sl}(2) \oplus kv_0 \oplus kt_0$ with center $C(\mathfrak{g}) = kv_0 \oplus kt_0$ and p-map given by

$$e^{[p]} := 0; \; h^{[p]} := h + v_0; \; f^{[p]} := t_0; \; v_0^{[p]} := 0; \; t_0^{[p]} := t_0.$$

According to Theorem 6.15, the principal block $\mathcal{B}_0(\mathfrak{g})$ is tame. By the results of [51, §8], the algebra $U_0(\mathfrak{g})$ is wild.

6.5.2 Infinitesimal groups

Our above example seems to portend trouble. Since subgroups of tame infinitesimal groups are not necessarily tame, it is a priori not clear how the results on Lie algebras may be employed in studying infinitesimal groups. In fact, the general theory of algebraic groups tells us that the Lie algebra $\mathrm{Lie}(\mathcal{G})$ usually does not capture much information of $\mathcal{G}$. All Frobenius kernels $(\mathcal{G}_r)_{r\geq 1}$ of an algebraic group $\mathcal{G}$ have the same Lie algebra.

However, our situation is rather special. Passage to subgroups does at least "preserve" the complexity (see Proposition 2.4), so that the tameness of $\mathcal{B}_0(\mathcal{G})$ still implies

$$\dim \hat{\mathcal{V}}_{\mathrm{Lie}(\mathcal{G})} \leq 2.$$

If $\mathcal{G}$ is semi-simple, then Theorem 6.12 yields $\mathrm{Lie}(\mathcal{G}) \cong \mathfrak{sl}(2)$, so that we obtain:

- $\mathcal{G} \subseteq \mathrm{SL}(2)$ is an infinitesimal subgroup of $\mathrm{SL}(2)$, and
- $\mathcal{G}_1 = \mathrm{SL}(2)_1$.

To illustrate the utility of the foregoing observations, we elaborate on the following result, which also exemplifies some of the methods relevant for the later developments. We let $T \subseteq \mathrm{SL}(2)$ be the standard maximal torus of diagonal matrices.

Proposition 6.16 ([49]). *Let $\mathcal{G}$ be a semi-simple infinitesimal group such that $\mathcal{B}_0(\mathcal{G})$ is tame. Then there exists $r \in \mathbb{N}$ with $\mathcal{G} \cong \mathrm{SL}(2)_1 T_r$.*

Proof. By the above, we have $\mathcal{G} \subseteq \mathrm{SL}(2)$ and $\mathcal{G}_1 = \mathrm{SL}(2)_1$. There results an embedding

$$\mathcal{G}/\mathcal{G}_1 \hookrightarrow \mathrm{SL}(2)/\,\mathrm{SL}(2)_1.$$

As the group SL(2) is smooth, the Frobenius endomorphism of SL(2) induces an isomorphism $\mathrm{SL}(2)/\,\mathrm{SL}(2)_1 \xrightarrow{\sim} \mathrm{SL}(2)$, so that $\mathcal{G}/\mathcal{G}_1$ can also be considered a subgroup of SL(2). Since $\mathcal{B}_0(\mathcal{G})$ is tame, the principal block $\mathcal{B}_0(\mathcal{G}/\mathcal{G}_1)$ is tame or representation-finite. In the former case, our above observations imply $\mathrm{SL}(2)_2 \subset \mathcal{G}$, so that

$$\mathrm{cx}_{\mathcal{G}}(k) \geq \mathrm{cx}_{\mathrm{SL}(2)_2}(k) = 3.$$

Since this contradicts Theorem 3.15, the principal block $\mathcal{B}_0(\mathcal{G}/\mathcal{G}_1)$ has finite representation type and Proposition 6.7 can be brought to bear. A detailed analysis then shows that there exists $r \in \mathbb{N}$ with

$$\mathcal{G} \cong \mathcal{A}_{[r]} \quad \text{or} \quad \mathcal{G} \cong \mathrm{SL}(2)_1 T_r,$$

where $\mathcal{A}_{[r]}$ is given by

$$\mathcal{A}_{[r]}(R) := \left\{ \begin{pmatrix} a & b \\ c & d \end{pmatrix} \in \mathrm{SL}(2)(R);\ a^{p^r} = 1 = d^{p^r},\ b^{p^2} = 0 = c^p \right\}.$$

It thus remains to rule out the former group, so we assume $\mathcal{G} = \mathcal{A}_{[r]}$.

Since every simple $\mathrm{SL}(2)_1$-module has the structure of an SL(2)-module, the simple $\mathrm{SL}(2)_1$-modules are restrictions of simple $\mathcal{A}_{[r]}$-modules, which we denote $L(0), \ldots, L(p-1)$. By Lemma 6.3, the simple $\mathcal{A}_{[r]}$-modules are therefore tensor products $L(i) \otimes_k M_j$, with the simple modules M_j of the factor group $\mathcal{G}' := \mathcal{G}/\mathcal{G}_1$. Direct computation shows that $\mathcal{G}' \cong \mathbb{G}_{a(1)} \rtimes \mu_{(p^{r-1})}$ is trigonalizable, so that $\mathcal{G}'$ has p^{r-1} simple modules, each having dimension 1. Using Lemma 6.4(1) one can now compute the Ext-quiver $Q_{\mathcal{B}_0(\mathcal{G})}$ of $\mathcal{B}_0(\mathcal{G})$. It has vertex set $V := \mathbb{Z}/(p^{r-1}) \times \mathbb{Z}/(2)$ and arrows

$$\alpha_{(i,j)} : (i,j) \mapsto (i+1,j); \quad \beta_{(i,j)} : (i,j) \mapsto (i+\ell, j+1),$$

where $\ell := \frac{p^{r-1}-1}{2}$. In particular, two arrows start and end at each vertex.

Let $\Lambda = k[Q_{\mathcal{B}_0(\mathcal{G})}]/I$ be the basic algebra of $\mathcal{B}_0(\mathcal{G})$. The Nakayama algebra $k\mathcal{G}'$ has a uniserial module of length 3 with trivial top. We consider its pull-back to $k\mathcal{G}$, which turns out to be a $\mathcal{B}_0(\mathcal{G})$-module. Thus, Λ also possesses a uniserial module N of length 3, whose composition factors can be computed. Using this information, one now shows that neither $\alpha_{(1,0)}\alpha_{(0,0)}$ nor $\beta_{(1,0)}\alpha_{(0,0)}$ annihilates the Λ-module M corresponding to N. This then implies that these paths are not summands of relations of the bound quiver algebra Λ, and Lemma 5.7 shows that $\mathcal{B}_0(\mathcal{G})$ is wild. □

Note that the group $\mathcal{G} = \mathrm{SL}(2)_1 T_r$ satisfies the technical conditions given in Section 2. In particular, $\mathcal{G}/\mathcal{G}_1 \cong \mu_{(p^{r-1})}$ is diagonalizable and the Ext-quiver of $k\mathcal{G}$ is closely related to the McKay quivers of the linearly reductive subgroup $\mathcal{G}/\mathcal{G}_1 \subseteq \mathrm{SL}(2)$. This information suffices to compute the quiver and the relations of $\mathcal{B}_0(\mathcal{G})$ (cf. [49]). In [52] these results were extended to arbitrary infinitesimal groups by a detailed analysis of the solvable radical $\mathcal{R}(\mathcal{G})$ of $\mathcal{G}$ and techniques of Galois extensions (see [7]). Let me quote a result that is easy to state:

Theorem 6.17 ([52]). *Let $\mathcal{G}$ be an infinitesimal group of odd characteristic. Then the following statements are equivalent:*
(1) *The principal block $\mathcal{B}_0(\mathcal{G})$ is tame.*
(2) *$\mathcal{B}_0(\mathcal{G}_1)$ is tame and $\mathcal{G}/\mathcal{G}_1$ is diagonalizable.*
In either case, we have $\mathcal{B}_0(\mathcal{G}_1) = \mathcal{B}_0(\mathcal{G})^{X(\mathcal{G}/\mathcal{G}_1)}$. □

The invariants in the last part are given relative to the standard action of $X(\mathcal{G})$ on $k\mathcal{G}$, which associates to $\gamma \in X(\mathcal{G})$ the convolution

$$\mathrm{id}_{k\mathcal{G}} * \gamma : k\mathcal{G} \longrightarrow k\mathcal{G}; \quad h \mapsto \sum_{(h)} h_{(1)}\gamma(h_{(2)}).$$

Moreover, $X(\mathcal{G}/\mathcal{G}_1)$ is identified with those characters that vanish on the ideal of $k\mathcal{G}$ that is generated by $k\mathcal{G}_1^{\dagger}$.

The above result implies in particular, that tameness is preserved under passage to Frobenius kernels, thereby providing an important remedy for the failure of descent that we have observed earlier. In [52], the infinitesimal groups with tame principal blocks are classified modulo their multiplicative centers and the quiver and relations for the blocks are given. It turns out that all blocks are special biserial. More precisely, the basic algebras are generalizations of trivial extensions $T(\tilde{A}_{2p^{r-1}-1})$ of the radical square zero tame hereditary algebras of type $\tilde{A}_{2p^{r-1}-1}$ (see Theorem 6.15 for an example of the nature of these generalizations). Here $r = \mathrm{ht}(\mathcal{G}/\mathcal{M}(\mathcal{G}))$ is the height of the factor group $\mathcal{G}/\mathcal{M}(\mathcal{G})$ of $\mathcal{G}$ by its multiplicative center $\mathcal{M}(\mathcal{G})$. The effect of $\mathcal{M}(\mathcal{G})$ on arbitrary blocks of $k\mathcal{G}$ is investigated in [53].

6.5.3 Finite group schemes

Let $\mathcal{G}$ be an infinitesimal group. We have seen in Theorem 6.17 that the tameness of $\mathcal{B}_0(\mathcal{G})$ implies the diagonalizability of $\mathcal{G}/\mathcal{G}_1$ along with $\mathcal{G}_1$ being closely related to the first Frobenius kernel $\mathrm{SL}(2)_1$. The Galois correspondence established there heavily relies on the character group $X(\mathcal{G}/\mathcal{G}_1)$, which fully describes the quotient $\mathcal{G}/\mathcal{G}_1$. This will only work for diagonalizable group schemes, so that Galois covering techniques are not suitable for arbitrary finite group schemes.

We thus change our point of view and note that the quivers occurring for infinitesimal groups can be interpreted as McKay quivers. For this approach to be successful in the general case, we need a good supply of linearly reductive groups. Here is the relevant generalization of Theorem 6.17, whose proof relies on the identification of the quivers of minimal counter-examples:

Proposition 6.18 ([41]). *Let $\mathcal{G}$ be a finite group scheme with tame principal block. Then the following statements hold:*
(1) *The principal block $\mathcal{B}_0(\mathcal{G}_1)$ is tame.*

(2) *The group $\mathcal{G}/\mathcal{G}_1$ is linearly reductive.*
In particular, the order of the finite group $\mathcal{G}(k)$ is not divisible by p. □

To ease on the technical aspects we shall discuss our main result in a simplified context. We thus assume that our group $\mathcal{G}$ and its Lie algebra $\mathfrak{g} := \mathrm{Lie}(\mathcal{G})$ satisfy the following conditions

- $\mathcal{G}^0 = \mathcal{G}_1$, and
- $C(\mathfrak{g}) = (0)$.

Setting $G := \mathcal{G}(k)$, the first condition implies

$$k\mathcal{G} \cong U_0(\mathfrak{g})G,$$

while the second condition says that the center of $\mathcal{G}_1$ is trivial.

Now assume $\mathcal{B}_0(\mathcal{G})$ to be tame. Thanks to Proposition 6.18, the principal block of $U_0(\mathfrak{g})$ is tame. Since $C(\mathfrak{g}) = (0)$, Proposition 6.13 yields $\mathfrak{g} \cong \mathfrak{sl}(2)$. Consequently,

$$k\mathcal{G} \cong U_0(\mathfrak{sl}(2))G \quad \text{and} \quad p \nmid \mathrm{ord}(G).$$

Let $N \trianglelefteq G$ denote the kernel of the representation $\varrho : G \longrightarrow \mathrm{Aut}_k(U_0(\mathfrak{sl}(2)))$. Since N is linearly reductive, the principal block $\mathcal{B}_0(\mathcal{G})$ is isomorphic to $\mathcal{B}_0(U_0(\mathfrak{sl}(2)) \cdot (G/N))$. We thus simplify further by assuming $N = \{1\}$.

The group G acts on $U_0(\mathfrak{sl}(2))$ via automorphisms of Hopf algebras. As a result, ϱ defines an embedding

$$G \hookrightarrow \mathrm{Aut}_k(U_0(\mathfrak{sl}(2))) \cong \mathrm{Aut}_p(\mathfrak{sl}(2)) \cong \mathrm{PSL}(2)(k).$$

Passage to the double cover provides a linearly reductive group $\tilde{G} \subseteq \mathrm{SL}(2)(k)$ such that

$$\mathcal{B}_0(U_0(\mathfrak{sl}(2))G) \cong \mathcal{B}_0(U_0(\mathfrak{sl}(2))\tilde{G}).$$

Thus, we may assume that $G \subseteq \mathrm{SL}(2)(k)$ is a binary polyhedral group. These groups were classified by Klein around 1884 [82], and we have already noted in Theorem 6.11 that they are uniquely determined by their McKay graphs relative to the two-dimensional standard representation $L(1) = k^2$. Let us recall the "classical" list of groups up to conjugacy in $\mathrm{SL}(2)(k)$:

G	$\Theta_{L(1)}(G)$
$\mathbb{Z}/(n)$	$\tilde{A}_{n-1}$
Q_n	$\tilde{D}_{n+2}$
$\hat{T}$	$\tilde{E}_6$
$\hat{O}$	$\tilde{E}_7$
$\hat{I}$	$\tilde{E}_8$.

Here, each quiver is obtained from the given graph by replacing a bond $\bullet \!-\!\!-\! \bullet$ is by a pair of arrows $\bullet \leftrightarrows \bullet$. Thus, modulo our simplifications, we know the groups that can occur, i.e., we understand the Hopf algebra structure. Moreover, the affine quivers describing the tame hereditary algebras also appear. It remains to relate the Ext-quiver of $B_0(\mathcal{G})$ to the McKay quivers of G.

Let S, T be simple $\mathcal{G}$-modules. Then G acts on $\mathrm{Ext}^1_{U_0(\mathfrak{sl}(2))}(S,T)$, and since G is linearly reductive, we have

$$\mathrm{Ext}^1_{\mathcal{G}}(S,T) \cong \mathrm{Ext}^1_{U_0(\mathfrak{sl}(2))}(S,T)^G.$$

Every simple $U_0(\mathfrak{sl}(2))$-module is the restriction of a simple $\mathrm{SL}(2)(k)$-module. Consequently, the same holds for the canonical restriction map

$$\mathrm{Res} : \mathrm{mod}\,\mathcal{G} \longrightarrow \mathrm{mod}\,U_0(\mathfrak{sl}(2)).$$

Let $L(0), \ldots, L(p-1)$ be the corresponding simple $\mathcal{G}$-modules, $M_1, \ldots, M_n$ be the simple kG-modules, which we view as simple $k\mathcal{G}$-modules. According to Lemma 6.3, $(L(i) \otimes_k M_j)_{ij}$ is a complete set of representatives for the simple $\mathcal{G}$-modules, and Lemma 6.4(2) implies

$$\mathrm{Ext}^1_{\mathcal{G}}(L(i) \otimes_k M_r, L(j) \otimes_k M_s) \cong \mathrm{Hom}_G(M_r, \mathrm{Ext}^1_{U_0(\mathfrak{sl}(2))}(L(i), L(j)) \otimes_k M_s).$$

Accordingly, the Ext-groups count the multiplicity of M_r in $\mathrm{Ext}^1_{U_0(\mathfrak{sl}(2))}(L(i), L(j)) \otimes_k M_s$. By virtue of [96], the $\mathrm{SL}(2)(k)$-module $\mathrm{Ext}^1_{U_0(\mathfrak{sl}(2))}(L(i), L(j))$ has the following structure

$$\mathrm{Ext}^1_{U_0(\mathfrak{sl}(2))}(L(i), L(j)) \cong \begin{cases} L(1)^{[1]} & i+j = p-2, \\ (0) & \text{else}, \end{cases}$$

where $L(1)^{[1]}$ is the Frobenius twist of the standard module. There results a close relationship between the Ext-quiver $Q_{k\mathcal{G}}$ and the McKay quiver of $\Theta_{L(1)}(G)$ of G. In this fashion, one obtains complete information on the block structure of $k\mathcal{G}$:

Theorem 6.19 ([41]). *Let $\mathcal{G}$ be a finite algebraic group such that $\mathcal{B}_0(\mathcal{G})$ tame. Then the following statements hold:*

(1) *There exists a finite, linearly reductive group scheme $\tilde{\mathcal{G}} \subseteq \mathrm{SL}(2)$ such that the* Ext*-quiver of $\mathcal{B}_0(\mathcal{G})$ is isomorphic to the McKay quiver $\Theta_{L(1)}(\tilde{\mathcal{G}})$.*

(2) *The block $\mathcal{B}_0(\mathcal{G})$ is Morita equivalent to a generalization of $T(Q_{\mathcal{B}_0(\mathcal{G})})$.* □

The Morita equivalence given in (2) follows from the structure of the principal indecomposable $\mathcal{G}$-modules, provided in Corollary 6.5. Recall that $T(Q_{\mathcal{B}_0(\mathcal{G})})$ is the trivial extension of the radical square zero hereditary algebra of type $Q_{\mathcal{B}_0(\mathcal{G})}$. The generalization involves the "lengthening" of commutativity relations analogous to the procedure delineated in Theorem 6.15.

Example. Let us consider the case, where $G = Q_n$ is the generalized quaternion group of order $4n$ with $p \nmid n$. Thus, $\tilde{\mathcal{G}} = N_{(n)}$ is the reduced group with $\tilde{\mathcal{G}}(k) = Q_n$, and $\mathcal{G} = \mathrm{SL}(2)_1 \rtimes N_{(n)}$. Theorem 6.19 provides a Morita equivalence

$$\mathcal{B}_0(\mathcal{G}) \sim_M T(k[\tilde{D}_{n+2}]).$$

In particular, $\mathcal{B}_0(\mathcal{G})$ has $n+3$ simple modules, whose dimensions can be read off from the subadditive function d_C defined in Theorem 5.13. Since Q_n has 4 one-dimensional simple modules, located at the ends of $\tilde{D}_{n+2}$, and n–1 two-dimensional simple modules, Lemma 6.3, Corollary 6.5 yield the dimensions of the simple and principal indecomposable $\mathcal{B}(\mathcal{G})$-modules. For $n = 3$, $\mathcal{B}_0(\mathcal{G})$ has two 1-dimensional simple modules, two (p–1)-dimensional simple modules, one 2-dimensional simple module and one (2p–2)-dimensional simple module. By the same token, the block has two principal indecomposable modules of dimension $4p$ and four principal indecomposables of dimension $2p$. Consequently, $\dim_k \mathcal{B}_0(\mathcal{G}) = 8p^2$.

References

[1] J. Alperin, Periodicity in groups, Illinois J. Math. **21** (1977), 776–783.

[2] J. Alperin, Local Representation Theory, Cambridge Studies in Advanced Mathematics **11**, Cambridge University Press, Cambridge, 1986.

[3] J. Alperin and L. Evens, Representations, resolutions and Quillen's dimension theorem, J. Pure Appl. Algebra **22** (1981), 1–9.

[4] I. Assem, D. Simson and A. Skowroński, Elements of the Representation Theory of Associative Algebras. London Mathematical Society Student Texts **65**, Cambridge University Press, Cambridge, 2006.

[5] M. Auslander, Rational singularities and almost split sequences, Trans. Amer. Math. Soc. **293** (1986), 511–531.

[6] M. Auslander and I. Reiten, McKay quivers and extended Dynkin diagrams, Trans. Amer. Math. Soc. **293** (1986), 293–301.

[7] M. Auslander, I. Reiten and S. Smalø, Galois actions on rings and finite Galois coverings, Math. Scand. **65** (1989), 5–32.

[8] M. Auslander, I. Reiten and S. Smalø, Representation Theory of Artin Algebras, Cambridge Studies in Advanced Mathematics **36**, Cambridge University Press, Cambridge, 1995.

[9] G. Avrunin and L. Scott, Quillen stratification for modules, Invent. math. **66** (1982), 277–286.

[10] J. Ballard, Injective modules for restricted enveloping algebras, Math. Z. **163** (1978), 57–63.

[11] G. Benkart and J. Osborn, Representations of rank one Lie algebras of characteristic p, Lecture Notes in Mathematics **933**, 1–37, Springer, Berlin, 1982.

[12] D. Benson, Representations and Cohomology, I, Cambridge Studies in Advanced Mathematics **30**, Cambridge University Press, Cambridge, 1991.

[13] D. Benson, Representations and Cohomology, II, Cambridge Studies in Advanced Mathematics **31**, Cambridge University Press, Cambridge, 1991.

[14] I. Bernstein, I. Gel'fand and V. Ponomarev, Coxeter functors and Gabriel's theorem, Russ. Math. Surv. **28** (1973), 17–32.

[15] V. Bondarenko and Yu. Drozd, Representation type of finite groups, J. Soviet Math. **20** (1982), 2515–2528.

[16] R. Brauer, A note on theorems of Burnside and Blichfeldt, Proc. Amer. Math. Soc. **15** (1964), 31–34.

[17] R. Brauer, On finite groups with cyclic Sylow subgroups, I, J. Algebra **40** (1976), 556–584.

[18] R. Brauer, On finite groups with cyclic Sylow subgroups, II, J. Algebra **58** (1979), 291–318.

[19] W. Burnside, Theory of Groups of Finite Order, Cambridge University Press, Cambridge, 1911.

[20] J. Carlson, The complexity and varieties of modules, Lecture Notes in Mathematics **882**, 415–422, Springer, Berlin, 1980.

[21] J. Carlson, Varieties and the cohomology ring of a module, J. Algebra **85** (1983), 104–143.

[22] J. Carlson, Modules and Group Algebras, Lectures in Mathematics, ETH Zürich, Birkhäuser Verlag, Basel, 1996.

[23] J. Carlson and E. Friedlander, Exact category of modules of constant Jordan type, Manin Festschrift, Progress in Mathematics **269**, 267–290, Birkhäuser-Verlag Basel, 2010.

[24] J. Carlson, E. Friedlander and J. Pevtsova, Modules of constant Jordan type, J. reine angew. Math. **614** (2008), 191–234.

[25] H. Cartan and S. Eilenberg, Homological Algebra, Princeton Mathematical Series **19**, Princeton University Press, Princeton, 1956.

[26] L. Chouinard, Projectivity and relative projectivity over group rings, J. Pure Appl. Algebra **7** (1976), 278–302.

[27] D. Collingwood and W. McGovern, Nilpotent Orbits in Semisimple Lie Algebras, Van Nostrand Reinhold, New York, 1993.

[28] W. Crawley-Boevey, On tame algebras and bocses, Proc. London Math. Soc. **56** (1988), 451–483.

[29] W. Crawley-Boevey, Tameness of biserial algebras, Arch. Math. **65** (1995), 399–407.

[30] E. Dade, Endo-permutation modules over p-groups, II, Ann. of Math. **108** (1978), 317–346.

[31] M. Demazure and P. Gabriel, Groupes Algébriques I, Masson/North Holland, 1970.

[32] P. Donovan and M. Freislich, The Representation Theory of Graphs and Algebras, Carleton Lecture Notes **5**. Ottawa, 1973.

[33] Yu. Drozd, Tame and wild matrix problems, In: Representation Theory II, Lecture Notes in Mathematics **832**, 242–258, Springer, Berlin, 1980.

[34] Yu. Drozd, On the representations of the Lie algebra sl_2, Visn. Kiiv. Mat. Mekh. **25** (1983), 70–77.

[35] D. Eisenbud, Commutative Algebra, Graduate Texts in Mathematics **150**, Springer Verlag, New York, 1996.

[36] K. Erdmann, Blocks of Tame Representation Type and Related Algebras, Lecture Notes in Mathematics **1428**, Springer, Berlin, 1990.

[37] L. Evens, The cohomology ring of a finite group, Trans. Amer. Math. Soc. **101** (1961), 224–239.

[38] R. Farnsteiner, Periodicity and representation type of modular Lie algebras,

J. reine angew. Math. **464** (1995), 47–65.

[39] R. Farnsteiner, Varieties of tori and Cartan subalgebras of restricted Lie algebras, Trans. Amer. Math. Soc. **356** (2004), 4181–4236.

[40] R. Farnsteiner, Block representation type of Frobenius kernels of smooth groups, J. reine angew. Math. **586** (2005), 45–69.

[41] R. Farnsteiner, Polyhedral groups, McKay quivers, and the finite algebraic groups with tame principal blocks, Invent. math. **166** (2006), 27–94.

[42] R. Farnsteiner, Tameness and complexity of finite group schemes, Bull. London Math. Soc. **39** (2007), 63–70.

[43] R. Farnsteiner, Support spaces and Auslander-Reiten components, Contemp. Math. **442** (2007), 61–87.

[44] R. Farnsteiner, G. Röhrle and D. Voigt, Infinitesimal unipotent group schemes of complexity 1, Colloq. Math. **89** (2001), 179–192.

[45] R. Farnsteiner and H. Strade, Shapiro's lemma and its consequences in the cohomology theory of modular Lie algebras, Math. Z. **206** (1991), 153–168.

[46] R. Farnsteiner and D. Voigt, Modules of solvable infinitesimal groups and the structure of representation-finite cocommutative Hopf algebras, Math. Proc. Cambridge Philos. Soc. **127** (1999), 441–459.

[47] R. Farnsteiner and D. Voigt, On cocommutative Hopf algebras of finite representation type, Adv. in Math. **155** (2000), 1–22.

[48] R. Farnsteiner and D. Voigt, Schemes of tori and the structure of tame restricted Lie algebras, J. London Math. Soc. **63** (2001), 553–570.

[49] R. Farnsteiner and D. Voigt, On infinitesimal groups of tame representation type, Math. Z. **244** (2003), 479–513.

[50] R. Farnsteiner and D. Voigt, Representations of infinitesimal groups of characteristic 2, Arch. Math. **78** (2002), 184–188.

[51] R. Farnsteiner and A. Skowroński, Classification of restricted Lie algebras with tame principal block, J. reine angew. Math. **546** (2002), 1–45.

[52] R. Farnsteiner and A. Skowroński, The tame infinitesimal groups of odd characteristic, Adv. in Math. **205** (2006), 229–274.

[53] R. Farnsteiner and A. Skowroński, Galois actions and blocks of tame infinitesimal group schemes, Trans. Amer. Math. Soc. **359** (2007), 5867–5898.

[54] J. Feldvoss and H. Strade, Restricted Lie algebras with bounded cohomology and related classes of algebras, Manuscr. math. **74** (1992), 47–67.

[55] G. Fischer, Darstellungstheorie des ersten Frobeniuskerns der SL_2, Dissertation Universität Bielefeld, Bielefeld, 1982.

[56] D. Fischman, S. Montgomery and H. Schneider, Frobenius extensions of subalgebras of Hopf algebras, Trans. Amer. Math. Soc. **349** (1996), 4857–4895.

[57] E. Friedlander and B. Parshall, Support varieties for restricted Lie algebras, Invent. math. **86** (1986), 553–562.

[58] E. Friedlander and B. Parshall, Cohomology of Lie algebras and algebraic groups, Amer. J. Math. **108** (1986), 235–253.

[59] E. Friedlander and B. Parshall, Geometry of p-unipotent Lie algebras, J. Algebra **109** (1987), 25–45.

[60] E. Friedlander and J. Pevtsova, Representation-theoretic support spaces

for finite group schemes, Amer. J. Math. **127** (2005), 379–420. Erratum: Amer. J. Math. **128** (2006), 1067–1068.
[61] E. Friedlander and J. Pevtsova, Π-Supports for modules for finite group schemes, Duke Math. J. **139** (2007), 317–368.
[62] E. Friedlander, J. Pevtsova and A. Suslin, Generic and maximal Jordan types, Invent. math. **168** (2007), 485–522.
[63] E. Friedlander and A. Suslin, Cohomology of finite group schemes over a field, Invent. math. **127** (1997), 209–270.
[64] P. Gabriel, Unzerlegbare Darstellungen, I, Manuscripta math. **6** (1972), 71–103.
[65] I. Gel'fand and V. Ponomarev, Indecomposable representations of the Lorentz group, Usp. Math. Nauk **23** (1968), 1–58.
[66] D. Happel, Triangulated Categories in the Representation Theory of Finite Dimensional Algebras, LMS Lecture Notes **119**, Cambridge University Press, Cambridge, 1988.
[67] D. Happel, U. Preiser and C. Ringel, Binary polyhedral groups and Euclidean diagrams, Manuscr. math. **31** (1980), 317–329.
[68] A. Heller, Indecomposable representations and the loop space operation, Proc. Amer. Math. Soc. **12** (1961), 640–643.
[69] D. Higman, Indecomposable representations at characteristic p, Duke Math. J. **21** (1954), 377–381.
[70] G. Hochschild, Cohomology of restricted Lie algebras, Amer. J. Math. **76** (1954), 555–580.
[71] G. Hochschild, Representations of restricted Lie algebras of characteristic p, Proc. Amer. Math. Soc. **5** (1954), 603–605.
[72] H. Hopf, Über die Topologie der Gruppen-Mannigfaltigkeiten und ihre Verallgemeinerungen, Ann. of Math. **42** (1941), 22–52.
[73] J. Humphreys, Algebraic Groups and Modular Lie Algebras, Mem. Amer. Math. Soc. **71**. Providence, 1967.
[74] J. Humphreys, Modular Representations of Finite Groups of Lie Type, London Math. Soc. Lecture Notes **326**. Cambridge University Press, Cambridge, 2005.
[75] N. Jacobson, Restricted Lie algebras of characteristic p, Trans. Amer. Math. Soc. **50** (1941), 15–25.
[76] N. Jacobson, Galois theory of purely inseparable fields of exponent 1, Amer. J. Math. **66** (1944), 645–648.
[77] N. Jacobson, Lie Algebras, Dover Publications Inc., New York, 1979.
[78] J. Jantzen, Darstellungen halbeinfacher Gruppen und ihrer Frobeniuskerne, J. reine angew. Math. **317** (1980), 157–199.
[79] J. Jantzen, Kohomologie von p-Lie-Algebren und nilpotente Elemente, Abh. Math. Sem. Univ. Hamburg **56** (1986), 191–219.
[80] J. Jantzen, Representations of Algebraic Groups, Mathematical Surveys and Monographs **107**, American Mathematical Society, Providence, 2003.
[81] C. Kassel, Quantum Groups, Graduate Texts in Mathematics **155**, Springer, New York, 1995.
[82] F. Klein, Vorlesungen über das Ikosaeder und die Auflösungen der Gle-

ichungen vom fünften Grade, Teubner-Verlag, Leipzig, 1884
[83] A. Koch, The Hopf algebra of a uniserial group, Pacific J. Math. **215** (2004), 347–356.
[84] H. Kupisch, Beträge zur Theorie nichthalbeinfacher Ringe mit Minimalbedingung, J. reine angew. Math. **201** (1959), 100–112.
[85] R. Larson and M. Sweedler, An associative orthogonal bilinear form for Hopf algebras, Amer. J. Math. **91** (1969), 75–94.
[86] H. Maschke, Ueber den arithmetischen Charakter der Coefficienten der Substitutionen endlicher linearer Substitutionsgruppen, Math. Ann. **50** (1898), 492–498.
[87] J. McKay, Graphs, singularities, and finite groups, Proc. Sympos. Pure Math. **37** (1980), 183–186.
[88] J. McKay, Cartan matrices, finite groups of quaternions, and Kleinian singularities, Proc. Amer. Math. Soc. **81** (1981), 153–154.
[89] S. Montgomery, Hopf Algebras and their Actions on Rings, CBMS **82**, Amer. Math. Soc., Providence, RI.
[90] L. Nazarova, Representations of quivers of infinite type, Izv. Akad. Nauk. SSSR, Ser. Mat. **37** (1973), 752–791.
[91] W. Nichols and B. Zoeller, A Hopf algebra freeness theorem, Amer. J. Math. **111** (1989), 381–385.
[92] U. Oberst and H. Schneider, Über Untergruppen endlicher algebraischer Gruppen, Manuscripta math. **8** (1973), 217–241.
[93] B. Pareigis, Kohomologie von p-Lie-Algebren, Math. Z. **104** (1968), 281–336.
[94] D. Passman and D. Quinn, Burnside's theorem for Hopf algebras, Proc. Amer. Math. Soc. **123** (1995), 327–333.
[95] W. Pfautsch, Die Köcher der Frobeniuskerne in der SL_2, Dissertation, Universität Bielefeld, Bielefeld, 1983.
[96] W. Pfautsch, Ext^1 for the Frobenius kernels of SL_2, Comm. Algebra **13** (1985), 169–179.
[97] W. Pfautsch and D. Voigt, The representation-finite algebraic groups of dimension zero, C.R. Acad. Sci. Paris **306** (1988), 685–689.
[98] R. Pollack, Restricted Lie algebras of bounded type, Bull. Amer. Math. Soc. **74** (1968), 326–331.
[99] A. Premet, The Green ring of a simple three-dimensional Lie p-algebra, Soviet Math. (Iz. VUZ.) **35** (1991), 51–60.
[100] D. Quillen, The spectrum of an equivariant cohomology ring, I, Ann. of Math. **94** (1971), 549–572.
[101] D. Quillen, The spectrum of an equivariant cohomology ring, II, Ann. of Math. **94** (1971), 573–602.
[102] D. Radford, The order of the antipode of a finite dimensional Hopf algebra is finite, Amer. J. Math. **98** (1976), 333–355.
[103] J. Rickard, The representation type of self-injective algebras, Bull. London Math. Soc. **22** (1990), 540–546.
[104] M. Rieffel, Burnside's theorem for representations of Hopf algebras, J. Algebra **6** (1967), 123–130.

[105] C. Ringel, The indecomposable modules of the dihedral 2-groups, Math. Ann. **214** (1975), 19–34.

[106] C. Ringel, The representation type of local algebras, Lecture Notes in Mathematics **488**, 282–305, Springer, Berlin, 1975.

[107] C. Ringel, Tame Algebras and Integral Quadratic Forms, Lecture Notes in Mathematics **1099**, Springer Verlag, Berlin, 1984.

[108] J. Rotman, An Introduction to Homological Algebra, Pure and Applied Mathematics **85**, Academic Press Inc., Orlando, 1979.

[109] A. Rudakov, Reducible p-representations of a simple three-dimensional Lie-p-algebra, Moscow Univ. Math. Bull. **37** (1982), 51–56.

[110] J. Schue, Symmetry for the enveloping algebra of a restricted Lie algebra, Proc. Amer. Math. Soc. **16** (1965), 1123–1124.

[111] A. Skowroński and J. Waschbüsch, Representation-finite biserial algebras, J. reine angew. Math. **345** (1983), 172–181.

[112] T. Springer, Invariant Theory, Lecture Notes in Mathematics **585**, Springer, Berlin, 1977.

[113] T. Springer, Linear Algebraic Groups, Progress in Mathematics **9**, Birkhäuser-Verlag, Basel, 1981.

[114] R. Steinberg, Complete sets of representations of algebras, Proc. Amer. Math. Soc. **13** (1962), 746–747.

[115] H. Strade and R. Farnsteiner, Modular Lie Algebras and their Representations, Pure and Applied Mathematics **116**. Marcel Dekker, New York, 1988.

[116] A. Suslin, E. Friedlander and C. Bendel, Infinitesimal 1-parameter subgroups and cohomology, J. Amer. Math. Soc. **10** (1997), 693–728.

[117] A. Suslin, E. Friedlander and C. Bendel, Support varieties for infinitesimal group schemes, J. Amer. Math. Soc. **10** (1997), 729–759.

[118] M. Sweedler, Integrals for Hopf algebras, Ann. of Math. **89** (1969), 323–335.

[119] H. Tachikawa, Representations of trivial extensions of hereditary algebras, In: Representation theory II. Lecture Notes in Mathematics **832**, 579–599, Springer-Verlag, 1980.

[120] L. Unger, The concealed algebras of the minimal wild, hereditary algebras, Bayreuth. Math. Schr. **99** (1989), 1–9.

[121] B. Venkov, Cohomology algebras for some classifying spaces, Dokl. Akad. Nauk SSSR **127** (1959), 943–944.

[122] D. Voigt, Induzierte Darstellungen in der Theorie der endlichen, algebraischen Gruppen, Lecture Notes in Mathematics **592**, Springer-Verlag, 1977.

[123] W. Waterhouse, Introduction to Affine Group Schemes, Graduate Texts in Mathematics **66**, Springer-Verlag, 1979.

Lie Theory and
Representation Theory
SMM 2, pp. 151–176

Cohomology of Algebraic Groups, Finite Groups, and Lie Algebras: Interactions and Connections

Daniel K. Nakano*

Abstract

In this article we survey recent developments in the representation theory and cohomology theory of reductive algebraic groups, their Frobenius kernels and their associated finite groups of Lie type. This includes work of the author with various collaborators on relating projectivity and support varieties between the Frobenius kernels and finite Chevalley groups, in addition to using the geometry of the flag variety to relate their cohomology theories.

2000 Mathematics Subject Classification: 20J06, 20G10.

Keywords and Phrases: Algebraic groups, finite groups, cohomology, support varieties.

1 Overview

1.1 Motivation

Let G be a simple, simply connected algebraic group scheme defined over a finite field $\mathbb{F}_p$, and $F^r : G \to G$ be the rth iteration of the Frobenius morphism. Let G_r be the scheme theoretic kernel of F^r, and $G(\mathbb{F}_q)$ be the finite Chevalley group obtained by looking at the fixed points under F^r. For a given rational G-module M one can consider the restriction of the action of M to the infinitesimal Frobenius kernel G_r and to the finite group $G(\mathbb{F}_q)$.

For over 40 years, representation theorists have attempted to understand the relationship between these three module categories. Two basic questions we will address in this paper are:

*Department of Mathematics, University of Georgia, Athens, GA 30602, USA. The author was supported in part by NSF grants DMS-0654169 and DMS-1002135. E-mail: nakano@math.uga.edu

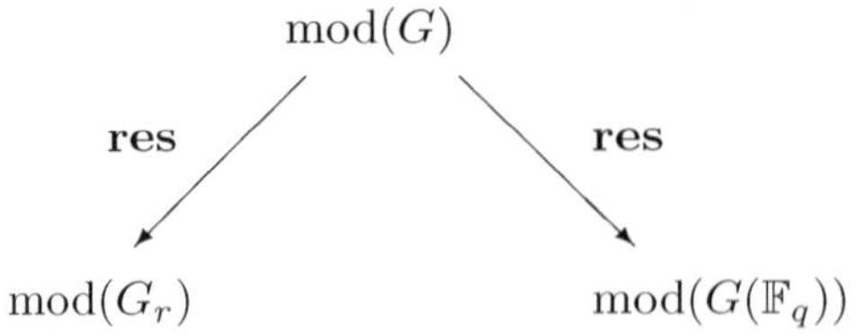

Figure 1

(1.1.1) What is the relationship between the representation theories of these categories?

(1.1.2) Is there a relationship between the cohomology theories of these categories?

Even though there is no direct functorial relationship between $\mathrm{mod}(G_r)$ and $\mathrm{mod}(G(\mathbb{F}_q))$ these categories have very similar features. For example, both categories correspond to module categories for finite dimensional cocommutative Hopf algebras. Curtis proved in the 1960s there exists a one-to-one correspondence between the simple G_r-modules and simple $G(\mathbb{F}_q)$-modules. This correspondence is given by simply restricting natural classes of simple G-modules. Furthermore, Steinberg's Tensor Product Theorem allows one to easily transfer the questions pertaining to the computation of irreducible characters between these three categories. The theme of lifting G_r-modules to G-structures was further studied by Ballard and Jantzen for projective G_r-modules. For a general guide to the history and results in this area we refer the reader to [Hu2].

The aim of this paper is to present new methods in relating these three categories which have been developed over the past 10 years. Even though the group $G(\mathbb{F}_p)$ is finite, we will see that there are ambient geometric structures which govern its representation theory and cohomology. One theme in these new approaches has been to use nilpotent orbit theory for semisimple Lie algebras and the geometry of the flag variety G/B to calculate extensions of modules over the group $G(\mathbb{F}_p)$. Often times the use of such tools can reduce difficult group cohomology calculations to combinatorics within the given root system.

1.2 Outline

This paper is organized as follows. In Section 2, we review the basic facts about the representation theory of G, G_r and $G(\mathbb{F}_q)$. We give a parametrization of the simple modules for each of these cases and describe how the simple modules are related. The projective modules for G_r and $G(\mathbb{F}_q)$ are discussed in relation to proving the existence of G-structures. In the following section (Section 3), we give the fundamental definitions and properties of the complexity and support varieties of a module for a finite dimensional cocommutative Hopf algebra (i.e., finite group scheme). In Section 4, we discuss the relationships between the complexity and support varieties of a given finite dimensional G-module when restricted to

$\text{mod}(G_1)$ and $\text{mod}(G(\mathbb{F}_p))$. These results were motivated by questions posed by Parshall in the 1980s. In Section 5, we investigate the connections between the cohomology groups in $\text{mod}(G)$, $\text{mod}(G_r)$ and $\text{mod}(G(\mathbb{F}_q))$ using techniques developed by Bendel, Pillen and the author. Formulas for Ext^1-groups will be exhibited, in addition to results pertaining to the existence of self-extensions. Finally, in Section 6, we present new developments in the computation of the cohomology groups $\text{H}^\bullet(G(\mathbb{F}_q), k)$. In the case when $r = 1$ and p is larger than the Coxeter number, we provide an upper bound for the dimension of these groups via Kostant's Partition Functions. Moreover, our methods allow us to give precise information about the existence of the first non-trivial cohomology when the root system is of type A or C.

The author would like to acknowledge his many discussions with and the important contributions to the subject in recent years by Henning Andersen, Christopher Bendel, Jon Carlson, Eric Friedlander, Zongzhu Lin, Brian Parshall, Cornelius Pillen, and Leonard Scott. The author would also like to thank Bin Shu and his staff for their efforts in hosting and organizing a vibrant summer school program at East China Normal University in Summer 2009 when the results in this paper were presented.

2 Representation theory

2.1 Notation

Throughout this paper, we will follow the basic Lie theoretic conventions used in [Jan1]. Let G be a simple simply connected algebraic group scheme which is defined and split over the finite field $\mathbb{F}_p$ with p elements. Moreover, let k denote an algebraically closed field of characteristic p. Let T be a maximal split torus and Φ be the root system associated to (G,T). The positive (resp. negative) roots will be denoted by Φ^+ (resp. Φ^-), with Δ the collection of simple roots. Let B be a Borel subgroup containing T corresponding to the negative roots and U be the unipotent radical of B. For a given root system of rank n, the simple roots in Δ will be denoted by $\alpha_1, \alpha_2, \dots, \alpha_n$. Our convention will be to employ the Bourbaki ordering of simple roots. In particular, for type B_n, α_n denotes the unique short simple root and for type C_n, α_n denotes the unique long simple root. The highest short root is labelled by α_0.

Let $\mathbb{E}$ be the Euclidean space associated with Φ, and the inner product on $\mathbb{E}$ will be denoted by $\langle\ ,\ \rangle$. Set $\alpha^\vee = 2\alpha/\langle\alpha,\alpha\rangle$ (the coroot corresponding to $\alpha \in \Phi$). The fundamental weights (basis dual to $\alpha_1^\vee, \alpha_2^\vee, \dots, \alpha_n^\vee$) will be denoted by ω_1, ω_2, $\dots$, ω_n. Let W denote the Weyl group associated to Φ and for $w \in W$, let $\ell(w)$ denote the length of w.

Let $X(T)$ be the integral weight lattice spanned by the fundamental weights. For $\lambda, \mu \in X(T)$, $\lambda \geq \mu$ if and only if $\lambda - \mu$ is a positive integral linear combination of simple roots. The set of dominant integral weights is defined by

$$X(T)_+ := \{\lambda \in X(T) :\ 0 \leq \langle\lambda, \alpha^\vee\rangle, \text{ for all } \alpha \in \Delta\}.$$

For a weight $\lambda \in X(T)$, set $\lambda^* := -w_0\lambda$ where w_0 is the longest word in the Weyl group W. The dot action is defined by $w \cdot \lambda := w(\lambda + \rho) - \rho$ on $X(T)$ where ρ is the half-sum of the positive roots. For $\alpha \in \Delta$, $s_\alpha \in W$ denotes the reflection in the hyperplane determined by α. The Coxeter number h equals $\langle \rho, \alpha_0^\vee \rangle + 1$.

Let M be a T-module. Then $M = \oplus M_\lambda$ where M_λ are the weight spaces of M. The formal character of M is $\text{ch } M = \sum_{\lambda \in X(T)} \dim M_\lambda \; e(\lambda)$. An important class of G-modules whose formal character is known are the induced or Weyl modules. For $\lambda \in X(T)_+$, let

$$H^0(\lambda) := \text{ind}_B^G \lambda = \{f : G \to \lambda : \; f(g \cdot b) = \lambda(b^{-1})f(g), \quad \text{for all } b \in B, \, g \in G\}$$

be the induced module. These G-modules can be interpreted geometrically as the global sections of the line bundle $\mathcal{L}(\lambda)$ over the flag variety G/B. The Weyl module of highest weight λ is defined by $V(\lambda) := H^0(\lambda^*)^*$. The following theorem provides basic properties of the induced/Weyl modules.

Theorem 2.1.1. (1) $H^0(\lambda) \neq 0$ *if and only if* $\lambda \in X(T)_+$;
(2) $\dim H^0(\lambda)_\lambda = 1$, $H^0(\lambda)_\mu \neq 0$ *implies that* $w_0\lambda \leq \mu \leq \lambda$;
(3) $\dim H^0(\lambda) = \prod_{\alpha \in \Phi^+} \frac{\langle \lambda + \rho, \alpha^\vee \rangle}{\langle \rho, \alpha^\vee \rangle}$.

If H is an algebraic group scheme defined over $\mathbb{F}_p$ then there exists a Frobenius morphism $F : H \to H$. In the case when $H = GL_n(-)$, F takes a matrix (a_{ij}) to (a_{ij}^p). Let F^r be the r-th iteration of the F with itself. For a rational H-module M, let $M^{(r)}$ be the module obtained by composing the underlying representation for M with F^r. Moreover, let M^* denote the dual module.

We will be interested in understanding the representation theory and cohomology of infinitesimal Frobenius kernels for reductive algebraic groups G. For $r \geq 1$, let $G_r := \ker F^r$ be the rth Frobenius kernel of G and $G(\mathbb{F}_q)$ be the associated finite Chevalley group obtained by looking at the $\mathbb{F}_q$ rational points of G. We note that in the case when $r = 1$, $\text{Mod}(G_1)$ is equivalent to $\text{Mod}(u(\mathfrak{g}))$ where $\mathfrak{g} = \text{Lie } G$ and $u(\mathfrak{g})$ is the restricted enveloping algebra of $\mathfrak{g}$.

2.2 Simple modules

In this section we will describe the classification of simple modules in $\text{mod}(G)$, $\text{mod}(G_r)$ and $\text{mod}(G(\mathbb{F}_q))$. All modules will be defined over the field k. The details can be found in [Jan1, II. Chapters 2–3].

First for $\lambda \in X(T)_+$, the socle of $H^0(\lambda)$ (largest semisimple submodule) is simple. Set $L(\lambda) = \text{soc}_G \; H^0(\lambda)$. Then the simple finite-dimensional modules are given by: $\{L(\lambda) : \; \lambda \in X(T)_+\}$. Note that when $k = \mathbb{C}$ then $L(\lambda) \cong H^0(\lambda)$ for all $\lambda \in X(T)_+$.

In order to classify the simple modules in $\text{mod}(G_r)$ and $\text{mod}(G(\mathbb{F}_q))$ we need to define the set of p^r-restricted weights. Let

$$X_r(T) = \{\lambda \in X(T)_+ : \; 0 \leq \langle \lambda, \alpha^\vee \rangle \leq p^r - 1, \; \text{for all } \alpha \in \Delta\}.$$

If $\lambda \in X_r(T)$ then $L(\lambda)|_{G_r}$ remain simple, and

(2.2.1) {simple finite-dimensional modules for G_r} $\leftrightarrow$ $\{L(\lambda)\ |_{G_r}\colon\ \lambda \in X_r(T)\}$ (one to one correspondence).

On the other hand, Curtis proved that one can restrict $L(\lambda)$ to $G(\mathbb{F}_q)$ and also get a classification of simple $G(\mathbb{F}_q)$-modules:

(2.2.2) {simple finite-dimensional modules for $G(\mathbb{F}_q)$} $\leftrightarrow \{L(\lambda)\ |_{G(\mathbb{F}_q)}\colon \lambda \in X_r(T)\}$ (one to one correspondence).

Now by a beautiful theorem of Steinberg (known as Steinberg's Tensor Product Theorem), one can describe all simple G-modules by tensoring by Frobenius twists of simple G_1-modules:

Theorem 2.2.1. *Let $\lambda \in X(T)_+$, $\lambda = \lambda_0 + \lambda_1 p + \cdots + \lambda_s p^s$, where $\lambda_i \in X_1(T)$. Then*

$$L(\lambda) \cong L(\lambda_0) \otimes L(\lambda_1)^{(1)} \otimes \cdots \otimes L(\lambda_s)^{(s)}.$$

If G is semisimple then $X_1(T)$ is finite. As a consequence of the Steinberg Tensor Product Theorem, if we know the characters of the simple G_1-modules $L(\lambda)$ where $\lambda \in X_1(T)$, we can compute the characters of all simple finite-dimensional G-modules. The characters of the simple G-modules are still undetermined. For $p > h$ there is a conjecture due to Lusztig which gives a recursive formula for the characters of simple modules via Kazhdan-Lusztig polynomials (cf. [Jan1, Chapter C]).

2.3 Projective modules

For $r \geq 1$, the infinitesimal Frobenius kernel G_r is a finite group scheme whose module category is equivalent to Mod(Dist(G_r)) where Dist(G_r) is a finite-dimen sional cocommutative Hopf algebra. On the other hand, the category of $G(\mathbb{F}_q)$ modules that we are considering is equivalent to Mod($kG(\mathbb{F}_q)$) where $kG(\mathbb{F}_q)$ is the group algebra of $G(\mathbb{F}_q)$ (also a finite-dimensional cocommutative Hopf algebra). In both instances the distribution algebra Dist(G_r) and $kG(\mathbb{F}_q)$ are symmetric algebras. This means that these algebras are self-injective and the projective cover of a simple module is isomorphic to its injective hull.

If $L(\lambda)$ ($\lambda \in X_r(T)$) is a simple module for G_r then let $Q_r(\lambda)$ be its injective hull (and projective cover). One natural question to ask is whether we can lift the G_r-structure on $Q_r(\lambda)$ to a compatible G-module structure. Ballard showed that this holds when $p \geq 3(h-1)$, and Jantzen later lowered the bound to $p \geq 2(h-1)$.

Theorem 2.3.1. *Let $\lambda \in X_r(T)$ and $p \geq 2(h-1)$. Then $Q_r(\lambda)$ admits a G-structure.*

The idea behind the proof given in [Jan1, II 11.11] is to first realize $Q_r(\lambda)$ as a G_r-direct summand of the G-module $\mathrm{St}_r \otimes L$, where L is an irreducible G-module and St_r is the rth Steinberg module. The proof entails showing that $Q_r(\lambda)$ is in fact a G-direct summand when $p \geq 2(h-1)$.

The Steinberg module St_r is projective upon restriction to G_r and $G(\mathbb{F}_q)$. Therefore, $\mathrm{St}_r \otimes L$ is projective upon restriction to $G(\mathbb{F}_q)$. Consequently, for $p \geq 2(h-1)$, one can lift $Q_r(\lambda)$ to a G-structure and upon restriction to $G(\mathbb{F}_q)$

this module will split into a direct summand of projective indecomposable $kG(\mathbb{F}_q)$-modules. For $\lambda \in X_r(T)$, let $U_r(\lambda)$ be the projective cover of $L(\lambda)$ in $\text{mod}(kG(\mathbb{F}_q))$. Chastkofsky [Ch] proved that

$$[Q_r(\lambda)|_{G(\mathbb{F}_q)} : U_r(\mu)] = \sum_{\nu \in \Gamma} [L(\mu) \otimes L(\nu) : L(\lambda) \otimes L(\nu)^{(r)}]_G.$$

Here $\Gamma = \{\nu \in X(T)_+ : \langle \nu, \alpha_0^\vee \rangle < h\}$. We shall see later that similar factors appear in Ext^1-formulas relating the extensions of simple modules for $G(\mathbb{F}_q)$ to simple modules for G.

3 Homological algebra

3.1 Complexity

Let $\mathcal{V} = \{V_j : j \in \mathbb{N}\}$ be a sequence of finite-dimensional k-vector spaces. The rate of growth of $\mathcal{V}$, denoted by $r(\mathcal{V})$, is the smallest positive integer c such that $\dim_k V_n \leq K \cdot n^{c-1}$ for some constant $K > 0$.

Given a finite-dimensional algebra A, Alperin [Al] introduced the notion of the complexity of a module in 1977 as a generalization of the notion of periodicity in group cohomology. The complexity is defined as follows. Let $M \in \text{mod}(A)$ and

$$\cdots \to P_2 \to P_1 \to P_0 \to M \to 0$$

be a minimal projective resolution of M. This means that the exact sequence above is a projective resolution of M and the kernels have no projective summands. The complexity of M in $\text{mod}(A)$ is $r(\{P_t : t \in \mathbb{N}\})$.

Let $\text{Ext}^j_A(M, -)$ be the jth right derived functor of $\text{Hom}_A(M, -)$ for M in $\text{Mod}(A)$. If A is an augmented algebra (i.e., A-admits the trivial module k), and N is in $\text{Mod}(A)$ set $\text{H}^j(A, N) = \text{Ext}^j_A(k, N)$. One can relate the complexity of N in terms of the rate of growth of extension groups.

Theorem 3.1.1. *Let A be a finite-dimensional algebra and $S_1, S_2, \ldots, S_u$ be a complete set of simple objects in* $\text{mod}(A)$ *then*

$$c_A(M) = r(\{\text{Ext}^t_A(\oplus_{j=1}^u S_j, N) : t \in \mathbb{N}\}).$$

The complexity becomes a finer invariant when one considers the case when A is self-injective, in particular when A is a finite-dimensional Hopf algebra. In this setting a module being injective is equivalent to it being projective, and the complexity is an invariant of projectivity.

Proposition 3.1.2. *Let A be a self injective algebra. Then $c_A(M) = 0$ if and only if M is projective.*

Proof. This can be shown as follows. Suppose $c_A(M) = 0$, then

$$0 \to P_s \to \cdots \to P_1 \to P_0 \to M \to 0$$

is a minimal projective resolution of M. Furthermore, P_s is injective so this sequence splits, hence M is projective. On the other hand if M is projective one can construct a resolution of M, $0 \to M \to M \to 0$, thus $c_A(M) = 0$. □

3.2 Support varieties

Support varieties were introduced in the pioneering work of Carlson [Ca1, Ca2] nearly 30 years ago as a method to study complexes and resolutions of modules over group algebras. Since that time the theory of support varieties has been extended to restricted Lie algebras by Friedlander and Parshall [FP1], to finite-dimensional sub Hopf algebras of the Steenrod algebra by Palmieri and the author [NPal], to infinitesimal group schemes by Suslin, Friedlander and Bendel [SFB1, SFB2], and to arbitrary finite-dimensional cocommutative Hopf algebras by Friedlander and Pevtsova [FPe]. Further attempts to generalize the theory have been made to finite-dimensional algebras by Solberg and Snashall [SS] via Hochschild cohomology and Hecke algebras by Erdmann and Holloway [EH], Lie superalgebras [BKN1, BKN2, BKN3, BaKN], quantum groups [Ost, BNPP], and quantum complete intersections [BE]. As a testament to the dissemination of this theory, Dmitry Rumynin wrote in a Mathematical Review (MR: 2003b:20063) that "support varieties have proved to be an indispensable tool in the arsenal of a modern representation theorist".

Friedlander and Suslin [FS] proved that for a finite-dimensional cocommutative Hopf algebra A, the even degree cohomology ring $R = \mathrm{H}^{2\bullet}(A,k)$ is a commutative finitely generated algebra, and $\mathrm{H}^{\bullet}(A,M)$ is a finitely generated R-module for any M in $\mathrm{mod}(A)$. Set $\mathcal{V}_A(k) = \mathrm{Maxspec}(R)$. For any finitely generated A-module M, one can assign a conical subvariety $\mathcal{V}_A(M)$ inside the spectrum of the cohomology ring $\mathcal{V}_A(k)$ by letting

$$\mathcal{V}_A(M) = \mathrm{Maxspec}(R/J_M)$$

where $J_M = \mathrm{Ann}_R \mathrm{Ext}_A^{\bullet}(k, M \otimes M^*)$. Here M^* is the dual module of M.

Support varieties provide a method to introduce the geometry of the spectrum of R into the representation theory of A. These varieties have natural geometric properties and provide a geometric interpretation of the complexity of M:

(3.2.1) $c_A(M) = \dim \mathcal{V}_A(M)$.

(3.2.2) $\mathcal{V}_A(M \oplus N) = \mathcal{V}_A(M) \cup \mathcal{V}_A(N)$.

(3.2.3) $\mathcal{V}_A(M \otimes N) = \mathcal{V}_A(M) \cap \mathcal{V}_A(N)$.

(3.2.4) If W is a conical subvariety of $\mathcal{V}_A(k)$ there exists M in $\mathrm{mod}(A)$ such that $\mathcal{V}_A(M) = W$.

(3.2.5) If $M \in \mathrm{mod}(A)$ is indecomposable then $\mathrm{Proj}(\mathcal{V}_A(M))$ is connected.

For finite groups a description of the spectrum of the cohomology ring was determined by Quillen [Q1, Q2] using elementary abelian subgroups. Avrunin and Scott [AS] proved a more general stratification theory for $\mathcal{V}_A(M)$. Support varieties become very transparent when one considers the case when $\mathfrak{g}$ is a restricted Lie algebra over a field k of characterstic $p > 0$ and A is the restricted universal enveloping algebra $u(\mathfrak{g})$. In this situation the spectrum of the cohomology ring is homeomorphic to the restricted nullcone

$$\mathcal{N}_1(\mathfrak{g}) = \{x \in \mathfrak{g} : x^{[p]} = 0\}.$$

Moreover, Friedlander and Parshall showed that

$$\mathcal{V}_{G_1}(M) \cong \{x \in \mathfrak{g} :\ x^{[p]} = 0,\ M|_{u(\langle x \rangle)} \text{ is not free}\} \cup \{0\}.$$

If G is a reductive algebraic group over k and $\mathfrak{g} = \text{Lie } G$ then $\mathcal{N}_1(\mathfrak{g})$ is a G-stable conical subvariety inside the cone of nilpotent elements (nullcone) $\mathcal{N} := \mathcal{N}(\mathfrak{g})$. In the special case when $G = GL_n(k)$,

$$\mathcal{N}_1(\mathfrak{g}) = \{A \in \mathfrak{gl}_n(k) :\ A^p = 0\}.$$

The group $GL_n(k)$ acts on $\mathcal{N}$ via conjugation and has finitely many orbits (indexed by partitions of n). Note that $\mathcal{N}_1(\mathfrak{g})$ is G-stable. For general reductive groups G, the nullcone $\mathcal{N}$ has been well studied (see [Car, CM, Hu2]) because of its beautiful geometric properties with deep connections to representation theory. The group G acts on $\mathcal{N}$ via the adjoint representation and $\mathcal{N}$ has finitely many G-orbit (which are classified). In this setting support varieties behave well with respect to the G-action.

(3.2.6) If M is a G-module, then $\mathcal{V}_A(M)$ is G-invariant.

(3.2.7) If $M \in \text{mod}(G)$, $\mathcal{V}_A(M) = \overline{G \cdot x_1} \cup \overline{G \cdot x_2} \cup \cdots \cup \overline{G \cdot x_s}$. There are only finitely many possibilities for $\mathcal{V}_A(M)$ in this case because there are finitely many nilpotent G-orbits.

4 Relating support varieties

4.1 Motivation

When I was a graduate student at Yale, my Ph.D advisor George Seligman told me a story about the Ph.D thesis of his former student David Pollack (circa 1966). Seligman said that he proposed the following problem to Pollack for his thesis: to determine the representation type for the restricted enveloping algebras $u(\mathfrak{g})$ where $\mathfrak{g} = \text{Lie } G$ and G is a reductive algebraic group. More specifically, at the time the question was to determine whether these algebras have finite or infinite representation type.

For group algebras this problem was settled earlier: a group algebra over a field of characteristic $p > 0$ has finite representation type if and only if its p-Sylow subgroups are cyclic. Therefore, when $G = SL_2(k)$, and $A = kG(\mathbb{F}_p)$, all Sylow subgroups are cyclic and conjugate to the subgroup of unipotent upper triangular matrices. Consequently, $A = kSL_2(\mathbb{F}_p)$ has finite representation type.

One of the first challenges was for Pollack to determine the representation type of $u(\mathfrak{sl}_2)$. At the time Walter Feit thought by analogy that this algebra should be of finite representation type. But, after some efforts in trying to prove this, Pollack constructed the projective indecomposables for the algebra, and using these concrete realizations he constructed infinitely many indecomposable modules for $u(\mathfrak{sl}_2)$. Moreover, Pollack was able to show that $u(\mathfrak{g})$ where $\mathfrak{g} = \text{Lie } G$ (G is a reductive algebraic group) has infinite representation type.

After hearing this story, I thought that there should be direct connections between understanding representation type between finite Chevalley groups and their associated Lie algebras. The pursuit in finding such connections was in part a motivating factor in the development of the results outlined in this section.

4.2 Parshall conjecture

In 1987, Parshall (another former student of Seligman) asked the following question in his article in the Arcata Conference Proceedings. Let G be a reductive algebraic group and M be in $\text{mod}(G)$. If M is projective as a G_1-module then is M projective over $G(\mathbb{F}_p)$? The affirmative version to this question has become known as the "Parshall Conjecture". As we have seen in Section 2.3 the Parshall Conjecture holds if $p \geq 2(h-1)$.

For this section, we will need some additional notion. Let B be a Borel subgroup of G, U be the unipotent radical of B. Furthermore, let $G(\mathbb{F}_p)$, $B(\mathbb{F}_p)$, and $U(\mathbb{F}_p)$ be the corresponding finite groups, and $\mathfrak{g} = \text{Lie } G$, $\mathfrak{b} = \text{Lie } B$, and $\mathfrak{n} = \text{Lie } U$ be the corresponding Lie algebras.

We will now describe a process which allows us to pass from $\text{mod}(G_1)$ to $\text{mod}(G(\mathbb{F}_p))$ without lifting to $\text{mod}(G)$. First consider

$$U(\mathbb{F}_p) = \Gamma_1 \supset \Gamma_2 \supset \cdots \supset \Gamma_n = \{1\}$$

the lower central series where $\Gamma_i = [\Gamma_{i-1}, U(\mathbb{F}_p)]$. Then the associated graded object $\text{gr } U(\mathbb{F}_p) = \bigoplus_{n\geq 1} \Gamma_n/\Gamma_{n+1}$ is a restricted Lie algebra. We can identify the associated graded Lie algebra with $\mathfrak{n}$ (for most primes). That is, $\text{gr } U(\mathbb{F}_p) \cong \mathfrak{n}$.

Now consider filtering the group algebra of $U(\mathbb{F}_p)$ with powers of its augmentation ideal:

$$kU(\mathbb{F}_p) = I_0 \supseteq I_1 \supseteq \cdots \supseteq I_t = \{0\}$$

where $I_1 = \text{rad } kU(\mathbb{F}_p)$. One can now form the associated graded algebra $\text{gr } kU(\mathbb{F}_p) = \bigoplus_{n\geq 0} I_n/I_{n+1}$. By using results of Quillen and Jennings, we have the following isomorphism of algebras:

$$\text{gr } kU(\mathbb{F}_p) \cong u(\text{gr } U(\mathbb{F}_p)) \otimes_{\mathbb{F}_p} k \cong u(\mathfrak{n}).$$

The filtration on the group algebra of $U(\mathbb{F}_p)$ can be used to construct the May spectral sequence. For M in $\text{mod}(U(\mathbb{F}_p))$,

$$E_1^{i,j} = \text{H}^{i+j}(u(\mathfrak{n}), \text{gr } M) \Rightarrow \text{H}^{i+j}(U(\mathbb{F}_p), M).$$

Here we are using the identification: $\text{gr } kU(\mathbb{F}_p) = u(\mathfrak{n})$. If M in $\text{mod}(B)$, then $\text{gr } M = M|_{U_1}$ and we can rewrite the spectral sequence as

$$E_1^{i,j} = \text{H}^{i+j}(U_1, M) \Rightarrow \text{H}^{i+j}(U(\mathbb{F}_p), M).$$

This shows that if M in $\text{mod}(B)$ then $c_{U_1}(M) \geq c_{U(\mathbb{F}_p)}(M)$. Moreover, one obtains the following theorem (cf. [LN1, Thm. 3.4]).

Theorem 4.2.1. *Let M be in* mod(G)*. Then* $c_{G(\mathbb{F}_p)}(M) \leq \frac{1}{2}c_{G_1}(M)$.

Proof. (Sketch) First $c_{G(\mathbb{F}_p)}(M) = c_{U(\mathbb{F}_p)}(M)$ because $U(\mathbb{F}_p)$ is a Sylow subgroup of $G(\mathbb{F}_p)$. Next $c_{U_1}(M) = c_{B_1}(M) = \frac{1}{2}c_{G_1}(M)$ by using a result of Spaltenstein which says that the intersection of a G-orbit inside the nilpotent cone with $\mathfrak{n}$ is a union of B-stable sets having one-half the dimension of the original orbit. Finally, one can apply the inequality $c_{U_1}(M) \geq c_{U(\mathbb{F}_p)}(M)$. □

As a corollary, one can obtain a proof of the Parshall Conjecture.

Corollary 4.2.2. *Let M be in* mod(G)*. If M is a projective G_1-module then M is a projective $kG(\mathbb{F}_p)$-module.*

Proof. If M is projective as a G_1-module then $c_{G_1}(M)=0$. Therefore, $c_{G(\mathbb{F}_p)}(M) = 0$ by using the complexity bound, thus M is a projective $kG(\mathbb{F}_p)$-module. □

Let us consider these results in the context of the representation type question in Section 4.1. Since $kSL_2(\mathbb{F}_p)$ has finite representation type we have $c_{SL_2(\mathbb{F}_p)}(k) = 1$. Therefore, $c_{(SL_2)_1}(k) \geq 2$ which demonstrates immediately that $(SL_2)_1$ has infinite representation type because the syzygies $\Omega^n(k)$ are indecomposable modules and must grow in dimension at least linearly in n.

4.3 Partial converse to the Parshall conjecture

A natural question to ask is whether the converse to the Parshall Conjecture holds, that is if M is in mod(G), and M is projective over $kG(\mathbb{F}_p)$, is it projective over G_1?

The immediate answer is no. Let $M = St^{(1)} = L((p-1)\rho)^{(1)}$. Then M is projective as $G(\mathbb{F}_p)$-module, and $M \simeq \oplus k$ as G_1-module. Let $\mathcal{C}_p$ be the full subcategory of G-modules whose composition factors have highest weight λ, satisfying $\langle \lambda, \tilde{\alpha}^\vee \rangle \leq p(p-1)$ where $\tilde{\alpha}$ is the highest root. The converse is valid as long as we restrict the collection of G-modules under consideration (cf. [LN1, Thm. 4.4]).

Theorem 4.3.1. *Let $M \in \mathcal{C}_p$. If M is projective over $kG(\mathbb{F}_p)$ then M is projective over G_1.*

Friedlander and Parshall proved that can one check projectivity over G_1 for modules in mod(G) by considering the restriction to the one-dimensional Lie algebra spanned by a long root vector (cf. [FP1, (2.4b) Prop.]).

Theorem 4.3.2. *Let $\alpha \in \Phi$ be long root and M be in* mod(G)*. Then M is projective over G_1 if and only if $M\mid_{u(\langle x_\alpha \rangle)}$ is projective.*

We can now apply Friedlander and Parshall's result above along with the validity of the Parshall Conjecture to prove a projectivity criterion over $G(\mathbb{F}_p)$ for modules in $\mathcal{C}_p$ (cf. [LN1, Thm 4.5]).

Corollary 4.3.3. *Let M be in $\mathcal{C}_p$. The following are equivalent:*

(1) *M is projective over $kG(\mathbb{F}_p)$;*

(2) $M\mid_{x_\alpha(\mathbb{F}_p)}$ *is projective for all* $\alpha \in \Phi$;
(3) $M\mid_{x_\beta(\mathbb{F}_p)}$ *is projective for* $\beta \in \Phi$ *long.*

Proof. The implications (1) $\Rightarrow$ (2) $\Rightarrow$ (3) are clear. We need to show that (3) $\Rightarrow$ (1). Suppose that $M\mid_{x_\beta(\mathbb{F}_p)}$ is projective. Then $M\mid_{SL_2(\mathbb{F}_p)_\beta}$ is projective where $SL_2(\mathbb{F}_p)_\beta$ is the semisimple part of the subgroup generated by $x_\beta(\mathbb{F}_p)$ and $x_{-\beta}(\mathbb{F}_p)$. By the partial converse to the Parshall Conjecture, $M\mid_{u(\mathfrak{sl}_2)_\beta}$ is projective. Therefore, $M\mid_{u(\langle x_\beta\rangle)}$ is projective. Friedlander and Parshall's result implies that M is projective as G_1-module. The Parshall Conjecture can be employed to conclude that M is projective as $G(\mathbb{F}_p)$-module. □

4.4 A map between support varieties

Parshall also posed the problem in his Arcata Conference Proceedings article to find a relationship between the support varieties for G_1 and $G(\mathbb{F}_p)$. The results in this section are due to the Carlson, Lin and the author [CLN].

Let $\mathcal{V}_{G_1}(M)$ denote the support variety of M in mod(G_1). Recall that

$$\mathcal{V}_{G_1}(M) = \{x \in \mathfrak{g} : x^{[p]} = 0,\ M|_{u(\langle x\rangle)} \text{ is not free}\} \cup \{0\} \subseteq \mathcal{N}_1(g) \subseteq \mathcal{N}.$$

We will assume that p is good (i.e., if $\beta \in \Phi^+$ with $\beta = \sum_{\alpha\in\Delta} n_\alpha \alpha$ then $p \nmid n_\alpha$ for any α). For example, $\Phi = E_6$, p is good except when $p = 2, 3$. The following theorem [NPV, (6.2.1) Thm.] provides a description of $\mathcal{V}_{G_1}(k) \cong \mathcal{N}_1(\mathfrak{g})$ when p is good.

Theorem 4.4.1. *Let p be good. Then there exist $J \subseteq \Delta$ such that $\mathcal{N}_1(\mathfrak{g}) = G\cdot\mathfrak{u}_J$, where $\mathfrak{g} = \mathfrak{u}_J \oplus \mathfrak{l}_J \oplus \mathfrak{u}_J^+$ where $\mathfrak{l}_J$ is the corresponding Levi subalgebra. Furthermore, the nilpotence degree of $\mathfrak{u}_J < p$.*

Let $\mathcal{U}$ be the unipotent variety, which is the set of unipotent element in G, and set

$$\mathcal{U}_1 = \{u \in \mathcal{U} : u^p = 1\}.$$

When p is good, $\mathcal{U}_1 = G\cdot U_J$ where U_J is the unipotent radical of the parabolic subgroup associated to J (as in the preceding theorem). By using variations on the exponential map, Seitz was able to prove the following theorem (cf. [Sei, Prop. 5.1]).

Theorem 4.4.2. *There exists a P_J-equivariant isomorphism of algebraic k-varieties:*

$$\exp : \mathfrak{u}_J \to \mathcal{U}_J$$

which is defined over $\mathbb{Z}_{(p)}$.

Under suitable normality conditions, we were able to extend Seitz's map to a G-equivariant isomorphism of varieties [CLN, Thm. 3].

Theorem 4.4.3. *Let $\mathcal{N}_1(g)$ be normal. Then exp extends to a G-equivariant isomorphism:*

$$\exp : G\cdot\mathfrak{u}_J \to G\cdot U_J.$$

From now on let us define

$$\log : \mathcal{U}_1 \to \mathcal{N}_1(\mathfrak{g})$$

as the inverse of exp. For M a $G(\mathbb{F}_p)$-module, let $\mathcal{V}_{G(\mathbb{F}_p)}(M)$ denote the support variety of M. One can describe the support variety of the trivial module as

$$\mathcal{V}_{G(\mathbb{F}_p)}(k) = \lim_{\overrightarrow{E}} \mathcal{V}_E(k),$$

where E ranges over the elementary abelian p-subgroups. Now one can identify $\mathcal{V}_E(k)$ with a rank variety $\mathcal{V}_E^{rank}(k)$ which allows one to construct a map $\mathcal{V}_E(k) \hookrightarrow \mathcal{U}_1$. By putting all these maps together we have

$$\mathcal{V}_{G(\mathbb{F}_p)}(k) \hookrightarrow \mathcal{U}_1/G(\mathbb{F}_p) \to \mathcal{N}_1(\mathfrak{g})/G(\mathbb{F}_p) \cong \mathcal{V}_{G_1}(k)/G(\mathbb{F}_p).$$

Therefore, we have defined a map of support varieties:

$$\psi : \mathcal{V}_{G(\mathbb{F}_p)}(k) \to \mathcal{V}_{G_1}(k)/G(\mathbb{F}_p).$$

Now define the category $\mathcal{D}$ as follows. Let $l = \text{rank}(\Phi)$ and $(b_{ij}) = (\langle \alpha_i, \alpha_j^\vee \rangle)^{-1}$. Let $\mathcal{D}$ be the full subcategory of G-module whose composition factors have high weight satisfying

$$\sum_{i=1}^{l} \sum_{j=1}^{l} \langle \lambda, \alpha_i^\vee \rangle b_{ij} < \frac{p(p-1)}{2}.$$

The following theorem gives an explicit description of the image of ψ on $\mathcal{V}_{G(\mathbb{F}_p)}(M)$ (cf. [CLN, Section 3, Cor. 1,2]).

Theorem 4.4.4. *Let G be a simple algebra group scheme defined over $\mathbb{F}_p$. Moreover, let $\mathcal{N}_1(g)$ be normal and $M \in \mathcal{D}$, then*

$$\psi(\mathcal{V}_{G(\mathbb{F}_p)}(M)) \subseteq \mathcal{V}_{G_1}(M)/G(\mathbb{F}_p).$$

More precisely:

$$\psi(\mathcal{V}_{G(\mathbb{F}_p)}(M)) = \{x \in \mathcal{N}_1^{\mathbb{F}_p} : \ x^{[p]} = 0, \ M|_{u(\langle x \rangle)} \textit{is not free}\} \cup \{0\}$$

where $\mathcal{N}_1^{\mathbb{F}_p}$ is the set of $\mathbb{F}_p$-expressible elements.

4.5 Applications

The results in the preceding section can be used to understand projectivity for finite groups of Lie type.

Theorem 4.5.1. *Let G be a simple algebraic group, M be in $\mathcal{C}_p$ which is not projective over $kG(\mathbb{F}_p)$. Then*

(1) *If M is indecomposable in* mod(kG) *and the rank of Φ is greater than or equal to 2 then* $\dim \mathcal{V}_{G(\mathbb{F}_p)}(M) \geq 2$;

(2) *If* $\Phi = A_n$ *then* $\dim \mathcal{V}_{GL_n(\mathbb{F}_p)}(M) \geq n$.

We can also determine the dimension of support varieties for simple modules for rank 2 groups.

Theorem 4.5.2. *Let* G *be a simple algebraic group where* Φ *has rank* 2 *(i.e.,* $\Phi = A_2$ $(p > 2)$, B_2 $(p > 7)$, G_2 $(p > 19)$, *and let* $\lambda \in X_1(T)$.

(1) *If* $\lambda = (p-1)\rho$ *then* $\mathcal{V}_{G(\mathbb{F}_p)}(L(\lambda)) = \{0\}$.
(2) *If* $\lambda \neq (p-1)\rho$ *then*

$$\dim \mathcal{V}_{G(\mathbb{F}_p)}(L(\lambda)) = \begin{cases} 2 & \Phi = A_2, \\ 3 & \Phi = B_2, G_2. \end{cases}$$

4.6 Generalizations

Friedlander [F2] has recently found an appropriate generalization of many of the aforementioned results for $G(\mathbb{F}_q)$ where $r \geq 1$. The idea involves using base changes on the Lie algebra level and to view $g_{\mathbb{F}_q}$ as an $\mathbb{F}_p$-Lie algebra. Set $A_q = u(g_{\mathbb{F}_q} \otimes_{\mathbb{F}_p} k)$. In particular, Friedlander proves a generalization of the Parshall Conjecture [F2, Cor. 4.4].

Theorem 4.6.1. *Let* $M \in \text{mod}(G)$. *If* $M|_{A_q}$ *is projective then* $M|_{G(\mathbb{F}_q)}$ *is projective.*

Furthermore, by using the Π-point theory developed by Friedlander and Pevtsova [FPe], he constructs a map

$$\psi : \mathcal{V}_{G(\mathbb{F}_q)}(k) \hookrightarrow \mathcal{V}_{A_q}(k)/G(\mathbb{F}_q)$$

(cf. [F2, Corollary 3.6]).

5 Relating cohomology

5.1 Spectral sequences via truncation

We will next investigate questions which involve relating extensions in the three categories $\text{Mod}(G(\mathbb{F}_q))$, $\text{Mod}(G)$, and $\text{Mod}(G_r)$. The results were obtained in [BNP1, BNP2, BNP3] with generalizations to the twisted cases in [BNP5, BNP6].

The main idea is to first construct spectral sequences which connect the cohomology theories of $\text{Mod}(G(\mathbb{F}_q))$ and $\text{Mod}(G)$. The basic problem is that $\text{Mod}(G)$ does not have enough projective objects. Therefore, it is better to work with "truncated categories" obtained via saturated sets of weights. These truncated categories are highest weight categories and Morita equivalent to the module category for some quasi-hereditary algebra (cf. [CPS]). The cohomology for these quasi-hereditary algebras provides a better approximation of the homological behavior in $\text{Mod}(G(\mathbb{F}_q))$. Once we provide a connection to the G-category, we apply the Lyndon-Hochschild-Serre (LHS) spectral sequence and use information about the cohomology in $\text{Mod}(G_r)$, in addition to the "geometry of the flag variety G/B" to obtain results about the cohomology for the finite Chevalley groups.

Let $\Pi = \{\lambda \in X(T)_+ : \langle \lambda+\rho, \alpha_0^\vee \rangle \leq 2p^r \langle \rho, \alpha_0^\vee \rangle\}$ and $\mathcal{C}$ be the full subcategory of modules in $\text{Mod}(G)$ whose composition factors have highest weight in Π. Let $\mathcal{T} : \text{Mod}(G) \to \mathcal{C}$ be the functor defined where $\mathcal{T}(M)$ is the largest submodule of M contained in $\mathcal{C}$. The functor $\mathcal{T}$ is left exact.

Now define the functor $\mathcal{G} : \text{Mod}(G(\mathbb{F}_q)) \to \mathcal{C}$ as

$$\mathcal{G}(N) = \mathcal{T}(\text{ind}_{G(\mathbb{F}_q)}^{G} N)$$

where $N \in \text{Mod}(G(\mathbb{F}_q))$. Since $G/G(\mathbb{F}_q)$ is affine, the induction functor is exact. Therefore, $\mathcal{G}$ is a left exact functor and admits higher right derived functors $R^j\mathcal{G}(-)$ for $j \geq 0$.

The following theorem provides a spectral sequence which relates extensions in $\text{Mod}(G(\mathbb{F}_q))$ and $\mathcal{C}$.

Theorem 5.1.1. *Let M in $\mathcal{C}$ and N in $\text{Mod}(G(\mathbb{F}_q))$. Then there exists a first quadrant spectral sequence:*

$$E_2^{i,j} = \text{Ext}^i_{\mathcal{C}}(M, R^j\mathcal{G}(N)) \Rightarrow \text{Ext}^{i+j}_{G(\mathbb{F}_q)}(M, N).$$

We note that since $\mathcal{C}$ is obtained from a saturated set of weight, we have

$$\text{Ext}^i_{\mathcal{C}}(M_1, M_2) = \text{Ext}^i_G(M_1, M_2)$$

for all M_1, M_2 in $\mathcal{C}$ and $i \geq 0$. One can apply this spectral sequence along with the lifting property of projective G_r-modules to obtain the following corollary.

Corollary 5.1.2. *Let $p \geq 2(h-1)$ with $\lambda, \mu \in X_r(T)$*

(1) $\text{Ext}^1_{G(\mathbb{F}_q)}(L(\lambda), L(\mu)) \cong \text{Ext}^1_G(L(\lambda), \mathcal{G}(L(\mu)))$;

(2) $\text{Ext}^2_G(L(\lambda), \mathcal{G}(L(\mu))) \hookrightarrow \text{Ext}^2_{G(\mathbb{F}_q)}(L(\lambda), L(\mu))$.

Proof. Let $M = L(\lambda)$, $N = L(\mu)$ in the spectral sequence. The spectral sequence yields a five term exact sequence:

$$0 \to E_2^{1,0} \to E_1 \to E_2^{0,1} \to E_2^{2,0} \to E_2.$$

In order to prove the result we need to show that $E_2^{0,1} = 0$. It suffices to prove that $R^1\mathcal{G}(L(\mu))$ has no composition factors isomorphic to $L(\lambda)$. When $p \geq 2(h-1)$, the projective cover $P(\lambda)$ of $L(\lambda)$ in $\mathcal{C}$ is isomorphic to $Q_r(\lambda)$. This fact uses the lifting property of projective G_r-modules. Moreover, one can apply the spectral sequence with $M = P(\lambda)$ and $N = L(\mu)$. The spectral sequence then collapses and yields:

$$\text{Hom}_{\mathcal{C}}(P(\lambda), R^1\mathcal{G}(L(\mu))) \cong \text{Ext}^1_{G(\mathbb{F}_q)}(P(\lambda), L(\mu)).$$

By putting all of this information together, we have

$$\begin{aligned} [R^1\mathcal{G}(L(\mu)) : L(\lambda)] &= \dim \text{Hom}_{\mathcal{C}}(P(\lambda), R^1\mathcal{G}(L(\mu))) \\ &= \dim \text{Ext}^1_{G(\mathbb{F}_q)}(P(\lambda), L(\mu)) \\ &= \dim \text{Ext}^1_{G(\mathbb{F}_q)}(Q_r(\lambda), L(\mu)) \\ &= 0. \end{aligned}$$

□

5.2 Extensions between simple modules

Our knowledge about extensions involving two simple modules in $\text{Mod}(G)$, $\text{Mod}(G_r)$ and $\text{Mod}(G(\mathbb{F}_q))$ is very minimal. We can state this in terms of three open problems:

(5.2.1) Determine $\text{Ext}^j_G(L(\sigma_1), L(\sigma_2))$ for $\sigma_1, \sigma_2 \in X(T)_+$, $j \geq 0$.

(5.2.2) Determine $\text{Ext}^j_{G_r}(L(\sigma_1), L(\sigma_2))$ for $\sigma_1, \sigma_2 \in X_r(T)$, $j \geq 0$.

(5.2.3) Determine $\text{Ext}^j_{G(\mathbb{F}_q)}(L(\sigma_1), L(\sigma_2))$ for $\sigma_1, \sigma_2 \in X_r(T)$, $j \geq 0$.

The Lusztig Conjecture predicts the characters of the simple G-modules when $p \geq h$. If the Lusztig Conjecture holds then (5.2.1) and (5.2.2) can be solved for regular weights. Moreover, if there are certain vanishing/non-vanishing conditions that hold when $j = 1$ for the extension groups in (5.2.1) and (5.2.2) then the Lusztig Conjecture is valid.

It is interesting to consider the case when $L(\sigma_1) \cong L(\sigma_2) \cong k$. Then $\text{H}^j(G, k) = 0$ for $j > 0$ and is isomorphic to k when $j = 0$. For the Frobenius kernels, $\text{H}^{2\bullet}(G_1, k) = k[\mathcal{N}]$ where $\mathcal{N}$ is the nilpotent cone, and $\text{H}^{2\bullet+1}(G_1, k) = 0$ when $p > h$ (cf. [AJ, FP2]). The cohomology for G_r is not known in general for $r > 1$. Finally, very little is known about the cohomology groups $\text{H}^\bullet(G(\mathbb{F}_q), k)$. We will investigate questions about these groups later in Section 6.

In the case when $\sigma = \sigma_1 = \sigma_2$ and $j = 1$ one is looking at the case of self-extensions. By standard arguments [Jan1, II 2.12], one can show that

$$\text{Ext}^1_G(L(\sigma), L(\sigma)) = 0$$

for $\sigma \in X(T)_+$. In 1984, Andersen [And1, Theorem 4.5] proved the following result about self extensions for G_r:

Theorem 5.2.1. *Suppose $\Phi \neq C_n$, when $p = 2$. Then $\text{Ext}^1_{G_r}(L(\sigma), L(\sigma)) = 0$ for all $\sigma \in X_r(T)$.*

Andersen used information about the structure of the higher line bundle cohomology groups in his calculation. The use of these geometric techniques will be a dominant theme in our examination of self-extensions and cohomology for finite Chevalley groups

5.3 Ext^1-formulas for simple $G(\mathbb{F}_q)$-modules

We will now run through a series of steps which will lead us to produce Ext^1-formulas for finite Chevalley groups.

(1) Let $\lambda, \mu \in X_r(T)$, then

$$\text{Ext}^1_{G(\mathbb{F}_q)}(L(\lambda), L(\mu)) \cong \text{Ext}^1_G(L(\lambda), L(\mu) \otimes \mathcal{G}(k))$$

for all $\lambda, \mu \in X_r(T)$ [BNP3, Theorem 2.2].

(2) We can explicitly describe $\mathcal{G}(k)$ as a G-module. Set $\Gamma_{2h-1} = \{\nu \in X(T)_+ : \langle \nu, \alpha_0^\vee \rangle < 2h-1\}$. Then a miracle happens! For $p \geq 3(h-1)$, $\mathcal{G}(k)$ is semisimple [BNP1, Theorem 7.4]. Moreover,

$$\begin{aligned}\mathcal{G}(k) &\cong \bigoplus_{\nu \in \Gamma_{2h-1}} L(\nu) \otimes [L(\nu)^{(r)}]^* \\ &\cong \bigoplus_{\nu \in \Gamma_{2h-1}} L(v) \otimes L(v^*)^{(r)}.\end{aligned}$$

(3) Set $\Gamma = \{\nu \in X(T)_+ : \langle \nu, \alpha_0^\vee \rangle < h-1\}$. By using the decomposition of $\mathcal{G}(k)$ along with some additional information about G-extensions, we obtain the following formula.

Theorem 5.3.1. *Let $p \geq 3(h-1)$ and $\lambda, \mu \in X_r(T)$. Then*

$$\operatorname{Ext}^1_{G(\mathbb{F}_q)}(L(\lambda), L(\mu)) \cong \bigoplus_{\nu \in \Gamma} \operatorname{Ext}^1_G(L(\lambda) \otimes L(\nu)^{(r)}, L(\mu) \otimes L(\nu)).$$

We can now refine the aforementioned Ext^1-formula when $r \geq 2$ and $r = 1$. First let $r \geq 2$. For $\sigma \in X_r(T)$ with $\sigma = \sigma_0 + \sigma_1 p + \cdots + \sigma_{r-2}p^{r-2} + \sigma_{r-1}p^{r-1}$ set $\widehat{\sigma} = \sigma_0 + \sigma_1 p + \cdots + \sigma_{r-2}p^{r-2}$. We can apply the LHS spectral sequence twice with $G_{r-1} \trianglelefteq G$, and $G_1 \trianglelefteq G$ to obtain the following theorem.

Theorem 5.3.2. *Let $p \geq 3(h-1)$, $\lambda, \mu \in X_r(T)$, $r \geq 2$. Then*

$$\operatorname{Ext}^1_{G(\mathbb{F}_q)}(L(\lambda), L(\mu)) = \operatorname{Ext}^1_G(L(\lambda), L(\mu)) \oplus R$$

where $R = \bigoplus_{\nu \in \Gamma - \{0\}} \operatorname{Hom}_G(L(\nu), \operatorname{Ext}^1_{G_1}(L(\lambda_{r-1}), L(\mu_{r-1}))) \otimes \operatorname{Hom}_G(L(\widehat{\lambda}), L(\widehat{\mu}) \otimes L(\nu))$.

One can also produce a formula when $r = 1$. In these formulas one should observe how the information from the Frobenius kernels is playing a role in the determination of cohomology for the finite Chevalley groups.

Theorem 5.3.3. *Let $p \geq 3(h-1)$ and $\lambda, \mu \in X_1(T)$. Then*

$$\operatorname{Ext}^1_{G(\mathbb{F}_p)}(L(\lambda), L(\mu)) \cong \operatorname{Ext}^1_G(L(\lambda), L(\mu)) \oplus R$$

where

$$R = \bigoplus_{\nu \in \Gamma - \{0\}} \operatorname{Hom}_G(L(\nu), \operatorname{Ext}^1_{G_1}(L(\lambda), L(\mu) \otimes L(\nu))^{(-1)}).$$

5.4 Self extensions

For $r \geq 2$, the groups $G(\mathbb{F}_q)$ do not admit self extensions when $p \geq 3(h-1)$.

Theorem 5.4.1. *Let $p \geq 3(h-1)$, $r \geq 2$ and $\lambda \in X_r(T)$. Then*

$$\mathrm{Ext}^1_{G(\mathbb{F}_q)}(L(\lambda), L(\lambda)) = 0.$$

Proof. As we have seen $\mathrm{Ext}^1_G(L(\lambda), L(\lambda)) = 0$. Therefore, we need to prove that $R = 0$ in Theorem 5.3.2. But, this follows by applying Andersen's result on self-extensions for G_1. □

For $r = 1$, Humphreys proved the existence of self extensions when the root system is of type C_n for general p. One can use a more detailed analysis using the cohomology of line bundles to obtain the following result (cf. [BNP3, Section 4]).

Theorem 5.4.2. *Let $p \geq 3(h-1)$ and $\lambda \in X_1(T)$. If either*

(1) *G does not have underlying root system of type A_1 or C_n or*
(2) *$\langle \lambda, \alpha_n^\vee \rangle \neq \frac{p-2-c}{2}$, where α_n is the unique long simple root and c is odd with $0 < |c| \leq h-1$,*

then $\mathrm{Ext}^1_{G(\mathbb{F}_p)}(L(\lambda), L(\lambda)) = 0$.

5.5 Applications

Our results involving Ext^1 for finite Chevalley group enabled us to address several open questions.

(1) Smith asked the following question (Question 5) in his 1985 paper [Sm]:

Question 5: Let V be a $G(\mathbb{F}_q)$-module satisfying Hypothesis A with all composition factors isomorphic. Is V completely reducible?

Our self extension results stated in Section 5.4 answer Smith's question independent of "Hypothesis A".

(2) Cline, Parshall, Scott, and vanderKallen's [CPSvdK] famous result on rational and generic cohomology can be stated as follows. Let V in $\mathrm{Mod}(G)$. For fixed $i \geq 0$, consider the restriction map in cohomology:

$$\mathrm{H}^i(G, V^{(s)}) \xrightarrow[\text{res}]{} \mathrm{H}^i(G(\mathbb{F}_q), V^{(s)}).$$

When r and s are sufficiently large, the map res is an isomorphism.

We can see the phenomenon in our Ext^1-formulas. Let $p \geq 3(h-1)$. Fix $\lambda, \mu \in X_r(T)$. As r get large $\lambda_{r-1} = \mu_{r-1} = 0$, and $\mathrm{Ext}^1_{G_1}(k,k) = 0$. Therefore, by Theorem 5.3.2, $R = 0$ and

$$\mathrm{Ext}^1_G(L(\lambda), L(\mu)) \xrightarrow[\text{res}]{} \mathrm{Ext}^1_{G(\mathbb{F}_q)}(L(\lambda), L(\mu))$$

is an isomorphism. Observe that in this instance no Frobenius twists are necessary.

(3) Jantzen [Jan4] proved the following result which insured that the restriction map in cohomology is an isomorphism given upper bounds on the highest weight of the composition factors of V.

Theorem 5.5.1. *Let* $V \in \text{Mod}(G)$ *where the composition factors* $L(\mu)$ *satisfy*

$$\langle \mu, \alpha_0^\vee \rangle \leq \begin{cases} p^r - 3p^{r-1} - 3 & \Phi = G_2, \\ p^r - 2p^{r-1} - 2 & \Phi \neq G_2. \end{cases}$$

Then res *is an isomorphism.*

For large primes, Andersen [And2] gives uniform conditions on the high weight on the composition factors to insure that the restriction map is an isomorphism.

Theorem 5.5.2. *Let* $p \geq 3(h-1)$ *and suppose that the composition factors of* V *satisfy*

$$\langle \mu, \alpha_0^\vee \rangle \leq p^r - p^{r-1} - 2 \quad (\Phi \neq A_1).$$

Then res *is an isomorphism.*

Our results [BNP2, Theorem 4.8(A)] enabled us to give the best possible bounds to insure that the restriction map is an isomorphism when $r \geq 2$. The optimal bounds when $r = 1$ are still unknown.

Theorem 5.5.3. *Let* G *be a simple, simply connected algebraic group with* $p \geq 3(h-1)$, $r \geq 2$. *If* V *has composition factors* $L(\mu)$ *satisfying*

$$\langle \mu, \alpha_0^\vee \rangle \leq \begin{cases} p^r - 2p^{r-1} & \Phi = A_1, \\ p^r - p^{r-1} & \Phi = A_n, \\ p^r & \Phi = B_n,\ C_n,\ D_n,\ E_6,\ E_7, \\ 2p^r - p^{r-1} + 1 & \Phi = E_8,\ F_4,\ G_2. \end{cases}$$

Then res *is an isomorphism.*

6 Computing cohomology for finite groups of Lie type

6.1 Vanishing ranges

In 2005 there was a Conference on the Cohomology of Finite Groups at Oberwolfach. At the beginning of Eric Friedlander's talk he started by stating that we know very little about the cohomology $\text{H}^\bullet(G(\mathbb{F}_q), k)$ where char $k = p > 0$ other than Quillen's description of the maximal ideal spectrum. Friedlander went on to say that it would be nice to know when the cohomology starts (i.e., in which degree the first non-trivial cohomology lives). There are two main aspects to this problem.

(6.1.1) Vanishing Ranges: Locating $D > 0$ such that the cohomology groups $\text{H}^i(G(\mathbb{F}_q), k) = 0$ for $0 < i < D$.

(6.1.2) Determining the First Non-Trivial Cohomology Class: Find a D such that $\text{H}^i(G(\mathbb{F}_q), k) = 0$ for $0 < i < D$ and $\text{H}^D(G(\mathbb{F}_q), k) \neq 0$. A D satisfying this property will be called a *sharp bound*.

Quillen computed the cohomology of $G(\mathbb{F}_q)$ in the non-describing characterisitic and provided a vanishing range (6.1.1) in the case of $GL_n(\mathbb{F}_q)$ in the describing characteristic case (cf. [Q3]). Later Friedlander [F1] and Hiller [H] discovered vanishing ranges for other groups of Lie type.

The aim of this section is to present some recent work of the author with Bendel and Pillen. We solve (6.1.2) for groups of types A_n and C_n where $p > h$ and $r = 1$. In order to obtain these results we first employ a variation on the ideas presented in Section 5 by investigating the induction $\mathcal{G}_r(k) = \text{ind}_{G(\mathbb{F}_q)}^G k$. We indicate that $\mathcal{G}_r(k)$ has a natural filtration as a $G \times G$-module. This allows us to reduce our problem to looking at sections of the filtration (i.e., modules of the form $H^0(\lambda) \otimes H^0(\lambda^*)^{(r)}$, $\lambda \in X(T)_+$). The next step is to apply the LHS spectral sequence in the case when $r = 1$. The spectral sequence collapses in the case when $p > h$ by using results due to Kumar, Lauritzen and Thompsen [KLT]. This in turn provides a method for giving an upper bound for the cohomology $\text{H}^\bullet(G(\mathbb{F}_p), k)$ using the combinatorics of the nilpotent cone via Kostant's Partition Functions. The steps are presented in the following diagram. The results and details in this section can be found in [BNP8, BNP9].

$\text{H}^i(G(\mathbb{F}_q), k)$

$\downarrow$ Induction Functor

$\text{H}^i(G, \mathcal{G}_r(k))$

$\downarrow$ Filtrations

$\text{H}^i(G, H^0(\lambda) \otimes H(\lambda^*)^{(r)}) \quad \longrightarrow \quad \text{H}^i(G_1, H^0(\lambda)) \quad \longrightarrow \quad$ Root Combinatorics.

(LHS Spectral Sequences) (Kostant Partition Functions)

6.2 Induction and filtrations

Let $\mathcal{G}_r = \text{ind}_{G(\mathbb{F}_q)}^G(-)$. This functor is exact and all its higher right derived functors vanish. Therefore, we can apply Frobenius reciprocity to obtain the following isomorphism of extension groups.

Proposition 6.2.1. *Let M, N be in* Mod(G)*. Then, for all $i \geq 0$,*

$$\text{Ext}^i_{G(\mathbb{F}_q)}(M, N) \cong \text{Ext}^i_G(M, N \otimes \mathcal{G}_r(k)).$$

As a $G \times G$-module the coordinate algebra $k[G]$ has a filtration with sections of the form $H^0(\lambda) \otimes H^0(\lambda^*)$ where $\lambda \in X(T)_+$. Each section of this form appears exactly one time. By using the Lang map, we were able to show that $\mathcal{G}_r(k) = k[G/G(\mathbb{F}_q)]$ admits a natural filtration as a $G \times G$-module.

Proposition 6.2.2. *The $G \times G$-module $\mathcal{G}_r(k)$ has a filtration with factors of the form $H^0(\lambda) \otimes H^0(\lambda^*)^{(r)}$ with multiplicity one for each $\lambda \in X(T)_+$.*

6.3 Isomorphisms

The existence of a filtration on $\mathcal{G}_r(k)$ with sections of the form $H^0(\lambda)\otimes H^0(\lambda^*)^{(r)}$, $\lambda \in X(T)_+$, in addition to the isomorphism in Proposition 6.2.1 allows us to deduce the next result on determining vanishing ranges.

Proposition 6.3.1. *Let m be the least positive integer such that there exists $\lambda \in X(T)_+$ with $\mathrm{H}^m(G, H^0(\lambda)\otimes H^0(\lambda^*)^{(r)}) \neq 0$. Then $\mathrm{H}^i(G(\mathbb{F}_q), k) \cong \mathrm{H}^i(G, \mathcal{G}_r(k)) = 0$ for $0 < i < m$.*

A more detailed analysis is necessary to address (6.1.2) using this aforementioned filtration (cf. [BNP7, Section 2.7]). Our results from this section are summarized below.

Theorem 6.3.2. *Let m be the least positive integer such that there exists $\nu \in X(T)_+$ with $\mathrm{H}^m(G, H^0(\nu)\otimes H^0(\nu^*)^{(r)}) \neq 0$. Let $\lambda \in X(T)_+$ be such that*

$$\mathrm{H}^m(G, H^0(\lambda)\otimes H^0(\lambda^*)^{(r)}) \neq 0.$$

Suppose $\mathrm{H}^{m+1}(G, H^0(\nu)\otimes H^0(\nu^*)^{(r)}) = 0$ for all $\nu < \lambda$ that are linked to λ. Then

(i) $\mathrm{H}^i(G(\mathbb{F}_q), k) = 0$ *for* $0 < i < m$;
(ii) $\mathrm{H}^m(G(\mathbb{F}_q), k) \neq 0$;
(iii) *if, in addition, $\mathrm{H}^m(G, H^0(\nu)\otimes H^0(\nu^*)^{(r)}) = 0$ for all $\nu \in X(T)_+$ with $\nu \neq \lambda$, then*

$$\mathrm{H}^m(G(\mathbb{F}_q), k) \cong \mathrm{H}^m(G, H^0(\lambda)\otimes H^0(\lambda^*)^{(r)}).$$

6.4 Good filtrations on cohomology

For $\lambda \in X(T)_+$ an open problem is to determine if $\mathrm{H}^i(G, H^0(\lambda))^{(-1)}$ admits a good filtration. When $i = 0, 1, 2$ this holds for all primes [Jan3, BNP4, W]. For $p > h$, from [AJ] and [KLT], we have

$$\mathrm{H}^i(G_1, H^0(\nu))^{(-1)} = \begin{cases} \mathrm{ind}_B^G(S^{\frac{i-\ell(w)}{2}}(\mathfrak{u}^*)\otimes\mu) & \text{if } \nu = w\cdot 0 + p\mu, \\ 0 & \text{otherwise,} \end{cases} \tag{6.4.1}$$

where $\mathfrak{u} = \mathrm{Lie}(U)$. Note also that since $p > h$ and ν is dominant, μ must also be dominant. This explicit description of the cohomology shows that in this case the cohomology groups $\mathrm{H}^i(G, H^0(\lambda))^{(-1)}$ admit a good filtration.

For $n > 0$, let $P_n(\nu)$ denote the number of times that ν can be expressed as a sum of exactly n positive roots. Note that $P_0(0) = 1$. The function P_n is often referred to as *Kostant's Partition Function.* By using work of [AJ] and the fact that the cohomology has a good filtration one can show the following result.

Proposition 6.4.1. *Assume $p > h$. Let $\lambda = p\mu + w\cdot 0 \in X(T)_+$. Then*

$$\dim \mathrm{H}^i(G, H^0(\lambda)\otimes H^0(\lambda^*)^{(1)}) = \sum_{u\in W}(-1)^{\ell(u)}P_{\frac{i-\ell(w)}{2}}(u\cdot\lambda - \mu).$$

6.5 Upper bound on cohomology:

Now we can use the existence of the filtration on $\mathcal{G}(k) = k[G/G(\mathbb{F}_q)]$ and the preceding proposition to provide an upper bound on the cohomology group $\mathrm{H}^i(G(\mathbb{F}_p), k)$.

Theorem 6.5.1. *Assume $p > h$.*

$$\dim \mathrm{H}^i(G(\mathbb{F}_p), k) \leq \sum_{\{w\in W\,|\,\ell(w)\equiv i \mod 2\}} \sum_{\mu\in X(T)_+} \sum_{u\in W} (-1)^{\ell(u)} P_{\frac{i-\ell(w)}{2}}(u\cdot(p\mu + w\cdot 0) - \mu).$$

6.6 Combinatorics

We would like to indicate how to address the question of vanishing ranges (6.1.1) by using root combinatorics and Proposition 6.3.1.

Proposition 6.6.1. *Suppose that $\Phi \neq G_2$. Let $p > h$, $\lambda = w\cdot 0 + p\mu$, $\mu \neq 0$ and $\mathrm{H}^i(G, H^0(\lambda)\otimes H^0(\lambda^*)^{(1)}) \neq 0$,*

(1) *For $\sigma \in \Phi^+$, $(p-1)\langle \mu, \sigma^\vee\rangle + \ell(w) + \langle w\cdot 0, \sigma^\vee\rangle \leq i$.*
(2) *For $\tilde{\alpha}$ the highest root, $(p-1)\langle \mu, \tilde{\alpha}^\vee\rangle - 1 \leq i$.*

Proof. (1): First we have

$$\begin{aligned} 0 \neq \mathrm{H}^i(G, H^0(\lambda)\otimes H^0(\lambda^*)^{(1)}) &\cong \mathrm{Hom}_G(V(\lambda)^{(1)}, \mathrm{H}^i(G_1, H^0(\lambda))) \\ &= \mathrm{Hom}_G(V(\lambda), \mathrm{ind}_B^G S^{\frac{i-\ell(w)}{2}}(\mathfrak{u}^*)\otimes \mu) \\ &= \mathrm{Hom}_B(V(\lambda), S^{\frac{i-\ell(w)}{2}}(\mathfrak{u}^*)\otimes \mu). \end{aligned}$$

Observe that $\lambda - \mu$ must be a weight of $S^{\frac{i-\ell(w)}{2}}(\mathfrak{u}^*)$. Consequently, $\lambda - \mu$ is expressable as $\frac{i-\ell(w)}{2}$ positive roots.

Next note that for $\sigma_1,\ \sigma_2 \in \Phi$, $\langle \sigma_1, \sigma_2^\vee\rangle \leq 2$ ($\Phi \neq G_2$). Therefore,

$$\langle \lambda - \mu, \sigma^\vee\rangle \leq (\frac{i-\ell(w)}{2})\cdot 2 = i - \ell(w).$$

By substituting $\lambda = p\mu + w\cdot 0$ one has

$$\langle p\mu + w\cdot 0 - \mu, \sigma^\vee\rangle \leq i - \ell(w)$$

and

$$(p-1)\langle \mu, \sigma^\vee\rangle + \ell(w) + \langle w\cdot 0, \sigma^\vee\rangle \leq i.$$

(2) The weight $-w\cdot 0$ can be expressed uniquely as $l(w)$ distinct positive roots. Since at most one can be $\tilde{\alpha}$ and $\langle \tilde{\alpha}, \tilde{\alpha}^\vee\rangle = 2$, it follows that

$$\langle -w\cdot 0, \tilde{\alpha}^\vee\rangle \leq (\ell(w)-1) + 2 = \ell(w) + 1.$$

Thus, $\langle w\cdot 0, \tilde{\alpha}^\vee\rangle \geq -l(w) - 1$. Now apply Part (1) to obtain the result. □

6.7 Applications

By applying Proposition 6.3.1 and Theorem 6.3.2 with the LHS spectral sequence we are able to deduce the following theorem which addresses (6.1.1) and (6.1.2) when $\Phi = C_n$ and A_n.

Theorem 6.7.1 (A). *Suppose Φ is of type C_n with $p > 2n$. Then*

(1) $\mathrm{H}^i(G(\mathbb{F}_q), k) = 0$ *for* $0 < i < r(p-2)$;
(2) $\mathrm{H}^{p-2}(G(\mathbb{F}_p), k) \cong k$;
(3) $\mathrm{H}^{r(p-2)}(G(\mathbb{F}_q), k) \neq 0$.

Theorem 6.7.2 (B). *Suppose Φ is of type A_n with $n \geq 2$. Suppose further that $p > n + 1$.*

(1) *(Generic case) If $p > n + 2$ and $n > 3$, then*
  (i) $\mathrm{H}^i(G(\mathbb{F}_p), k) = 0$ *for* $0 < i < 2p - 3$;
  (ii) $\mathrm{H}^{2p-3}(G(\mathbb{F}_p), k) = k$.
(2) *If $p = n + 2$, then*
  (i) $\mathrm{H}^i(G(\mathbb{F}_p), k) = 0$ *for* $0 < i < p - 2$;
  (ii) $\mathrm{H}^{p-2}(G(\mathbb{F}_p), k) = k \oplus k$.
(3) *If $n = 2$ and 3 divides $p - 1$, then*
  (i) $\mathrm{H}^i(G(\mathbb{F}_p), k) = 0$ *for* $0 < i < 2p - 6$;
  (ii) $\mathrm{H}^{2p-6}(G(\mathbb{F}_p), k) = k \oplus k$.
(4) *If $n = 2$ and 3 does not divide $p - 1$, then*
  (i) $\mathrm{H}^i(G(\mathbb{F}_p), k) = 0$ *for* $0 < i < 2p - 3$;
  (ii) $\mathrm{H}^{2p-3}(G(\mathbb{F}_p), k) = k$.
(5) *If $n = 3$ and $p > 5$, then*
  (i) $\mathrm{H}^i(G(\mathbb{F}_p), k) = 0$ *for* $0 < i < 2p - 6$;
  (ii) $\mathrm{H}^{2p-6}(G(\mathbb{F}_p), k) = k$.

For the type C_n case when $r = 1$, we have seen that for $p > h = 2n$, $\mathrm{H}^i(G(\mathbb{F}_q), k) = 0$ for $0 < i < (p-2)$. We will now outline how to to construct λ such that $\mathrm{H}^{p-2}(G, H^0(\lambda) \otimes H^0(\lambda^*)^{(1)}) \neq 0$. Let $w = s_1 s_2 \cdots s_{n-1} s_n s_{n-1} \cdots s_1 \in W$. Then

$$-w \cdot 0 = n\tilde{\alpha} = 2n\omega_1 = \text{sum of all positive roots containing } \alpha_1.$$

Furthermore, $\lambda = p\omega_1 + w \cdot 0 = (p - 2n)\omega_1$, thus $\lambda - \omega_1 = (\frac{p-1}{2} - n)\tilde{\alpha}$.

Next one shows that

$$P_{\frac{p-1}{2}-n}(u \cdot \lambda - \mu) = \begin{cases} 1 & \mu = 1, \\ 0 & \mu \neq 1. \end{cases}$$

This implies $\mathrm{H}^{p-2}(G, H^0(\lambda) \otimes H^0(\lambda^*)) \neq 0$. Finally, one has to also prove that $\lambda \in X(T)_+$ (as above) is the only such weight in $X(T)_+$ such that $\mathrm{H}^{p-2}(G, H^0(\lambda) \otimes H^0(\lambda^*)^{(1)}) \neq 0$ (see [BNP8, Section 5.3]).

We remark that (6.1.1) and (6.1.2) for other types (when G is simply connected) are addressed further in [BNP8]. Moreover, we exhibit uniform bounds when G is of adjoint type.

References

[AM] A. Adem, R.J. Milgram, Cohomology of Finite Groups, 2nd ed., Grundlehren der Mathematischen Wissenschaften **309**, Springer, Berlin, 2004.

[Al] J.L. Alperin, Periodicity in groups, Illinois J. Math. 21 (1977), 776–783.

[And1] H.H. Andersen, Extensions of modules for algebraic groups, American J. Math. 106 (1984), 489–504.

[And2] H.H. Andersen, Extensions of simple modules for finite Chevalley groups, J. Algebra 111 (1987), 388–403.

[AJ] H.H. Andersen, J.C. Jantzen, Cohomology of induced representations for algebraic groups, Math. Ann. 269 (1984), 487–525.

[AS] G.S. Arvunin, L.L. Scott, Quillen stratification for modules, Invent. Math. 66 (1982), 277–286.

[BaKN] I. Bagci, J. Kujawa, D.K. Nakano, Cohomology and support varieties for the Lie superalgebra $W(n)$, International Math. Research Notices, doi:10.1093/imrn/rrn115, (2008).

[Ben] D.J. Benson, Representations and cohomology, Volumes I and II, Cambridge University Press, Cambridge, 1991.

[BNP1] C.P. Bendel, D.K. Nakano, C. Pillen, On comparing the cohomology of algebraic groups, finite Chevalley groups and Frobenius kernels, J. Pure and Applied Algebra 163 (2001), 119–146.

[BNP2] C.P. Bendel, D.K. Nakano, C. Pillen, Extensions for finite Chevalley groups II, Transactions of the AMS 354, (2002), 4421–4454.

[BNP3] C.P. Bendel, D.K. Nakano, C. Pillen, Extensions for finite Chevalley groups I, Advances in Math. 183 (2004), 380–408.

[BNP4] C.P. Bendel, D.K. Nakano, C. Pillen, Extensions for Frobenius kernels, J. Algebra 272 (2004), 476–511.

[BNP5] C.P. Bendel, D.K. Nakano, C. Pillen, Extensions for finite groups of Lie type: Twisted Groups, Finite Groups 2003 (held at the University of Florida), De Gruyter, New York, 2004, 29–46.

[BNP6] C.P. Bendel, D.K. Nakano, C. Pillen, Extensions for finite groups of Lie type II: Filtering the truncated induction functor, Cont. Math. 413 (2006), 1–23.

[BNP7] C.P. Bendel, D.K. Nakano, C. Pillen, Second cohomology for Frobenius kernels and related structures, Advances in Math. 209 (2007), 162–197.

[BNP8] C.P. Bendel, D.K. Nakano, C. Pillen, On the vanishing ranges for the cohomology of finite groups of Lie type, preprint.

[BNP9] C.P. Bendel, D.K. Nakano, C. Pillen, On the vanishing ranges for the cohomology of finite groups of Lie type II, in preparation.

[BNPP] C.P. Bendel, D.K. Nakano, B.J. Parshall, C. Pillen, Quantum group cohomology via the geometry of the nullcone, preprint.

[BE] P.A. Bergh, K. Erdmann, The Avrunin-Scott theorem for quantum complete intersections, J. Algebra 322 (2009), 479–488.

[BKN1] B.D. Boe, J. Kujawa, D.K. Nakano, Cohomology and support varieties

for Lie superalgebras, Transactions of the AMS 362 (2010), 6551–6590.
[BKN2] B.D. Boe, J. Kujawa, D.K. Nakano, Cohomology and support varieties for Lie superalgebras II, Proceedings of the London Math. Soc. 98 (2009), 19–44.
[BKN3] B.D. Boe, J. Kujawa, D.K. Nakano, Complexity and module varieties for classical Lie superalgebras, International Math. Research Notices, doi:10.1093/imrn/rnq090, (2010).
[Ca1] J.F. Carlson, The varieties and cohomology ring of a module, J. Algebra 85 (1983), 104–143.
[Ca2] J.F. Carlson, The variety of an indecomposable module is connected, Invent. Math. 77 (1984), 291–299.
[Ca3] J.F. Carlson, Module varieties and cohomology rings of finite groups, Vorlesungen aus dem Fachbereich Mathematick der Universitat Essen, 1985.
[CLN] J.F. Carlson, Z. Lin, D.K. Nakano, Support varieties for modules over Chevalley groups and classical Lie algebras, Transactions of the AMS 360 (2008), 1879–1906.
[CLNP] J.F. Carlson, Z. Lin, D.K. Nakano, B.J. Parshall, The restricted nullcone, Cont. Math. 325 (2003), 51–73.
[CMN] J.F. Carlson, N. Mazza, D.K. Nakano, Endotrivial groups for finite Chevalley groups, J. Reine Angew. Math. 595 (2006), 93–120.
[Car] R.W. Carter, Finite groups of Lie type, Wiley-Interscience, Chichester, 1985.
[Ch] L. Chastkofsky, Characters of projective indecomposable modules for finite Chevalley groups, Proc. Sympos. Pure Math. 37 (1980), 359–362.
[CPS] E.T. Cline, B.J. Parshall, L.L. Scott, Finite dimensional algebras and highest weight categories, J. Reine Angew. Math. 391, (1988), 85–99.
[CPSvdK] E. Cline, B. Parshall, L. Scott, W. van der Kallen, Rational and generic cohomology, Invent. Math. 39 (1977), 143-163.
[CM] D.H. Collingwood, W.M. McGovern, Nilpotent Orbits in Semisimple Lie Algebras, Van Nostrand Reinhold, New York, 1993.
[EH] K. Erdmann, M. Holloway, Rank varieties and projectivity for a class of local algebras, Math. Zeitschrift 247 (2004), 67–87.
[F1] E.M. Friedlander, Computations of K-theories of finite fields, Topology 15 (1976), 87–109.
[F2] E.M. Friedlander, Weil restriction and support varieties, preprint.
[FP1] E.M. Friedlander, B.J. Parshall, Support varieties for restricted Lie algebras, Invent. Math. 86 (1986), 553–562.
[FP2] E.M. Friedlander, B.J. Parshall, Cohomology of Lie algebras and algebraic groups, American J. Math. 108 (1986), 235–253.
[FP3] E.M. Friedlander, B.J. Parshall, Cohomology of infinitesimal and discrete groups, Math. Ann. 273 (1986), 353–374.
[FPe] E.M. Friedlander, J. Pevtsova, Representation theoretic support spaces for finite group schemes, American J. Math. 127 (2005), 379–420.
[FS] E.M. Friedlander, A. Suslin, Cohomology of finite group schemes over a field, Invent. Math. 127 (1997), no. 2, 209–270.

[GH] R.M. Guralnick, C. Hoffman, The first cohomology group and generation of simple groups, Groups and geometries (Siena, 1996), 81–89, Trends Math., Birkhauser, Basel, 1998.

[GT] R.M. Guralnick, P.H. Tiep, First cohomology groups of Chevalley groups in cross characteristic, preprint.

[H] H.L. Hiller, Cohomology of Chevalley groups over finite fields, J. Pure Appl. Alg. 16 (1980), 259–263.

[Hu1] J.E. Humphreys, Conjugacy Classes in Semisimple Algebraic Groups, Math. Surveys Monographs, Vol. 43, Amer. Math. Soc., Providence, RI, 1995.

[Hu2] J.E. Humphreys, Modular Representations of Finite Groups of Lie Type, London Math. Soc. Lecture Note Series 326 Cambridge University Press, Cambridge, 2005.

[Jan1] J.C. Jantzen, Representations of Algebraic Groups, American Mathematical Society, Mathematical Surveys and Monographs, Vol. 203, 2003.

[Jan2] J.C. Jantzen, Support varieties of Weyl modules, Bull of London Math. Soc. 19 (1987), 238–244.

[Jan3] J.C. Jantzen, First cohomology groups for classical Lie algebras, Progress in Mathematics 95, Birkhauser, 1991, 289–315.

[Jan4] J.C. Jantzen, Low dimensional representations of reductive groups are semisimple, Algebraic Groups and Lie Groups, ed. G.I. Lehrer, Cambridge University Press, 1997, 255–266.

[Jan5] J.C. Jantzen, Nilpotent orbits in representation theory, Lie theory, Progr. Math. 228, 1–211, Birkhuser, Boston, MA, 2004.

[KLT] S. Kumar, N. Lauritzen, J. Thomsen, Frobenius splitting of cotangent bundles of flag varieties, Invent. Math. 136 (1999), 603–621.

[LN1] Z. Lin, D.K. Nakano, Complexity for modules over finite Chevalley groups and classical Lie algebras, Invent. Math. 138 (1999), 85–101.

[LN2] Z. Lin, D.K. Nakano, Projective modules for Chevalley groups and Frobenius kernels, Bulletin of the London Math. Soc. 39 (2007), 1019–1028.

[McG] W. McGovern, Rings of regular functions on nilpotent orbits and their covers, Invent. Math. 97 (1989), 209–217.

[NPal] D.K. Nakano, J.H. Palmieri, Support varieties for the Steenrod algebra, Math. Zeit. 227 (1998), 663–684.

[NPV] D.K. Nakano, B.J. Parshall, D.C. Vella, Support varieties for algebraic groups, J. Reine Angew. Math. 547 (2002), 15–49.

[Ost] V. Ostrik, Support varieties for quantum groups, Funct. Anal. Appl. 32 (1998), 237–246.

[Par] B.J. Parshall, Cohomology of algebraic groups, Proc. Symp. Pure Math. 47 (1987), 233–248.

[Q1] D.G. Quillen, The spectrum of an equivariant cohomology ring I, Ann. Math. 94 (1971), 549–572.

[Q2] D.G. Quillen, The spectrum of an equivariant cohomology ring II, Ann. Math. 94 (1971), 573–602.

[Q3] D.G. Quillen, On the cohomology and K-theory of finite fields, Ann. Math. 96 (1972), 551–586.

[Sc] L.L. Scott, Some new examples in 1-cohomology, J. Algebra 260 (2003), 416–425.

[Sei] G. Seitz, Unipotent elements, tilting modules, and saturation, Invent. Math 141 (2000), 467–502.

[Sin1] P. Sin, Extensions of simple modules for $\mathrm{SP}_4(2^n)$ and $\mathrm{Suz}_3(2^m)$, Bulletin of the London Math. Soc. 24 (1992), 159–164.

[Sin2] P. Sin, Extensions of simple modules for $\mathrm{SL}_3(2^n)$ and $\mathrm{SU}_3(2^n)$, Proceedings of the London Math. Soc. 65 (1992), 265–296.

[Sin3] P. Sin, Extensions of simple modules for $G_2(3^n)$ and ${}^2G_2(3^m)$, Proceedings of the London Math. Soc. 66 (1993), 327–357.

[Sin4] P. Sin, On the 1-cohomology of the groups $G_2(2^n)$, Comm. Algebra 20 (1992), 2653–2662.

[Sin5] P. Sin, The cohomology in degree 1 of the group F_4 in characteristic 2 with coefficients in a simple modules, J. Algebra 164 (1994), 695–717.

[Sm] S.D. Smith, Sheaf homology and complete reducibility, J. Algebra 95 (1985), 72–80.

[SS] N. Snashall, O Solberg, Support varieties and Hochschild cohomology rings, Proceedings of the London Math. Soc. 88 (2004), 705–732.

[SFB1] A. Suslin, E.M. Friedlander, C.P. Bendel, Infinitesimal one-parameter subgroups and cohomology, Journal of the AMS 10 (1997), no. 3, 693–728.

[SFB2] A. Suslin, E.M. Friedlander, C.P. Bendel, Support varieties for infinitesimal group schemes, Journal of the AMS 10 (1997), no. 3, 729–759.

[UGA1] University of Georgia VIGRE Algebra Group, Varieties of nilpotent elements for simple Lie algebras I: Good Primes, J. Algebra 280 (2004), 719–737.

[UGA2] University of Georgia VIGRE Algebra Group, Varieties of nilpotent elements for simple Lie algebras II: Bad Primes, J. Algebra 292 (2005), 65–99.

[UGA3] University of Georgia VIGRE Algebra Group, Support varieties for Weyl modules over bad primes, J. Algebra 312 (2007), 602–633.

[W] C.B. Wright, Second cohomology groups for Frobenius kernels, Ph.D Thesis, University of Georgia, 2008.

Lie Theory and
Representation Theory
SMM 2, pp. 177–219

D-modules and Representation Theory

Toshiyuki Tanisaki*

Abstract

These are notes for a series of lectures given at "Summer school on Lie theory and representation theory II", held at East China Normal University in 2009 July.[1] We present an overview of the theory of D-modules and its application to representations of Lie algebras. Brief accounts of the sheaf theory and the homological algebra are also included so that readers who are not yet familiar with those notions could enjoy the essence of the theory.

2000 Mathematics Subject Classification: 14F10, 32C38, 20G05.

Keywords and Phrases: D-modules, representation theory.

1 Motivation

1.1 Differential equations and D-modules (Sato's Philosophy)

We first explain the original motivation of the D-module theory due to Mikio Sato.

For simplicity we consider the case where the base space X is $\mathbb{C}^n$. We denote by $\mathcal{O}$ the ring of holomorphic functions on X. Let D be the set of differential operators acting on $\mathcal{O}$. Namely, D consists of the elements of $\mathrm{End}_{\mathbb{C}}(\mathcal{O})$ of the form

$$\sum_{i_1,\dots,i_n\geq 0} f_{i_1,\dots,i_n}(x_1,\dots,x_n)\left(\frac{\partial}{\partial x_1}\right)^{i_1}\cdots\left(\frac{\partial}{\partial x_n}\right)^{i_n} \qquad \text{(finite sum)},$$

*Department of Mathematics, Osaka City University, 3-3-138, Sugimoto, Sumiyoshi-ku, Osaka, 558-8585 Japan. The author was partially supported by the Grants-in-Aid for Scientific Research, Scientific Research (B) No. 19340010 and Challenging Exploratory Research No. 21654005 from Japan Society for the Promotion of Science.

[1]The notes are also based on the author's lectures on the same subjects on various occasions at Osaka University, Nagoya University, Kumamoto University, Algebra camp of Seoul National University in Gyeongju, Chinese Academy of Sciences in Beijing, Osaka City University, Seoul National University.

where $(x_1, \dots, x_n)$ is the coordinate of X. In fact D is a subalgebra of $\mathrm{End}_{\mathbb{C}}(\mathcal{O})$. We thus have a non-commutative ring D and a left D-module $\mathcal{O}$.

Consider the differential equation

$$Pu = 0 \tag{1.1}$$

for an unknown function u, where P is a given element of D. According to Sato we should consider the left D-module

$$M = D/DP \tag{1.2}$$

correspondingly. The following observation is crucial.

Observation 1.1. *The vector space consisting of holomorphic solutions to the differential equation* (1.1) *is naturally isomorphic to* $\mathrm{Hom}_D(M, \mathcal{O})$.

Proof. Note

$$\mathrm{Hom}_D(M, \mathcal{O}) = \mathrm{Hom}_D(D/DP, \mathcal{O}) \cong \{\varphi \in \mathrm{Hom}_D(D, \mathcal{O}) \mid \varphi(P) = 0\}.$$

Note also

$$\mathrm{Hom}_D(D, \mathcal{O}) \cong \mathcal{O} \qquad (\psi \leftrightarrow \psi(1)).$$

For $\varphi \in \mathrm{Hom}_D(D, \mathcal{O})$ we have $\varphi(P) = P(\varphi(1))$, and hence we obtain

$$\mathrm{Hom}_D(M, \mathcal{O}) \cong \{f \in \mathcal{O} \mid P(f) = 0\}.$$ □

More generally we have also the following whose proof is left to the readers.

Observation 1.2. *Consider the system*

$$\sum_{j=1}^{\ell} P_{ij} u_j = 0 \qquad (i = 1, \dots, k) \tag{1.3}$$

of linear differential equations for unknown functions $u_1, \dots, u_\ell$, *where* P_{ij} *are given differential operators. Correspondingly, define a left D-module M by*

$$M = D^\ell / \sum_{i=1}^{k} D(P_{i1}, \dots, P_{i\ell}). \tag{1.4}$$

Here

$$D^\ell = \{(Q_1, \dots, Q_\ell) \mid Q_i \in D\},$$

and the left D-module structure of D^ℓ *is given by*

$$R(Q_1, \dots, Q_\ell) = (RQ_1, \dots, RQ_\ell).$$

Then the vector space consisting of holomorphic solutions to the system (1.3) *of differential equations is naturally isomorphic to* $\mathrm{Hom}_D(M, \mathcal{O})$.

By the above observation we can say that the left D-module M in (1.4) contains all information on the system of linear differential equations (1.3). Since k and ℓ in Observation 1.2 are finite, left D-modules M which correspond to some system of differential equations are finitely presented. Namely, there exists an exact sequence

$$D^k \to D^\ell \to M \to 0$$

of left D-modules.

We have seen that the theory of systems of linear differential equations is nothing but the investigation of the contravariant functor

$$\boxed{\begin{array}{c}\{\text{finitely presented left } D\text{-module}\} \to \{\text{complex vector space}\} \\ (M \mapsto \mathrm{Hom}_D(M, \mathcal{O}))\end{array}} \tag{1.5}$$

between abelian categories.

1.2 Language of sheaves

In order to develop the idea explained in Section 1.1 we need to be more careful as seen from the following example.

Example 1.3. Fix $\lambda \in \mathbb{C}$, and consider the differential equation

$$x\frac{du}{dx} = \lambda u \tag{1.6}$$

on $X = \mathbb{C}$. The corresponding D-module is given by $M_\lambda = D/D(x\frac{d}{dx} - \lambda)$. We know that the solution to (1.6) is given by $\mathbb{C}x^\lambda$. For $\lambda \in \mathbb{Z}_{\geq 0}$ the solution x^λ is a holomorphic function on X, and for $\lambda \in \mathbb{Z}_{<0}$ x^λ is a meromorphic function on X with pole at 0. But if λ is not an integer, then x^λ is a multi-valued function given by

$$(re^{i\theta})^\lambda = e^{\lambda \log_e(re^{i\theta})} = e^{\lambda(\log_e(r)+i\theta)}.$$

By analytic continuation we have $(re^{i(\theta+2\pi)})^\lambda = e^{2\pi i\lambda}(re^{i\theta})^\lambda$. It follows that $\mathrm{Hom}_D(M_\lambda, \mathcal{O}) = \mathbb{C}$ for $\lambda \in \mathbb{Z}_{\geq 0}$, and $\mathrm{Hom}_D(M_\lambda, \mathcal{O}) = 0$ otherwise.

This example tells us that the scheme (1.5) is not sufficient to catch local solutions. More precisely, we should take into account of

- how solutions look like locally?

and

- how local solutions are patched together globally?

The language for this purpose is the theory of sheaves (see, for example, Kashiwara-Schapira [23] for a detailed treatment on the theory of sheaves). We consider the sheaf D_X of differential operators on X. Namely, for each open subset U of X we consider the ring $D_X(U)$ of differential operators defined on U. The family of rings $\{D_X(U) \mid U : \text{open subset of } X\}$ are glued together to form a sheaf D_X in the sense that for each pair $U \supset V$ of open subsets of X we are given a ring homomorphism $r_V^U : D_X(U) \to D_X(V)$ obtained by the restriction. They satisfy

(1) $r^V_W \circ r^U_V = r^U_W$ for $U \supset V \supset W$.
(2) If $U = \bigcup_{\lambda\in\Lambda} U_\lambda$ is an open covering of an open subset U of X, then the natural ring homomorphism

$$D_X(U) \to \{(P_\lambda)_{\lambda\in\Lambda} \in \prod_{\lambda\in\Lambda} D_X(U_\lambda) \mid r^{U_\lambda}_{U_\lambda\cap U_\mu}(P_\lambda) = r^{U_\mu}_{U_\lambda\cap U_\mu}(P_\mu)\ (\lambda,\mu\in\Lambda)\},$$
$$(P \mapsto (r^U_{U_\lambda}(P))_{\lambda\in\Lambda})$$

is an isomorphism.

Note that the condition (2) says that a collection of local information which are compatibly patched together determines uniquely a global information.

We will also frequently use the sheaves $\mathcal{O}_X$ and $\mathbb{C}_X$ of rings on X defined similarly from

$$\mathcal{O}_X(U) = \{\text{holomorphic functions on } U\},$$
$$\mathbb{C}_X(U) = \{\text{locally constant } \mathbb{C}\text{-valued functions on } U\}.$$

Definition 1.4. Assume that for each open subset U of X we are given a left $D_X(U)$-module $M(U)$. Assume also that for each pair $U \supset V$ of open subsets of X we are given a homomorphism $\rho^U_V : M(U) \to M(V)$ of $D_X(U)$-modules. We say that the family $M = (\{M(U)\}_U, \{\rho^U_V\}_{U\supset V})$ is a (left) D_X-module, or a sheaf of (left) D_X-modules, if it satisfies

(1) $\rho^V_W \circ \rho^U_V = \rho^U_W$ for $U \supset V \supset W$,
(2) If $U = \bigcup_{\lambda\in\Lambda} U_\lambda$ is an open covering of an open subset U of X, then the natural homomorphism

$$M(U) \to \{(m_\lambda)_{\lambda\in\Lambda} \in \prod_{\lambda\in\Lambda} M(U_\lambda) \mid \rho^{U_\lambda}_{U_\lambda\cap U_\mu}(m_\lambda) = \rho^{U_\mu}_{U_\lambda\cap U_\mu}(m_\mu)\ (\lambda,\mu\in\Lambda)\}$$

is an isomorphism.

Example 1.5. *$\mathcal{O}_X$ is naturally a D_X-module.*

We can similarly define the notion of right D_X-modules, $\mathcal{O}_X$-modules, $\mathbb{C}_X$-modules e.t.c.

Definition 1.6. Let M and N be D_X-modules. Assume that for each open subset U of X we are given $\varphi_U \in \mathrm{Hom}_{D_X(U)}(M(U), N(U))$. Then the collection $\varphi = (\varphi_U)_U$ is called a homomorphism of D_X-modules if the diagram

$$\begin{array}{ccc} M(U) & \xrightarrow{\varphi_U} & N(U) \\ {\scriptstyle\rho^U_V}\downarrow & & \downarrow{\scriptstyle\rho^U_V} \\ M(V) & \xrightarrow[\varphi_V]{} & N(V) \end{array}$$

is commutative for any pair $U \supset V$ of open subsets of X. We denote by $\mathrm{Hom}_{D_X}(M, N)$ the $\mathbb{C}$-module consisting of all homomorphism of D_X-modules from M to N.

By Definition 1.4 and Definition 1.6 we obtain an abelian category $\mathrm{Mod}(D_X)$ consisting of D_X-modules. More generally, for a sheaf $\mathcal{R}$ of rings we denote by $\mathrm{Mod}(\mathcal{R})$ the abelian category of $\mathcal{R}$-modules.

Definition 1.7. (1) We say that a left D_X-module M is finitely presented if there exists an exact sequence in $\mathrm{Mod}(D_X)$ of the form

$$D_X^k \to D_X^\ell \to M \to 0.$$

(2) We say that a left D_X-module M is locally finitely presented if for any $x \in X$ there exists an open neighborhood U of x such that the restriction $M|_U$ of M to U is a finitely presented D_U-module.

(3) We denote by $\mathrm{Mod}_{lfp}(D_X)$ the full subcategory of $\mathrm{Mod}(D_X)$ consisting of locally finitely presented D_X-modules.

We will see later that $\mathrm{Mod}_{lfp}(D_X)$ is an abelian category (see Lemma 4.5 below).

Definition 1.8. Let $M, N \in \mathrm{Mod}(D_X)$. We denote by $\mathcal{H}om_{D_X}(M,N)$ the $\mathbb{C}_X$-module given by

$$(\mathcal{H}om_{D_X}(M,N))\,(U) = \mathrm{Hom}_{D_U}(M|_U, N|_U)$$

for any open subset U of X. Here, $M|_U$ is a D_U-module given by $(M|_U)(V) = M(V)$ for an open subset V of U.

Remark 1.9. The canonical homomorphism

$$(\mathcal{H}om_{D_X}(M,N))\,(U) \to \mathrm{Hom}_{D_X(U)}(M(U), N(U))$$

is not an isomorphism in general.

More generally, for a sheaf $\mathcal{R}$ of rings on X and $M, N \in \mathrm{Mod}(\mathcal{R})$ we define $\mathcal{H}om_{\mathcal{R}}(M,N)$ similarly. It belongs to $\mathrm{Mod}(Z(\mathcal{R}))$, where $Z(\mathcal{R})$ is the center of $\mathcal{R}$.

For $M \in \mathrm{Mod}_{lfp}(D_X)$ we have the "solution sheaf"

$$\mathcal{H}om_{D_X}(M, \mathcal{O}_X) \in \mathrm{Mod}(\mathbb{C}_X),$$

and we obtain a contravariant functor

$$\boxed{\begin{array}{c}\mathrm{Mod}_{lfp}(D_X) \to \mathrm{Mod}(\mathbb{C}_X)\\ (M \mapsto \mathcal{H}om_{D_X}(M, \mathcal{O}_X))\end{array}} \tag{1.7}$$

between abelian categories as a modification of the scheme (1.5).

1.3 Necessity of homological algebra

We still have to point out that the modified scheme (1.7) is not yet satisfactory by the following reason. As a contravariant functor between abelian categories (1.7) is additive, but not exact. Namely, (1.7) does not preserve the exactness of sequences. The functor (1.7) is only left exact in the sense that for any short exact sequence

$$0 \to L \to M \to N \to 0, \tag{1.8}$$

the associated sequence

$$0 \to \mathcal{H}om_{D_X}(N, \mathcal{O}_X) \to \mathcal{H}om_{D_X}(M, \mathcal{O}_X) \to \mathcal{H}om_{D_X}(L, \mathcal{O}_X) \tag{1.9}$$

is exact (but the rightmost arrow is not surjective in general). This causes a trouble in developing the theory since properties of L and N are not directly reflected to M. We can overcome this difficulty by introducing the ideas of homological algebra. In our situation, for $M, N \in \mathrm{Mod}(D_X)$ we have objects

$$\mathcal{E}xt^i_{D_X}(M, N) \in \mathrm{Mod}(\mathbb{C}_X) \qquad (i = 0, 1, 2, \dots)$$

such that

$$\mathcal{E}xt^0_{D_X}(M, N) = \mathcal{H}om_{D_X}(M, N),$$

and (1.9) is extended to a long exact sequence

$$\begin{aligned} 0 &\to \mathcal{H}om_{D_X}(N, \mathcal{O}_X) \to \mathcal{H}om_{D_X}(M, \mathcal{O}_X) \to \mathcal{H}om_{D_X}(L, \mathcal{O}_X) \\ &\to \mathcal{E}xt^1_{D_X}(N, \mathcal{O}_X) \to \mathcal{E}xt^1_{D_X}(M, \mathcal{O}_X) \to \mathcal{E}xt^1_{D_X}(L, \mathcal{O}_X) \\ &\to \mathcal{E}xt^2_{D_X}(N, \mathcal{O}_X) \to \mathcal{E}xt^2_{D_X}(M, \mathcal{O}_X) \to \mathcal{E}xt^2_{D_X}(L, \mathcal{O}_X) \\ &\to \cdots \end{aligned} \tag{1.10}$$

in $\mathrm{Mod}(\mathbb{C}_X)$.

For $M \in \mathrm{Mod}_{lfp}(D_X)$ we have the "higher solution sheaves"

$$\mathcal{E}xt^i_{D_X}(M, \mathcal{O}_X) \in \mathrm{Mod}(\mathbb{C}_X) \qquad (i = 0, 1, 2, \dots),$$

and we should consider the family

$$\boxed{\begin{array}{c} \mathrm{Mod}_{lfp}(D_X) \to \mathrm{Mod}(\mathbb{C}_X) \\ (M \mapsto \mathcal{E}xt^i_{D_X}(M, \mathcal{O}_X)) \end{array}} \qquad (i = 0, 1, 2, \dots) \tag{1.11}$$

of contravariant functors between abelian categories as a modification of the scheme (1.7).

We finally point out that using a more sophisticated language of derived categories the scheme (1.11) is further replaced by a single contravariant functor

$$\boxed{\begin{array}{c} D^b_{lfp}(D_X) \to D^b(\mathbb{C}_X) \\ (M^\bullet \mapsto R\mathcal{H}om_{D_X}(M^\bullet, \mathcal{O}_X)) \end{array}} \tag{1.12}$$

between derived categories (see Section 3 below).

2 Basic concepts

2.1 Ring of differential operators

Let X be a complex manifold or a smooth algebraic variety over $\mathbb{C}$. In the analytic setting where X is a complex manifold we denote by $\mathcal{O}_X$, Θ_X, Ω^1_X, Ω_X the sheaves of holomorphic functions, holomorphic vector fields, holomorphic differential forms of degree 1, and holomorphic differential forms of the top degree respectively. When X is a smooth algebraic variety over $\mathbb{C}$, we denote by the same symbols the corresponding notions in the algebraic setting. In particular, $\mathcal{O}_X$ denotes the sheaf of regular functions. In the algebraic setting we have also the notion of a local coordinate in the following sense.

Fact 2.1. *Let X be a smooth algebraic variety of dimension n. Then for any $x \in X$ we can take an affine open neighborhood U of x, and functions $x_1, \dots, x_n \in \mathcal{O}_X(U)$ such that $\Omega^1_U = \bigoplus_{i=1}^n \mathcal{O}_U dx_i$ and the basis $\{\frac{\partial}{\partial x_i}\}_{i=1}^n$ of Θ_U, which is dual to $\{dx_i\}_{i=1}^n$, satisfies $[\frac{\partial}{\partial x_i}, \frac{\partial}{\partial x_j}] = 0$ $(i, j = 1, \dots, n)$.*

We call $(U, \{x_i\}_{i=1}^n)$ as above the local coordinate of X around x.

Remark 2.2. For a local coordinate $(U, \{x_i\}_{i=1}^n)$ of an algebraic variety X, the map $U \to \mathbb{C}^n$ is not necessarily injective. For example, consider the algebraic variety $X = \mathbb{C} \setminus \{0\} = \operatorname{Spec} \mathbb{C}[x, x^{-1}]$. Then for $m \geq 2$ we have a local coordinate (U, x^m) for which the canonical map $X \to \mathbb{C}$ is not injective.

Definition 2.3. (1st definition of D_X)
We define D_X to be the subring of $\mathcal{E}nd_{\mathbb{C}_X}(\mathcal{O}_X)$ generated by $\mathcal{O}_X$ and Θ_X.

Note that by definition we have

$$\Theta_X = \{\theta \in \mathcal{E}nd_{\mathbb{C}_X}(\mathcal{O}_X) \mid \theta(fg) = \theta(f)g + f\theta(g) \ (f, g \in \mathcal{O}_X)\}.$$

We have regarded $\mathcal{O}_X$ as a subring of $\mathcal{E}nd_{\mathbb{C}_X}(\mathcal{O}_X)$ by the embedding

$$\mathcal{O}_X \ni f \mapsto m_f = [g \to fg] \in \mathcal{E}nd_{\mathbb{C}_X}(\mathcal{O}_X).$$

In terms of a local coordinate $(U, \{x_i\}_{i=1}^n)$ we have

$$D_U = \bigoplus_{I \in \mathbb{Z}^n_{\geq 0}} \mathcal{O}_U \partial^I,$$

where for $I = (i_1, \dots, i_n) \in \mathbb{Z}^n_{\geq 0}$ we set $\partial^I = (\frac{\partial}{\partial x_1})^{i_1} \cdots (\frac{\partial}{\partial x_n})^{i_n}$. We define an increasing filtration

$$0 = \cdots = F_{-2}D_X = F_{-1}D_X \subset F_0 D_X \subset F_1 D_X \subset \cdots \subset D_X \tag{2.1}$$

of D_X by $F_p D_X = \{0\}$ for $p < 0$ and

$$F_p D_X = \{R \in D_X \mid [R, \mathcal{O}_X] \subset F_{p-1} D_X\}$$

for $p \geq 0$. In terms of a local coordinate we have

$$F_p D_U = \bigoplus_{|I| \leq p} \mathcal{O}_U \partial^I,$$

where for $I = (i_1, \dots, i_n) \in \mathbb{Z}^n_{\geq 0}$ we set $|I| = i_1 + \cdots + i_n$. In particular, we have

$$F_0 D_X = \mathcal{O}_X, \quad F_1 D_X = \mathcal{O}_X \oplus \Theta_X, \quad \bigcup_p F_p D_X = D_X.$$

By the last fact we can also define D_X as follows.

Definition 2.4. (2nd definition of D_X)
We define D_X to be the subring of $\mathcal{E}nd_{\mathbb{C}_X}(\mathcal{O}_X)$ defined inductively by

$$D_X = \bigcup_p F_p D_X, \qquad F_p D_X = \{0\} \quad (p < 0),$$

$$F_p D_X = \{R \in D_X \mid [R, \mathcal{O}_X] \subset F_{p-1} D_X\} \quad (p \geq 0).$$

Note that we have the relation

$$\theta f = f\theta + \theta(f) \qquad (f \in \mathcal{O}_X,\, \theta \in \Theta_X) \tag{2.2}$$

in D_X since

$$(\theta \circ m_f)(g) = \theta(fg) = f\theta(g) + \theta(f)g = (m_f \circ \theta + m_{\theta(f)})(g) \qquad (g \in \mathcal{O}_X).$$

Definition 2.5. (3rd definition of D_X)
We define D_X to be the sheaf of rings generated by the symbols $\tilde{f}$ $(f \in \mathcal{O}_X)$, $\tilde{\theta}$ $(\theta \in \Theta_X)$ satisfying the relations:

(1) $\tilde{f} + \tilde{g} = \widetilde{f + g}, \qquad \tilde{f}\tilde{g} = \widetilde{fg} \qquad (f, g \in \mathcal{O}_X)$,
(2) $\tilde{f}\tilde{\theta} = \widetilde{f\theta} \qquad (f \in \mathcal{O}_X, \theta \in \Theta_X)$,
(3) $\tilde{\theta}_1\tilde{\theta}_2 - \tilde{\theta}_2\tilde{\theta}_1 = \widetilde{[\theta_1, \theta_2]} \qquad (\theta_1, \theta_2 \in \Theta_X)$,
(4) $\tilde{\theta}\tilde{f} = \tilde{f}\tilde{\theta} + \widetilde{\theta(f)} \qquad (f \in \mathcal{O}_X, \theta \in \Theta_X)$.

We can check that the 3rd definition is equivalent to the 1st one using local coordinate.

2.2 Left and right D-modules

The following is obvious in view of the 3rd definition of D_X.

Lemma 2.6. *The following are equivalent.*

(1) *M is a left D_X-module.*
(2) *M is an $\mathcal{O}_X$-module equipped with a $\mathbb{C}_X$-linear map*

$$\nabla : \Theta_X \to \mathcal{E}nd_{\mathbb{C}_X}(M) \qquad (\theta \mapsto \nabla_\theta) \tag{2.3}$$

satisfying

(1) $\nabla_{f\theta}(m) = f\nabla_\theta(m) \qquad (f \in \mathcal{O}_X, \theta \in \Theta_X, m \in M)$,
(2) $\nabla_\theta(fm) = \theta(f)m + f\nabla_\theta(m) \qquad (f \in \mathcal{O}_X, \theta \in \Theta_X, m \in M)$,
(3) $\nabla_{[\theta_1,\theta_2]}(m) = [\nabla_{\theta_1}, \nabla_{\theta_2}](m) \qquad (\theta_1, \theta_2 \in \Theta_X, m \in M)$.

Recall that we have an equivalence

$$\{\text{vector bundle on } X\} \cong \{\text{locally free } \mathcal{O}_X\text{-module of finite rank}\} \tag{2.4}$$

of categories associating a vector bundle V on X with the $\mathcal{O}_X$-module $\mathcal{V}$ defined by $\mathcal{V}(U) = \{\text{section of } V \text{ over } U\}$. In the case M is a locally free $\mathcal{O}_X$-module of finite rank, a $\mathbb{C}_X$-linear map (2.3) satisfying (1) and (2) in Lemma 2.6 is called a connection of the corresponding vector bundle. If it also satisfies (3), then it is called an integrable connection (or a flat connection). Hence left D_X-modules are natural extension of the notion of vector bundles with integrable connections.

We have also the following analogue of Lemma 2.6 for right D-modules.

Lemma 2.7. *The following are equivalent.*

(1) *M is a right D_X-module.*
(2) *M is an $\mathcal{O}_X$-module equipped with a $\mathbb{C}_X$-linear map*

$$\nabla' : \Theta_X \to \mathcal{E}nd_{\mathbb{C}_X}(M) \qquad (\theta \mapsto \nabla'_\theta) \tag{2.5}$$

satisfying

(1) $\nabla'_{f\theta}(m) = \nabla'_\theta(fm) \qquad (f \in \mathcal{O}_X, \theta \in \Theta_X, m \in M)$,
(2) $\nabla'_\theta(fm) = \theta(f)m + f\nabla'_\theta(m) \qquad (f \in \mathcal{O}_X, \theta \in \Theta_X, m \in M)$,
(3) $\nabla'_{[\theta_1,\theta_2]}(m) = [\nabla'_{\theta_1}, \nabla'_{\theta_2}](m) \qquad (\theta_1, \theta_2 \in \Theta_X, m \in M)$.

If M is a right D_X-module, then we can regard it as an $\mathcal{O}_X$-module by $f \cdot m = mf$ for $f \in \mathcal{O}_X$, $m \in M$ (note that $\mathcal{O}_X$ is a commutative ring), and we have a $\mathbb{C}_X$-linear map (2.5) defined by $\nabla'_\theta(m) = -m\theta$ for $\theta \in \Theta_X$, $m \in M$.

By Lemma 2.7 we can easily check the following.

Proposition 2.8. *Ω_X is a right D_X-module by*

$$(\nabla'_\theta(\omega))(\theta_1, \dots, \theta_n) = \theta(\omega(\theta_1, \dots, \theta_n)) - \sum_{i=1}^{n} \omega(\theta_1, \dots, [\theta, \theta_i], \dots, \theta_n)$$
$$(\omega \in \Omega_X,\ \theta, \theta_1, \dots, \theta_n \in \Theta_X).$$

In terms of a local coordinate $(U, \{x_i\}_{i=1}^n)$ the right D_X-module structure of Ω_X is described by

$$(f dx_1 \wedge \dots \wedge dx_n) \frac{\partial}{\partial x_i} = -\frac{\partial f}{\partial x_i} dx_1 \wedge \dots \wedge dx_n \qquad (f \in \mathcal{O}_X).$$

If $\mathcal{V}$ is a locally free $\mathcal{O}_X$-module of finite rank, we have an obvious notion the ring of differential operators acting on $\mathcal{V}$. Namely, we have a subring $D_X^{\mathcal{V}}$ of $\mathcal{E}nd_{\mathbb{C}_X}(\mathcal{V})$ consisting of operators locally of the form

$$^t(f_1, \dots, f_r) \mapsto (P_{ij})\ ^t(f_1, \dots, f_r)$$

for some $(P_{ij}) \in M_r(D_X)$ (here, we have identified $\mathcal{V}$ with $\mathcal{O}_X^r$ locally). A more intrinsic description of $D_X^{\mathcal{V}}$ is as follows. We have the identification

$$D_X^{\mathcal{V}} = \mathcal{V} \otimes_{\mathcal{O}_X} D_X \otimes_{\mathcal{O}_X} \mathcal{V}^*,$$

where $\mathcal{V}^* = \mathcal{H}om_{\mathcal{O}_X}(\mathcal{V}, \mathcal{O}_X)$ is the dual of $\mathcal{V}$ (corresponding to the dual vector bundle), and the ring $\mathcal{O}_X$ acts on D_X from the left and the right by the left multiplication and the right multiplication in the ring D_X respectively. The multiplication of $D_X^{\mathcal{V}}$ is given by

$$(v \otimes P \otimes v^*)(w \otimes Q \otimes w^*) = v \otimes P\langle v^*, w\rangle Q \otimes w^* \qquad (v, w \in \mathcal{V}, v^*, w^* \in \mathcal{V}^*, P, Q \in D_X),$$

and the action of $D_X^{\mathcal{V}}$ on $\mathcal{V}$ is given by

$$(v \otimes P \otimes v^*)(w) = P(\langle v^*, w\rangle)v \qquad (v, w \in \mathcal{V}, v^* \in \mathcal{V}^*, P \in D_X).$$

In the case $\mathcal{V}$ is an invertible $\mathcal{O}_X$-module (i.e. locally free $\mathcal{O}_X$-module of rank one), $D_X^{\mathcal{V}}$ is a sheaf of rings on X which is locally isomorphic to D_X, and we call it a TDO (twisted differential operators).

In general for a ring (or a sheaf of rings) R we denote by R^{op} the opposite ring. Namely, we have a canonical bijection $R \leftrightarrow R^{\mathrm{op}}$ $(r \leftrightarrow r^\circ)$ such that $r_1^\circ + r_2^\circ = (r_1 + r_2)^\circ$ and $r_1^\circ r_2^\circ = (r_2 r_1)^\circ$. Then the category of right R-modules is identified with the category $\mathrm{Mod}(R^{\mathrm{op}})$ of left R^{op}-modules.

By our local description of the right D_X-module structure of Ω_X we see easily the following.

Proposition 2.9. *We have*

$$D_X^{\mathrm{op}} \cong D_X^{\Omega_X}$$

as a sheaf of rings on X.

From this we see easily the following.

Theorem 2.10. *We have an equivalence*

$$\mathrm{Mod}(D_X) \cong \mathrm{Mod}(D_X^{\mathrm{op}}) \qquad (M \mapsto \Omega_X \otimes_{\mathcal{O}_X} M)$$

of abelian categories.

In fact we have

$$\mathrm{Mod}(D_X) \cong \mathrm{Mod}(D_X^{\mathcal{V}}) \qquad (M \mapsto \mathcal{V} \otimes_{\mathcal{O}_X} M)$$

for any invertible $\mathcal{O}_X$-module $\mathcal{V}$. The inverse functor is given by

$$\mathrm{Mod}(D_X^{\mathcal{V}}) \cong \mathrm{Mod}(D_X) \qquad (M \mapsto \mathcal{V}^{\otimes -1} \otimes_{\mathcal{O}_X} M).$$

Remark 2.11. *For an invertible $\mathcal{O}_X$-module $\mathcal{V}$ we sometimes write $\mathcal{V}^{\otimes -1} := \mathcal{H}om_{\mathcal{O}_X}(\mathcal{V}, \mathcal{O}_X)$.*

2.3 Inverse images

For a topological space X we denote by $\mathrm{Sh}(X)$ the category of sheaves of abelian groups on X. It is naturally identified with $\mathrm{Mod}(\mathbb{Z}_X)$, where $\mathbb{Z}_X$ denotes the sheaf of rings on X such that

$$\mathbb{Z}_X(U) = \{\text{locally constant } \mathbb{Z}\text{-valued function on } U\}$$

for any open subset U of X. Let $f : X \to Y$ be a continuous map between topological spaces. For $K \in \mathrm{Sh}(Y)$ its inverse image $f^{-1}K \in \mathrm{Sh}(X)$ is defined as the sheaf associated to the presheaf on X

$$U \rightsquigarrow \varinjlim_{V \supset f(U)} K(V) \qquad (U : \text{open subset of } X).$$

This gives an exact functor

$$f^{-1} : \mathrm{Sh}(Y) \to \mathrm{Sh}(X), \tag{2.6}$$

called the inverse image functor. If R is a sheaf of rings on Y, then $f^{-1}R$ is naturally a sheaf of rings on X, and (2.6) induces

$$f^{-1} : \mathrm{Mod}(R) \to \mathrm{Mod}(f^{-1}R). \tag{2.7}$$

Now assume that $f : X \to Y$ is a morphism of complex manifolds or smooth algebraic varieties over $\mathbb{C}$. Then we have the inverse image functor

$$f^* : \mathrm{Mod}(\mathcal{O}_Y) \to \mathrm{Mod}(\mathcal{O}_X) \tag{2.8}$$

for $\mathcal{O}$-modules given by

$$f^*M = \mathcal{O}_X \otimes_{f^{-1}\mathcal{O}_Y} f^{-1}M.$$

Note that (2.8) is right exact, but not exact in general.

The inverse image functor for D-modules is defined in the following manner. Let $M \in \mathrm{Mod}(D_Y)$. By applying (2.8) to M regarded as an $\mathcal{O}_Y$-module we obtain an $\mathcal{O}_X$-module $f^*M = \mathcal{O}_X \otimes_{f^{-1}\mathcal{O}_Y} f^{-1}M$. We want to have an action of Θ_X on f^*M so that f^*M becomes a D_X-module. Note that we have a natural homomorphism

$$\Theta_X \to \mathcal{O}_X \otimes_{f^{-1}\mathcal{O}_Y} f^{-1}\Theta_Y \tag{2.9}$$

of $\mathcal{O}_X$-modules. In terms of local coordinates $(U, \{x_i\}_{i=1}^n)$ and $(V, \{y_j\}_{j=1}^m)$ of X and Y respectively, (2.9) is given by

$$\frac{\partial}{\partial x_i} \mapsto \sum_{j=1}^m \frac{\partial y_j}{\partial x_i} \otimes \frac{\partial}{\partial y_j}.$$

By using (2.9) we obtain the desired action of Θ_X on f^*M as follows. Let $\theta \in \Theta_X$, and assume that (2.9) sends θ to $\sum_j \psi_j \otimes \theta_j$ ($\psi_j \in \mathcal{O}_X, \theta_j \in \Theta_Y$). Then we set

$$\theta \cdot (\varphi \otimes m) = \theta(\varphi) \otimes m + \sum_j \varphi\psi_j \otimes \theta_j m \qquad (\varphi \in \mathcal{O}_X, m \in M).$$

By the 3rd definition of D_X it is easy to check that this action of Θ_X on f^*M gives rise to a D_X-module structure of f^*M. Hence we have obtained a right exact functor

$$f^* : \mathrm{Mod}(D_Y) \to \mathrm{Mod}(D_X). \tag{2.10}$$

Note that D_Y is a left D_Y-module by the left multiplication. Hence applying (2.10) to it we obtain a left D_X-module structure of $f^*D_Y = \mathcal{O}_X \otimes_{f^{-1}\mathcal{O}_Y} f^{-1}D_Y$. On the other hand the right D_Y-module structure of D_Y given by the right multiplication induces a right $f^{-1}D_Y$-module structure of $\mathcal{O}_X \otimes_{f^{-1}\mathcal{O}_Y} f^{-1}D_Y$:

$$(\varphi \otimes P) \cdot Q = \varphi \otimes PQ \qquad (\varphi \in \mathcal{O}_X, P, Q \in D_Y).$$

Hence we obtained a $(D_X, f^{-1}D_Y)$-bimodule

$$D_{X \to Y} := \mathcal{O}_X \otimes_{f^{-1}\mathcal{O}_Y} f^{-1}D_Y. \tag{2.11}$$

In terms of (2.11) the inverse image functor (2.10) can be described by

$$f^*M = D_{X \to Y} \otimes_{f^{-1}D_Y} f^{-1}M. \tag{2.12}$$

2.4 Direct images

Let $f : X \to Y$ be a continuous map between topological spaces. For $K \in \mathrm{Sh}(X)$ its direct image $f_*K \in \mathrm{Sh}(Y)$ is defined by

$$(f_*K)(V) = K(f^{-1}V).$$

This gives a left exact functor

$$f_* : \mathrm{Sh}(X) \to \mathrm{Sh}(Y) \tag{2.13}$$

called the direct image functor. If R is a sheaf of rings on Y, then (2.13) induces a left exact functor

$$f_* : \mathrm{Mod}(f^{-1}R) \to \mathrm{Mod}(R). \tag{2.14}$$

Now assume that $f : X \to Y$ is a morphism of complex manifolds or smooth algebraic varieties over $\mathbb{C}$. We want to construct a direct image functor $\mathrm{Mod}(D_X) \to \mathrm{Mod}(D_Y)$ for D-modules. In fact it is easier to construct such a functor for right D-module. Namely, we have a functor

$$\mathrm{Mod}(D_X^{\mathrm{op}}) \to \mathrm{Mod}(D_Y^{\mathrm{op}}) \qquad (M \rightsquigarrow f_*(M \otimes_{D_X} D_{X \to Y})). \tag{2.15}$$

This functor is a composite of the right exact functor $(\bullet) \otimes_{D_X} D_{X \to Y}$ and the left exact functor f_*. Hence it is neither left exact nor right exact in general. A method to overcome this disadvantage will be given later after introducing the notion of derived categories. Here, we continue naive consideration on the functor (2.15).

Let us define a direct image functor for left D-modules using the correspondence between left D-modules and right D-modules (Theorem 2.10). Namely, we define a functor

$$\mathrm{Mod}(D_X) \to \mathrm{Mod}(D_Y) \tag{2.16}$$

by the commutative diagram

$$\begin{array}{ccc} \mathrm{Mod}(D_X) & \longrightarrow & \mathrm{Mod}(D_Y) \\ {\scriptstyle \Omega_X \otimes_{\mathcal{O}_X} (\bullet)} \downarrow \cong & & \cong \downarrow {\scriptstyle \Omega_Y \otimes_{\mathcal{O}_Y} (\bullet)} \\ \mathrm{Mod}(D_X^{\mathrm{op}}) & \underset{(2.15)}{\longrightarrow} & \mathrm{Mod}(D_Y^{\mathrm{op}}). \end{array}$$

By definition the functor (2.16) sends $M \in \mathrm{Mod}(D_X)$ to

$$f_*((\Omega_X \otimes_{\mathcal{O}_X} M) \otimes_{D_X} D_{X\to Y}) \otimes_{\mathcal{O}_Y} \Omega_Y^{\otimes -1}.$$

Note that

$$\begin{aligned} & f_*((\Omega_X \otimes_{\mathcal{O}_X} M) \otimes_{D_X} D_{X\to Y}) \otimes_{\mathcal{O}_Y} \Omega_Y^{\otimes -1} \\ \cong & f_*((\Omega_X \otimes_{\mathcal{O}_X} M) \otimes_{D_X} D_{X\to Y} \otimes_{f^{-1}\mathcal{O}_Y} f^{-1}\Omega_Y^{\otimes -1}) \\ \cong & f_*((\Omega_X \otimes_{\mathcal{O}_X} D_{X\to Y} \otimes_{f^{-1}\mathcal{O}_Y} f^{-1}\Omega_Y^{\otimes -1}) \otimes_{D_X} M). \end{aligned}$$

Here we used the projection formula

$$f_*(K) \otimes_R N \cong f_*(K \otimes_{f^{-1}R} f^{-1}N)$$

for a locally free R-module N of finite rank in the first isomorphism and the formula

$$(\Omega_X \otimes_{\mathcal{O}_X} M) \otimes_{D_X} N \cong (\Omega_X \otimes_{\mathcal{O}_X} N) \otimes_{D_X} M \quad ((\omega \otimes m) \otimes n \leftrightarrow (\omega \otimes n) \otimes m)$$

for $M, N \in \mathrm{Mod}(D_X)$ in the second isomorphism. Therefore, the functor (2.16) is given by

$$M \rightsquigarrow f_*(D_{Y\leftarrow X} \otimes_{D_X} M), \tag{2.17}$$

where $D_{Y\leftarrow X}$ is an $(f^{-1}D_Y, D_X)$-bimodule given by

$$D_{Y\leftarrow X} = \Omega_X \otimes_{\mathcal{O}_X} D_{X\to Y} \otimes_{f^{-1}\mathcal{O}_Y} f^{-1}\Omega_Y^{\otimes -1}. \tag{2.18}$$

3 Derived category

3.1 Homological algebra

We recall some basics on homological algebra.

Definition 3.1. Let $\mathcal{A}$ and $\mathcal{B}$ be abelian categories and let $F : \mathcal{A} \to \mathcal{B}$ be an additive functor.

(1) We say that F is exact if it satisfies the following equivalent conditions.
   (a) F sends any exact sequence in $\mathcal{A}$ to an exact sequence in $\mathcal{B}$.
   (b) F sends any exact sequence in $\mathcal{A}$ of the form

$$0 \to L \to M \to N \to 0$$

   (a short exact sequence) to an exact sequence in $\mathcal{B}$.

(2) We say that F is left exact if it sends any exact sequence in $\mathcal{A}$ of the form

$$0 \to L \to M \to N$$

to an exact sequence in $\mathcal{B}$.

(3) We say that F is right exact if it sends any exact sequence in $\mathcal{A}$ of the form

$$L \to M \to N \to 0$$

to an exact sequence in $\mathcal{B}$.

Definition 3.2. Let $\mathcal{A}$ an abelian category.

(1) We say that $I \in \mathcal{A}$ is an injective object if the functor $\mathrm{Hom}_{\mathcal{A}}(\bullet, I) : \mathcal{A} \to \{\text{abelian groups}\}$ is exact.
(2) We say that $P \in \mathcal{A}$ is an projective object if the functor $\mathrm{Hom}_{\mathcal{A}}(P, \bullet) : \mathcal{A} \to \{\text{abelian groups}\}$ is exact.
(3) We say that $\mathcal{A}$ has enough injectives if for any $M \in \mathcal{A}$ there exists an injection $M \to I$ such that I is injective.
(4) We say that $\mathcal{A}$ has enough projectives if for any $M \in \mathcal{A}$ there exists an surjection $P \to M$ such that P is projective.

Let $\mathcal{A}$ and $\mathcal{B}$ be abelian categories and let $F : \mathcal{A} \to \mathcal{B}$ be an additive functor. Assume that F is left exact and $\mathcal{A}$ has enough injectives. Then we have a family of functors

$$R^k F : \mathcal{A} \to \mathcal{B} \qquad (k = 0, 1, 2, \ldots), \tag{3.1}$$

called the right derived functors of F, such that $R^0 F = F$, and for any short exact sequence

$$0 \to L \to M \to N \to 0$$

in $\mathcal{A}$ we have a long exact sequence

$$\begin{aligned} 0 &\to R^0F(L) \to R^0F(M) \to R^0F(N) \\ &\to R^1F(L) \to R^1F(M) \to R^1F(N) \\ &\to R^2F(L) \to R^2F(M) \to R^2F(N) \to \cdots \end{aligned} \tag{3.2}$$

in $\mathcal{B}$. (3.1) is defined as follows. Let $M \in \mathcal{A}$. Take an injective resolution $M \to I^\bullet$ of M. Namely, take an exact sequence of the form

$$0 \to M \to I^0 \to I^1 \to \cdots \tag{3.3}$$

in $\mathcal{A}$ such that I^j is injective for any j. Sending the complex

$$I^\bullet = [\cdots \to 0 \to 0 \to I^0 \to I^1 \to I^2 \to \cdots]$$

in $\mathcal{A}$ by F we obtain a complex

$$FI^\bullet = [\cdots \to 0 \to 0 \to FI^0 \to FI^1 \to FI^2 \to \cdots]$$

in $\mathcal{B}$. Then $R^kF(M)$ is defined as the k-th cohomology object

$$H^k(FI^\bullet) = \mathrm{Ker}(FI^k \to FI^{k+1})/\mathrm{Im}(FI^{k-1} \to FI^k)$$

of $FI^\bullet$.

3.2 Derived category

The notion of the derived category was invented in order to define derived functors on the level of the complexes. In this subsection we give a definition of the derived category of an abelian category $\mathcal{A}$ (see, for example, Kashiwara-Schapira [23], for a detailed theory of derived categories).

We start with the category $C(\mathcal{A})$ of complexes in $\mathcal{A}$. An object of $\mathcal{A}$ is a complex

$$M^\bullet = (\{M^i\}_{i\in\mathbb{Z}}, \{d^i_M\}_{i\in\mathbb{Z}})$$

with $M^i \in \mathcal{A}, d^i_M \in \mathrm{Hom}_{\mathcal{A}}(M^i, M^{i+1})$ and $d^{i+1}_M \circ d^i_M = 0$. For $M^\bullet, N^\bullet \in C(\mathcal{A})$ we define morphisms of $C(\mathcal{A})$ by

$$\begin{aligned}&\mathrm{Hom}_{C(\mathcal{A})}(M^\bullet, N^\bullet)\\ &= \{f^\bullet = \{f^i\}_{i\in\mathbb{Z}} \mid f^i \in \mathrm{Hom}_{\mathcal{A}}(M^i, N^i),\ d^i_N \circ f^i = f^{i+1} \circ d^i_M\}.\end{aligned}$$

For each $i \in \mathbb{Z}$ we have a functor

$$H^i : C(\mathcal{A}) \to \mathcal{A} \qquad (M^\bullet \rightsquigarrow \mathrm{Ker}(d^i_M)/\mathrm{Im}(d^{i-1}_M)). \tag{3.4}$$

Next we define the homotopy category $K(\mathcal{A})$ of $\mathcal{A}$ as follows. An object of $K(\mathcal{A})$ is a complex in $\mathcal{A}$. Namely, we have $\mathrm{Ob}(K(\mathcal{A})) = \mathrm{Ob}(C(\mathcal{A}))$. We define morphisms of $K(\mathcal{A})$ by

$$\mathrm{Hom}_{K(\mathcal{A})}(M^\bullet, N^\bullet) = \mathrm{Hom}_{C(\mathcal{A})}(M^\bullet, N^\bullet)/\sim,$$

where $\sim$ is an equivalence relation given by

$$f^\bullet \sim g^\bullet \iff \exists s^i \in \mathrm{Hom}_{\mathcal{A}}(M^i, N^{i-1}) \text{ s.t. } f^i - g^i = d^{i-1}_N \circ s_i + s^{i+1} \circ d^i_M.$$

Then $K(\mathcal{A})$ is naturally an additive category, and (3.4) descends to

$$H^i : K(\mathcal{A}) \to \mathcal{A}. \tag{3.5}$$

We finally define the derived category $D(\mathcal{A})$ of $\mathcal{A}$ as follows. An object of $D(\mathcal{A})$ is again a complex in $\mathcal{A}$. Namely, we have $\mathrm{Ob}(D(\mathcal{A})) = \mathrm{Ob}(K(\mathcal{A})) = \mathrm{Ob}(C(\mathcal{A}))$. Morphisms of $D(\mathcal{A})$ are defined by the localization with respect to the multiplicative system

$$Qis = \{s^\bullet : \text{a morphism of } K(\mathcal{A}) \mid H^i(s^\bullet) \text{ is an isomorphism in } \mathcal{A}\ (\forall i \in \mathbb{Z})\}$$

of morphisms of $K(\mathcal{A})$. Namely,

$$\mathrm{Hom}_{D(\mathcal{A})}(M^\bullet, N^\bullet) = \mathcal{C}(M, N)/\sim,$$

where $\mathcal{C}(M, N)$ consists of triples $(L^\bullet, s^\bullet, f^\bullet)$ such that

$$L^\bullet \in K(\mathcal{A}),\ s^\bullet \in \mathrm{Hom}_{K(\mathcal{A})}(L^\bullet, M^\bullet) \cap Qis,\ f^\bullet \in \mathrm{Hom}_{K(\mathcal{A})}(L^\bullet, N^\bullet),$$

and $\sim$ is the equivalence relation given by $(L^\bullet_1, s^\bullet_1, f^\bullet_1) \sim (L^\bullet_2, s^\bullet_2, f^\bullet_2)$ if there exists $(L^\bullet, s^\bullet, f^\bullet) \in \mathcal{C}(M, N)$, $g_1 : L^\bullet \to L^\bullet_1$, $g_2 : L^\bullet \to L^\bullet_2$ such that $s_1 \circ g_1 = s_2 \circ g_2 = s$

and $f_1 \circ g_1 = f_2 \circ g_2 = f$. Then $D(\mathcal{A})$ is naturally an additive category, and (3.5) descends to

$$H^i : D(\mathcal{A}) \to \mathcal{A}. \tag{3.6}$$

We can identify $\mathcal{A}$ with a subcategory of $D(\mathcal{A})$ by identifying $M \in \mathcal{A}$ with a complex $M^\bullet$ such that $M^0 = M$ and $M^i = 0$ for $i \neq 0$.

For $n \in \mathbb{Z}$ we define a shift functor

$$D(\mathcal{A}) \ni M^\bullet \rightsquigarrow M^\bullet[n] \in D(\mathcal{A}) \qquad (M^\bullet[n])^i = M^{i+n}).$$

The derived category $D(\mathcal{A})$ is not an abelian category, and we do not have the notion of an exact sequence in $D(\mathcal{A})$. Instead we have the notion of a distinguished triangle in $D(\mathcal{A})$ by which $D(\mathcal{A})$ is endowed with a structure of a triangulated category. A triangle in $D(\mathcal{A})$ consists of

$$L^\bullet \xrightarrow{u^\bullet} M^\bullet \xrightarrow{v^\bullet} N^\bullet \xrightarrow{w^\bullet} L^\bullet[1],$$

where $L^\bullet, M^\bullet, N^\bullet$ and $u^\bullet, v^\bullet, w^\bullet$ are objects and morphisms in $D(\mathcal{A})$ respectively. A special class of triangles called distinguished triangles is defined using "mapping cones". Here, we do not give its definition. We only give some properties below.

Fact 3.3. (1) *A short exact sequence*

$$0 \to L \to M \to N \to 0$$

in $\mathcal{A}$ yields a distinguished triangle

$$L \to M \to N \to L[1]$$

in $D(\mathcal{A})$.

(2) *A sequence $L^\bullet \to M^\bullet \to N^\bullet$ of morphisms in $C(\mathcal{A})$ such that*

$$0 \to L^i \to M^i \to N^i \to 0$$

is a short exact sequence in $\mathcal{A}$ for any i yields a distinguished triangle

$$L^\bullet \to M^\bullet \to N^\bullet \to L^\bullet[1]$$

in $D(\mathcal{A})$.

(3) *For a distinguished triangle*

$$L^\bullet \to M^\bullet \to N^\bullet \to L^\bullet[1]$$

in $D(\mathcal{A})$ we have an exact sequence

$$\begin{aligned} \cdots \to H^{i-1}(L^\bullet) &\to H^{i-1}(M^\bullet) \to H^{i-1}(N^\bullet) \\ \to H^i(L^\bullet) &\to H^i(M^\bullet) \to H^i(N^\bullet) \\ \to H^{i+1}(L^\bullet) &\to H^{i+1}(M^\bullet) \to H^{i+1}(N^\bullet) \to \cdots \end{aligned}$$

in $\mathcal{A}$.

We denote by $D^+(\mathcal{A})$ (resp. $D^-(\mathcal{A})$) the full subcategory of $D(\mathcal{A})$ consisting of $M^\bullet \in D(\mathcal{A})$ such that $M^j = 0$ for $j \ll 0$ (resp. $j \gg 0$). We also set $D^b(\mathcal{A}) = D^+(\mathcal{A}) \cap D^-(\mathcal{A})$.

A a full abelian subcategory $\mathcal{A}'$ of $\mathcal{A}$ is called a thick subcategory of $\mathcal{A}$ if it is closed under extensions in $\mathcal{A}$. For a thick subcategory $\mathcal{A}'$ of $\mathcal{A}$ we denote by $D^\sharp_{\mathcal{A}'}(\mathcal{A})$ ($\sharp = \emptyset, +, -, b$) the full subcategory of $D^\sharp(\mathcal{A})$ consisting of $M^\bullet$ such that $H^i(M^\bullet) \in \mathcal{A}'$ for any i.

3.3 Derived functor

Let $F : \mathcal{A} \to \mathcal{B}$ be a left exact functor between abelian categories. We assume that $\mathcal{A}$ has enough injectives. Then we have a functor

$$RF : D^+(\mathcal{A}) \to D^+(\mathcal{B}) \tag{3.7}$$

called the right derived functor of F. It is defined as follows. For $M^\bullet \in D^+(\mathcal{A})$ we can take an isomorphism $M^\bullet \cong I^\bullet$ in $D^+(\mathcal{A})$ such that I^j is an injective object for any $j \in \mathbb{Z}$. Then we set $RF(M^\bullet) := F(I^\bullet)$.

Remark 3.4. Let $M \in \mathcal{A}$. An injective resolution (3.3) of M yields a morphism $M \to I^\bullet$ in $C(\mathcal{A})$ such that $H^i(M) \to H^i(I^\bullet)$ is an isomorphism for any i. Hence, we have an isomorphism $M \to I^\bullet$ in $D^+(\mathcal{A})$. Therefore, we have

$$H^k(RF(M)) = R^kF(M).$$

Fact 3.5. *RF sends a distinguished triangle of $D^+(\mathcal{A})$ to a distinguished triangle of $D^+(\mathcal{B})$.*

Assume that we are given a short exact sequence

$$0 \to L \to M \to N \to 0$$

in $\mathcal{A}$. By Fact 3.3 (1) it yields a distinguished triangle

$$L \to M \to N \to L[1]$$

in $D^+(\mathcal{A})$. By Fact 3.5 we have a distinguished triangle

$$RF(L) \to RF(M) \to RF(N) \to RF(L)[1]$$

in $D^+(\mathcal{B})$. By Fact 3.3 (3) this gives the long exact sequence (3.2).

Remark 3.6. In defining RF we can replace {injective objects} with an F-injective subcategory of $\mathcal{A}$ defined as follows. A full additive subcategory $\mathcal{I}$ of $\mathcal{A}$ is called F-injective if it satisfies the following conditions:

(1) Any $M \in \mathcal{A}$ is a subobject of some $I \in \mathcal{I}$.
(2) Let $0 \to I_1 \to I_2 \to I_3 \to 0$ be a short exact sequence in $\mathcal{A}$. If $I_1, I_2 \in \mathcal{I}$, then we have $I_3 \in \mathcal{I}$.
(3) Let $0 \to I_1 \to I_2 \to I_3 \to 0$ be a short exact sequence in $\mathcal{A}$ such that $I_1, I_2, I_3 \in \mathcal{I}$. Then $0 \to F(I_1) \to F(I_2) \to F(I_3) \to 0$ is exact in $\mathcal{B}$.

Example 3.7. Let R be a sheaf of rings on a topological space X. It is well-known that $\mathrm{Mod}(R)$ has enough injectives. We set

$$D^\sharp(R) := D^\sharp(\mathrm{Mod}(R)) \qquad (\sharp = \emptyset, +, -, b). \tag{3.8}$$

Let $f : X \to Y$ be a continuous map between topological spaces, and let R be a sheaf of rings on Y. We have the right derived functor

$$Rf_* : D^+(f^{-1}R) \to D^+(R)$$

of the left exact functor $f_* : \mathrm{Mod}(f^{-1}R) \to \mathrm{Mod}(R)$. In computing Rf_* we can use the f_*-injective subcategory of $\mathrm{Mod}(f^{-1}R)$ consisting of flabby (flasque) $f^{-1}R$-modules.

Example 3.8. Let $\mathcal{A}$ be an abelian category. For $M, N \in \mathcal{A}$ we have left exact functors

$$\mathrm{Hom}_{\mathcal{A}}(\bullet, N) : \mathcal{A} \to \mathrm{Mod}(\mathbb{Z}), \tag{3.9}$$

$$\mathrm{Hom}_{\mathcal{A}}(M, \bullet) : \mathcal{A} \to \mathrm{Mod}(\mathbb{Z}). \tag{3.10}$$

Note that (3.10) is covariant, while (3.9) is contravariant. We say that an additive contravariant functor $F : \mathcal{A} \to \mathcal{B}$ between abelian categories is left (resp. right) exact if the corresponding covariant functor $\mathcal{A}^{\mathrm{op}} \to \mathcal{B}$ is left (resp. right) exact. Here, $\mathcal{A}^{\mathrm{op}}$ denotes the opposite category of $\mathcal{A}$ given by $\mathrm{Ob}(\mathcal{A}) = \mathrm{Ob}(\mathcal{A}^{\mathrm{op}})$ and $\mathrm{Hom}_{\mathcal{A}^{\mathrm{op}}}(M, N) = \mathrm{Hom}_{\mathcal{A}}(N, N)$. If $\mathcal{A}$ has enough projectives, then $\mathcal{A}^{\mathrm{op}}$ has enough injectives and (3.9) has a right derived functor

$$R^I \mathrm{Hom}_{\mathcal{A}}(\bullet, N) : D^+(\mathcal{A}^{\mathrm{op}})(\cong D^-(\mathcal{A})^{\mathrm{op}}) \to D^+(\mathbb{Z}). \tag{3.11}$$

If $\mathcal{A}$ has enough injectives, then (3.10) has a right derived functor

$$R^{II} \mathrm{Hom}_{\mathcal{A}}(M, \bullet) : D^+(\mathcal{A}) \to D^+(\mathbb{Z}). \tag{3.12}$$

If both exist, we have

$$R^I \mathrm{Hom}_{\mathcal{A}}(M, N) \cong R^{II} \mathrm{Hom}_{\mathcal{A}}(M, N).$$

More generally, if $\mathcal{A}$ has either enough injectives or enough projectives, then we have a bi-functor

$$R\mathrm{Hom}_{\mathcal{A}}(\bullet, \bullet) : D^-(\mathcal{A})^{\mathrm{op}} \times D^+(\mathcal{A}) \to D^+(\mathbb{Z}) \tag{3.13}$$

such that $R\mathrm{Hom}_{\mathcal{A}}(\bullet, N) \cong R^I \mathrm{Hom}_{\mathcal{A}}(\bullet, N)$ if $\mathcal{A}$ has enough projectives, and $R\mathrm{Hom}_{\mathcal{A}}(M, \bullet) \cong R^{II} \mathrm{Hom}_{\mathcal{A}}(M, \bullet)$ if $\mathcal{A}$ has enough injectives.

Example 3.9. Let R be a sheaf of rings on a topological space X. For $M, N \in \mathrm{Mod}(R)$ we have left exact functors

$$\mathcal{H}om_R(\bullet, N) : \mathrm{Mod}(R) \to \mathrm{Mod}(\mathbb{Z}_X), \tag{3.14}$$

$$\mathcal{H}om_R(M, \bullet) : \mathrm{Mod}(R) \to \mathrm{Mod}(\mathbb{Z}_X). \tag{3.15}$$

(3.15) has a right derived functor

$$R\mathcal{H}om_R(M,\bullet) : D^+(R) \to D^+(\mathbb{Z}_X). \tag{3.16}$$

More generally, we have a bi-functor

$$R\mathcal{H}om_R(\bullet,\bullet) : D^-(R)^{\mathrm{op}} \times D^+(R) \to D^+(\mathbb{Z}_X). \tag{3.17}$$

Let us also give an account of the left derived functors of right exact functors.

Let $F : \mathcal{A} \to \mathcal{B}$ be a right exact functor between abelian categories. We assume that $\mathcal{A}$ has enough projectives. Then we have a functor

$$LF : D^-(\mathcal{A}) \to D^-(\mathcal{B}) \tag{3.18}$$

called the right derived functor of F. For $M^\bullet \in D^-(\mathcal{A})$ we can take an isomorphism $P^\bullet \cong M^\bullet$ in $D^-(\mathcal{A})$ such that P^j is a projective object for any $j \in \mathbb{Z}$. Then we set $LF(M^\bullet) := F(P^\bullet)$.

Fact 3.10. *LF sends a distinguished triangle of $D^-(\mathcal{A})$ to a distinguished triangle of $D^-(\mathcal{B})$.*

Remark 3.11. In defining LF we can replace {projective objects} with an F-projective subcategory of $\mathcal{A}$ defined as follows. A full additive subcategory $\mathcal{P}$ of $\mathcal{A}$ is called F-projective if it satisfies the following conditions:

(1) Any $M \in \mathcal{A}$ is a quotient object of some $P \in \mathcal{P}$.
(2) Let $0 \to P_1 \to P_2 \to P_3 \to 0$ be a short exact sequence in $\mathcal{A}$. If $P_2, P_3 \in \mathcal{P}$, then we have $P_1 \in \mathcal{P}$.
(3) Let $0 \to P_1 \to P_2 \to P_3 \to 0$ be a short exact sequence in $\mathcal{A}$ such that $P_1, P_2, P_3 \in \mathcal{P}$. Then $0 \to F(P_1) \to F(P_2) \to F(P_3) \to 0$ is exact in $\mathcal{B}$.

Example 3.12. Let R be a ring. It is well-known that the category $\mathrm{Mod}(R)$ of left R-modules has enough projectives.

Example 3.13. Let R be a sheaf of rings on a topological space X. Then the category $\mathrm{Mod}(R)$ does not have enough projectives in general; however, in defining the left derived functor LF for a right exact functor F defined on $\mathrm{Mod}(R)$, one can often use F-projective resolutions instead of projective resolutions. For example take a right R-module K and consider the right exact functor

$$K \otimes_R (\bullet) : \mathrm{Mod}(R) \to \mathrm{Mod}(\mathbb{Z}_X).$$

Then {flat R-modules} is a $K \otimes_R (\bullet)$-projective subcategory, and hence we can define the left derived functor

$$K \otimes_R^L (\bullet) : D^-(R) \to D^-(\mathbb{Z}_X)$$

of $K \otimes_R (\bullet)$ using flat resolutions.

More generally, we have a bifunctor

$$(\bullet) \otimes_R^L (\bullet) : D^-(R^{\mathrm{op}}) \times D^-(R) \to D^-(\mathbb{Z}_X).$$

3.4 Derived categories for D-modules

Let $f : X \to Y$ be a morphism of complex manifolds or smooth algebraic varieties over $\mathbb{C}$.

We defined the inverse image functor for D-modules as the right exact functor

$$f^* : \mathrm{Mod}(D_Y) \to \mathrm{Mod}(D_X) \qquad (M \rightsquigarrow D_{X\to Y} \otimes_{f^{-1}D_Y} f^{-1}M) \tag{3.19}$$

between abelian categories, and the direct image functor for D-modules as the functor

$$\mathrm{Mod}(D_X) \to \mathrm{Mod}(D_Y) \qquad (M \rightsquigarrow f_*(D_{Y\leftarrow X} \otimes_{D_X} M)) \tag{3.20}$$

between abelian categories. But what are theoretically more important are the functors

$$Lf^* : D^-(D_Y) \to D^-(D_X) \qquad (M^\bullet \rightsquigarrow D_{X\to Y} \otimes^L_{f^{-1}D_Y} f^{-1}M^\bullet), \tag{3.21}$$

$$D_{Y\leftarrow X} \otimes^L_{D_X} (\bullet) : D^-(D_X) \to D^-(f^{-1}D_Y) \qquad (M^\bullet \rightsquigarrow D_{Y\leftarrow X} \otimes^L_{D_X} M^\bullet), \tag{3.22}$$

$$Rf_* : D^+(f^{-1}D_Y) \to D^+(D_Y) \qquad (N^\bullet \rightsquigarrow Rf_*(N^\bullet)) \tag{3.23}$$

between derived categories. It can be shown that (3.21), (3.22), (3.23) induce functors

$$Lf^* : D^b(D_Y) \to D^b(D_X), \tag{3.24}$$

$$\int_f : D^b(D_X) \to D^b(D_Y) \tag{3.25}$$

between bounded derived categories. In the rest of this note we call (3.24) and (3.25) the inverse image functor and the direct image functor for D-modules respectively.

4 Coherent D-modules

4.1 Coherent modules

Definition 4.1. Let R be a sheaf of rings on a topological space X. We say that $M \in \mathrm{Mod}(R)$ is coherent if it satisfies the following conditions:

(1) for any $x \in X$ there exists an open neighborhood U of X, $m < \infty$, and an exact sequence of the form $(R|_U)^m \to M|_U \to 0$ in $\mathrm{Mod}(R|_U)$,

(2) assume that we are given an open subset U of X, $m < \infty$, and a homomorphism $\varphi : (R|_U)^m \to M|_U$ in $\mathrm{Mod}(R|_U)$. Then for any $x \in U$ there exists an open neighborhood V of x in U and a homomorphism $\psi : (R|_V)^\ell \to (R|_V)^m$ in $\mathrm{Mod}(R|_V)$ such that $(R|_V)^\ell \to (R|_V)^m \to M|_V$ is an exact sequence.

Notation 4.2. For a sheaf of rings R on a topological space X we denote by $\mathrm{Mod}_c(R)$ the full subcategory of $\mathrm{Mod}(R)$ consisting of coherent R-modules.

Lemma 4.3. *For a sheaf of rings R on a topological space X $\mathrm{Mod}_c(R)$ is a thick subcategory of $\mathrm{Mod}(R)$.*

Definition 4.4. A sheaf of rings R on a topological space X is called coherent if $R \in \mathrm{Mod}_c(R)$.

Lemma 4.5. *Assume that R is a coherent sheaf of rings on a topological space X. Then the following conditions for $M \in \mathrm{Mod}(R)$ are equivalent.*

(1) $M \in \mathrm{Mod}_c(R)$.
(2) *M is locally finitely presented, i.e., for any $x \in X$ there exists an open neighborhood U of x, $m, n < \infty$, and an exact sequence of the form*

$$(R|_U)^n \to (R|_U)^m \to M|_U \to 0.$$

4.2 D_X is coherent

Let X be a complex manifold or a smooth algebraic variety over $\mathbb{C}$.

Theorem 4.6. *D_X is a coherent sheaf of rings.*

This is a consequence of the following.

Theorem 4.7. *$\mathcal{O}_X$ is a coherent sheaf of rings.*

Theorem 4.7 in the algebraic setting is not difficult; however, that in the analytic setting is a difficult result of K. Oka.

We set

$$D_c^b(D_X) := D^b_{\mathrm{Mod}_c(D_X)}(\mathrm{Mod}(D_X)). \tag{4.1}$$

4.3 Solution functor and de Rham functor

Definition 4.8. For a complex manifold X we define functors

$$\mathrm{Sol} : D_c^b(D_X) \to D^b(\mathbb{C}_X)^{\mathrm{op}}, \tag{4.2}$$
$$DR : D_c^b(D_X) \to D^b(\mathbb{C}_X) \tag{4.3}$$

by

$$\mathrm{Sol}(M^\bullet) = R\mathcal{H}om_{D_X}(M^\bullet, \mathcal{O}_X), \qquad DR(M^\bullet) = R\mathcal{H}om_{D_X}(\mathcal{O}_X, M^\bullet). \tag{4.4}$$

We call (4.2) and (4.3) the solution functor and the de Rham functor respectively.

In the algebraic setting where X is a smooth algebraic variety over $\mathbb{C}$, it is not a good idea to define the solution functor and the de Rham functor as in (4.4) since $\mathcal{H}om_{D_X}(M, \mathcal{O}_X)$ only takes solutions of the corresponding differential equation by the algebraic functions. For example, hypergeometric differential equation is defined algebraically, while its solution (hypergeometric function) is a non-algebraic holomorphic function. This example suggests us that even when we consider algebraic D-modules, the solutions should be taken in the analytic category. So we proceed as follows. Let X be a smooth algebraic variety over $\mathbb{C}$. We denote by X_{an} the associated complex manifold with classical topology. We have a continuous map $\iota : X_{\mathrm{an}} \to X$ and a ring homomorphism $\iota^{-1}\mathcal{O}_X \to \mathcal{O}_{X_{\mathrm{an}}}$. Then

$$D_{X_{\mathrm{an}}} = \mathcal{O}_{X_{\mathrm{an}}} \otimes_{\iota^{-1}\mathcal{O}_X} \iota^{-1}D_X = \iota^{-1}D_X \otimes_{\iota^{-1}\mathcal{O}_X} \mathcal{O}_{X_{\mathrm{an}}},$$

and hence we have a ring homomorphism $\iota^{-1}D_X \to D_{X_{\mathrm{an}}}$. The exact functor

$$\mathrm{Mod}(D_X) \to \mathrm{Mod}(D_{X_{\mathrm{an}}}) \qquad (M \rightsquigarrow M^{\mathrm{an}} := D_{X_{\mathrm{an}}} \otimes_{\iota^{-1}D_X} \iota^{-1}M) \tag{4.5}$$

induces a functor

$$D^b_c(D_X) \to D^b_c(D_{X_{\mathrm{an}}}) \qquad (M^\bullet \rightsquigarrow M^{\bullet\mathrm{an}} = D_{X_{\mathrm{an}}} \otimes_{\iota^{-1}D_X} \iota^{-1}M^\bullet) \tag{4.6}$$

between derived categories.

Definition 4.9. For a smooth algebraic variety X over $\mathbb{C}$ we define functors

$$\mathrm{Sol} : D^b_c(D_X) \to D^b(\mathbb{C}_{X_{\mathrm{an}}})^{\mathrm{op}}, \tag{4.7}$$
$$DR : D^b_c(D_X) \to D^b(\mathbb{C}_{X_{\mathrm{an}}}) \tag{4.8}$$

by

$$\mathrm{Sol}(M^\bullet) = \mathrm{Sol}(M^{\bullet\mathrm{an}}), \qquad DR(M^\bullet) = DR(M^{\bullet\mathrm{an}}). \tag{4.9}$$

4.4 Good filtrations and characteristic varieties

Recall that D_X has a canonical filtration (2.1). It satisfies

$$1 \in F_0D_X, \tag{4.10}$$
$$(F_pD_X)(F_qD_X) \subset F_{p+q}D_X \qquad (p, q \in \mathbb{Z}), \tag{4.11}$$
$$[F_pD_X, F_qD_X] \subset F_{p+q-1}D_X \qquad (p, q \in \mathbb{Z}). \tag{4.12}$$

Set

$$\mathrm{gr}D_X = \bigoplus_{p\in\mathbb{Z}} \mathrm{gr}_pD_X, \qquad \mathrm{gr}_pD_X = F_pD_X/F_{p-1}D_X, \tag{4.13}$$

and denote by $\sigma_p : F_pD_X \to \mathrm{gr}_pD_X$ the canonical morphism. By (4.10), (4.11), (4.12) $\mathrm{gr}D_X$ has a structure of a commutative sheaf of rings on X by $\sigma_p(P)\sigma_q(Q) = \sigma_{p+q}(PQ)$. Take a local coordinate $(U, \{x_i\}_{i=1}^n)$ of X. Then we have $D_U = \bigoplus_I \mathcal{O}_U\partial^I$, $F_pD_U = \bigoplus_{|I|\le p} \mathcal{O}_U\partial^I$, and hence

$$\mathrm{gr}D_U = \bigoplus_{|I|\le p} \mathcal{O}_U\xi^I = \mathcal{O}_U[\xi_1, \dots, \xi_n],$$

where $\xi_i = \sigma_1(\frac{\partial}{\partial x_i}) \in \mathrm{gr}_1D_U$, and $\xi^I = \xi_1^{i_1} \cdots \xi_n^{i_n}$ for $I = (i_1, \dots, i_n)$. Namely, $\mathrm{gr}D_U$ consists of functions on $U \times \mathbb{C}^n$, where $(\xi_1, \dots, \xi_n)$ is regarded as a coordinate of $\mathbb{C}^n$. In a more intrinsic language we can say that

"$\mathrm{gr}D_X$ consists of functions on the cotangent bundle T^*X." (4.14)

Namely, $(\xi_1, \dots, \xi_n)$ is naturally regarded as a coordinate of the fibers of the cotangent bundle (this is obvious from $\xi_i = \sigma_1(\frac{\partial}{\partial x_i})$). For

$$P = \sum_{|I| \le p} f_I(x)\partial^I \in F_p D_X$$

we call

$$\sigma_p(P) = \sum_{|I|=p} f_I(x)\xi^I \in \mathrm{gr}_p D_X,$$

which is a homogeneous function on the cotangent bundle T^*X, the principal symbol of the differential operator P. In general, if X is a complex manifold or a smooth algebraic variety over $\mathbb{C}$, we denote by $\pi_X : T^*X \to X$ the cotangent bundle of X. A sheaf theoretic restatement of (4.14) is the following.

Lemma 4.10. *In the algebraic setting we have* $\mathrm{gr} D_X \cong (\pi_X)_* \mathcal{O}_{T^*X}$, *and in the analytic setting we have* $\mathrm{gr} D_X \subset (\pi_X)_* \mathcal{O}_{T^*X}$,

Now let us consider filtrations of D-modules.

Definition 4.11. Let $M \in \mathrm{Mod}(D_X)$. A filtration

$$\cdots \subset F_{-2}M \subset F_{-1}M \subset F_0 M \subset F_1 M \subset F_2 M \subset \cdots \subset M$$

of M by $\mathcal{O}_X$-submodules is called a good filtration of M if it satisfies the following conditions.

(1) $(F_p D_X)(F_q M) \subset F_{p+q} M$,
(2) $F_p M = 0$ for $p \ll 0$,
(3) $\bigcup_p F_p M = M$,
(4) $(F_p D_X)(F_q M) = F_{p+q} M$ for $p \ge 0, q \gg 0$.

Let M be a coherent D_X-module. Then locally on X there exists a coherent $\mathcal{O}_X$-submodule M_0 of M such that $M = D_X M_0$. Then we obtain a good filtration of M by setting $F_p M = (F_p D_X) M_0$. It follows that any coherent D_X-module admits locally a good filtration. Conversely, we can show that if a D_X-module admits a good filtration, then it is coherent.

Remark 4.12. In the algebraic setting it is known that any coherent D_X-module admits a global good filtration.

For a coherent D_X-module M equipped with a good filtration $F = \{F_p M\}_{p \in \mathbb{Z}}$ we set

$$\mathrm{gr}^F M = \bigoplus_p \mathrm{gr}_p^F M, \qquad \mathrm{gr}_p^F M = F_p M / F_{p-1} M,$$

and denote by $\tau_p : F_p M \to \mathrm{gr}_p^F M$ the canonical morphism. Then $\mathrm{gr}^F M$ is naturally a $\mathrm{gr} D_X$-module by $\sigma_p(P)\tau_q(m) = \tau_{p+q}(Pm)$. Hence we obtain a coherent $\mathcal{O}_{T^*X}$-module

$$\widetilde{\mathrm{gr}}^F M := \mathcal{O}_{T^*X} \otimes_{\pi_X^{-1}\mathrm{gr} D_X} \pi_X^{-1} \mathrm{gr}^F M.$$

Fact 4.13. *For a coherent D_X-module M the support* $\mathrm{Supp}(\widetilde{\mathrm{gr}}^F M)$ *of* $\widetilde{\mathrm{gr}}^F M$ *does not depend on the choice of a good filtration F of M.*

Definition 4.14. For a coherent D_X-module M we define a closed conic (i.e. stable under the action of $\mathbb{C}^\times$ on fibers) subvariety $\mathrm{Ch}(M)$ of T^*X by

$$\mathrm{Ch}(M) = \mathrm{Supp}(\widetilde{\mathrm{gr}}^F M),$$

where F is a good filtration of M. We call $\mathrm{Ch}(M)$ the characteristic variety of M.

The above definition is well-defined by Fact 4.13 and the local existence of a good filtration.

Lemma 4.15. *For $M \in \mathrm{Mod}_c(D_X)$ we have $\mathrm{Ch}(M) = \emptyset$ if and only if $M = 0$.*

Lemma 4.16. *For an exact sequence*

$$0 \to M_1 \to M_2 \to M_3 \to 0$$

in $\mathrm{Mod}_c(D_X)$ we have $\mathrm{Ch}(M_2) = \mathrm{Ch}(M_1) \cup \mathrm{Ch}(M_3)$.

For a closed submanifold Y of X we denote by $T^*_Y X$ the conormal bundle of Y, i.e.,

$$T^*_Y X = \{(y, \xi) \in T^*X \mid y \in Y, \langle \xi, TY \rangle = \{0\}\}.$$

In particular, $T^*_X X$ denotes the zero-section of the cotangent bundle T^*X. We give some examples below.

Example 4.17. Assume that $M \in \mathrm{Mod}_c(D_X)$ is generated by a single section $u \in \Gamma(X, M)$. Then we have

$$M = D_X u \cong D_X / I,$$

where I is a locally finitely generated left ideal of D_X. We have a good filtration F of M given by $F_p M = (F_p D_X) u$, for which

$$\mathrm{gr}^F M = \mathrm{gr} D_X / \mathrm{gr}\, I, \qquad \mathrm{gr}\, I = \bigoplus_p \{\sigma_p(P) \mid P \in I \cap F_p D_X\} \subset \mathrm{gr} D_X.$$

Hence we have

$$\mathrm{Ch}(M) = \{y \in T^*X \mid \varphi(y) = 0 \ (\varphi \in \mathrm{gr}\, I)\}.$$

Assume that $I = \sum_{k=1}^r D_X P_k$ for $P_k \in F_{p_k} D_X \setminus F_{p_k - 1} D_X$. The we have obviously

$$\mathrm{gr}\, I \supset \sum_{k=1}^r (\mathrm{gr} D_X)\, \sigma_{p_k}(P_k),$$

and hence

$$\mathrm{Ch}(M) \subset \{y \in T^*X \mid (\sigma_{p_k}(P_k))(y) = 0 \ (k = 1, \dots, r)\}. \tag{4.15}$$

We have the equality in (4.15) in the case $r = 1$, but not in general. In conclusion, we can say that the characteristic variety of the coherent D-module corresponding to the differential equation

$$P_k(u) = 0 \qquad (k = 1, \cdots, r)$$

for a single unknown function u is "roughly" estimated by the common zero set of the principal symbols $\sigma_{p_k}(P_k)$ $(k = 1, \ldots, r)$.

Example 4.18. We consider the case $X = \mathbb{C}$ with canonical coordinate (x). Set $\partial = \frac{d}{dx}$ and $\xi = \sigma_1(\partial)$. Then we have

$$T^*X = \{(x, \xi) \mid x, \xi \in \mathbb{C}\} \cong \mathbb{C}^2.$$

We have $\mathcal{O}_X \cong D_X/D_X\partial$ $(1 \leftrightarrow \overline{1})$, and hence

$$\mathrm{Ch}(\mathcal{O}_X) = \{(x, \xi) \mid \xi = 0\} = T^*_X X \tag{4.16}$$

by Example 4.17. Regard $\mathcal{O}_X[x^{-1}]$ as a D_X-module via the natural action of differential operators on meromorphic functions. Then we have $\mathcal{O}_X[x^{-1}] \cong D_X/D_X(x\partial + 1)$ $(x^{-1} \leftrightarrow \overline{1})$. By $\sigma_1(x\partial + 1) = x\xi$ we have

$$\mathrm{Ch}(\mathcal{O}_X[x^{-1}]) = \{(x, \xi) \mid x\xi = 0\} = T^*_X X \cup T^*_{\{0\}}X \tag{4.17}$$

by Example 4.17. Set $\mathcal{B} = \mathcal{O}_X[x^{-1}]/\mathcal{O}_X$. This is a D_X-module supported on a single point $\{0\}$. We see easily that $\mathcal{B} \cong D_X/D_X x$ $(\overline{x^{-1}} \leftrightarrow \overline{1})$, and hence

$$\mathrm{Ch}(\mathcal{B}) = \{(x, \xi) \mid x = 0\} = T^*_{\{0\}}X \tag{4.18}$$

by Example 4.17. $\mathcal{B}$ is a coherent D_X-module corresponding to the differential equation $x\delta = 0$ satisfied by the delta function δ supported by the origin $\{0\}$. Note that we have also

$$\mathcal{B} \cong H^0\Big(\int_i \mathcal{O}_{\{0\}}\Big) \qquad (i : \{0\} \hookrightarrow X).$$

For a closed submanifold Z of X we define a D_X-module $\mathcal{B}_{Z|X}$ by

$$\mathcal{B}_{Z|X} = H^0\Big(\int_i \mathcal{O}_Z\Big) \qquad (i : Z \hookrightarrow X)$$

(we have $H^k(\int_i \mathcal{O}_Z) = 0$ for $k \neq 0$). In terms of a local coordinate $(U, \{x_i\}_{i=1}^n)$ such that $Z = \{x_{r+1} = \cdots = x_n = 0\}$ we have

$$(\mathcal{B}_{Z|X})|_U \cong D_U \Big/ \left(\sum_{i=1}^{r} D_U \frac{\partial}{\partial x_i} + \sum_{j=r+1}^{n} D_U x_j \right),$$

and hence $\mathcal{B}_{Z|X}$ is a coherent D_X-module corresponding to the differential equation satisfied by the delta function supported by Z.

Lemma 4.19. *For a closed submanifold Z of X we have* $\mathrm{Ch}(\mathcal{B}_{Z|X}) = T^*_Z X$.

4.5 Integrable connections

Proposition 4.20. *The following conditions on a coherent D_X-module M are equivalent.*

(1) *M is coherent over $\mathcal{O}_X$.*
(2) *M is locally free of finite rank over $\mathcal{O}_X$.*
(3) $\mathrm{Ch}(M) = T^*_X X$ *or* $\mathrm{Ch}(M) = \emptyset$.

Definition 4.21. We call a D_X-module M satisfying the conditions in Proposition 4.20 an integrable connection. We denote by $\mathrm{Conn}(X)$ the category of integrable connections on X.

In general, for a left D_X-module M, its $\mathcal{O}_X$-dual $\mathcal{H}om_{\mathcal{O}_X}(M, \mathcal{O}_X)$ is endowed with a structure of a left D_X-module by

$$(\theta \cdot \varphi)(m) = -\varphi(\theta m) \qquad (\varphi \in \mathcal{H}om_{\mathcal{O}_X}(M, \mathcal{O}_X), \theta \in \Theta_X, m \in M).$$

On $\mathrm{Conn}(X)$ this defines a contravariant exact functor

$$\mathrm{Conn}(X) \to \mathrm{Conn}(X) \qquad (M \rightsquigarrow M^* = \mathcal{H}om_{\mathcal{O}_X}(M, \mathcal{O}_X)) \tag{4.19}$$

satisfying $M^{**} \cong M$.

In general for left D_X-modules M, N their tensor product $M \otimes_{\mathcal{O}_X} N$ over $\mathcal{O}_X$ is endowed with a structure of a left D_X-module by

$$\theta \cdot (m \otimes n) = \theta m \otimes n + m \otimes \theta n, \qquad (\theta \in \Theta_X, m \in M, n \in N).$$

On $\mathrm{Conn}(X)$ this defines an exact bi-functor

$$(\bullet) \otimes_{\mathcal{O}_X} (\bullet) : \mathrm{Conn}(X) \times \mathrm{Conn}(X) \to \mathrm{Conn}(X). \tag{4.20}$$

Let X be a complex manifold. We consider $\mathrm{Sol}(M)$ and $DR(M)$ for $M \in \mathrm{Conn}(X)$. For $M \in \mathrm{Conn}(X)$ and $N \in \mathrm{Mod}(D_X)$ we have

$$\mathcal{H}om_{D_X}(M, N) \cong \mathcal{H}om_{D_X}(\mathcal{O}_X, M^* \otimes_{\mathcal{O}_X} N),$$

and hence

$$\mathrm{Sol}(M) = R\mathcal{H}om_{D_X}(M, \mathcal{O}_X) \cong R\mathcal{H}om_{D_X}(\mathcal{O}_X, M^*) = DR(M^*) \tag{4.21}$$

for $M \in \mathrm{Conn}(X)$. Hence we have only to consider $DR(M)$. Since M is a locally free $\mathcal{O}_X$-module of finite rank, we can take a local coordinate $(U, \{x_i\}_{i=1}^n)$ such that $M|_U \cong (\mathcal{O}_U)^r$ as an $\mathcal{O}_U$-module. Fix an $\mathcal{O}_U$-basis $u_1, \dots, u_r$ of $M|_U$. Write

$$\frac{\partial}{\partial x_i} u_\ell = \sum_{k=1}^r a^i_{kl} u_k \qquad (a^i_{k\ell} \in \mathcal{O}_U) \tag{4.22}$$

and set $A^i = (a^i_{k\ell}) \in M_r(\mathcal{O}_U)$. Note that

$$\begin{aligned}\frac{\partial}{\partial x_i}\frac{\partial}{\partial x_j} u_m &= \frac{\partial}{\partial x_i}\left(\sum_{\ell=1}^r a^j_{\ell m} u_\ell\right) = \sum_{\ell=1}^r \frac{\partial a^j_{\ell m}}{\partial x_i} u_\ell + \sum_{k,\ell=1}^r a^j_{\ell m} a^i_{kl} u_k \\ &= \sum_{\ell=1}^r \left(\frac{\partial a^j_{\ell m}}{\partial x_i} + \sum_{k=1}^m a^i_{\ell k} a^j_{km}\right) u_\ell.\end{aligned}$$

By $\frac{\partial}{\partial x_i}\frac{\partial}{\partial x_j}u_m = \frac{\partial}{\partial x_j}\frac{\partial}{\partial x_i}u_m$ we have

$$\frac{\partial a^j_{\ell m}}{\partial x_i} + \sum_{k=1}^{m} a^i_{\ell k} a^j_{km} = \frac{\partial a^i_{\ell m}}{\partial x_j} + \sum_{k=1}^{m} a^j_{\ell k} a^i_{km}.$$

In other words we have

$$[A^i, A^j] = \frac{\partial A^i}{\partial x_j} - \frac{\partial A^j}{\partial x_i} \qquad (i, j = 1, \dots, r). \tag{4.23}$$

Conversely, a free $\mathcal{O}_U$-module with basis $u_1, \dots, u_r$ is endowed with a structure of a D_U-module by (4.22) if (4.23) is satisfied. Now let us consider $H^0(DR(M|_U))$. We have

$$\begin{aligned} H^0(DR(M|_U)) &= \mathcal{H}om_{D_U}(\mathcal{O}_U, M|_U) \\ &\cong \{m \in M|_U \mid \frac{\partial}{\partial x_i} m = 0 \ (i = 1, \dots, n)\} \\ &\cong \{f = {}^t(f_1, \dots, f_r) \in (\mathcal{O}_U)^r \mid \frac{\partial}{\partial x_i} f = -Af \ (i = 1, \dots, n)\}. \end{aligned}$$

By (4.23) and the classical Frobenius theorem on Pfaff systems the local holomorphic solutions to the system of differential equation:

$$\frac{\partial}{\partial x_i} f = -Af \qquad (i = 1, \dots, n)$$

for $f = {}^t(f_1, \dots, f_r)$ on a connected simply-connected open subset V of U forms an r-dimensional $\mathbb{C}$-subspace W of $\mathcal{O}(V)^r$ such that $\mathcal{O}(V)^r = \mathcal{O}(V) \otimes_{\mathbb{C}} W$. In conclusion we have the following.

- Let $M \in \mathrm{Conn}(X)$ of rank r. Then $L = H^0(DR(M))$ is identified with a $\mathbb{C}_X$-submodule of M given by $\{m \in M \mid \Theta_X m = \{0\}\}$. Moreover, L is a local system of rank r satisfying $M = \mathcal{O}_X \otimes_{\mathbb{C}_X} L$.

Definition 4.22. Let X be a complex manifold. We call a locally free $\mathbb{C}_X$-module of rank $r < \infty$ a local system of rank r. We denote by $\mathrm{Loc}(X)$ the category of local systems on X.

Now let us give a sketch of the proof of $H^i(DR(M)) = 0$ for $i > 0$. By $M = \mathcal{O}_X \otimes_{\mathbb{C}_X} L$ we have

$$\begin{aligned} DR(M) =& R\mathcal{H}om_{D_X}(\mathcal{O}_X, M) = R\mathcal{H}om_{D_X}(\mathcal{O}_X, \mathcal{O}_X \otimes_{\mathbb{C}_X} L) \\ \cong& R\mathcal{H}om_{D_X}(\mathcal{O}_X, \mathcal{O}_X) \otimes_{\mathbb{C}_X} L = DR(\mathcal{O}_X) \otimes_{\mathbb{C}_X} L, \end{aligned}$$

and hence it is sufficient to show $H^i(DR(\mathcal{O}_X)) = 0$ for $i > 0$. Using a locally free resolution of the left D_X-module $\mathcal{O}_X$ of the form

$$0 \to D_X \otimes_{\mathcal{O}_X} \bigwedge^n \Theta_X \to \cdots \to D_X \otimes_{\mathcal{O}_X} \bigwedge^0 \Theta_X \to \mathcal{O}_X \to 0$$

we see that

$$\begin{aligned} R\mathcal{H}om_{D_X}(\mathcal{O}_X, \mathcal{O}_X) &\cong \mathcal{H}om_{D_X}(D_X \otimes_{\mathcal{O}_X} \bigwedge^{\bullet} \Theta_X, \mathcal{O}_X) \\ &\cong \mathcal{H}om_{\mathcal{O}_X}(\bigwedge^{\bullet} \Theta_X, \mathcal{O}_X) \cong \bigwedge^{\bullet} \Omega_X^1 \cong \mathbb{C}_X, \end{aligned}$$

where the last isomorphism is the holomorphic Poincaré lemma. By the above argument we have the following result.

Proposition 4.23. *Let X be a complex manifold. Then we have an equivalence*

$$\mathrm{Conn}(X) \cong \mathrm{Loc}(X) \qquad (M \rightsquigarrow DR(M)). \tag{4.24}$$

The inverse of (4.24) is given by $L \rightsquigarrow \mathcal{O}_X \otimes_{\mathbb{C}_X} L$.
We define the duality functor on $\mathrm{Loc}(X)$ by

$$\mathrm{Loc}(X) \to \mathrm{Loc}(X) \qquad (L \rightsquigarrow L^* = \mathcal{H}om_{\mathbb{C}_X}(L, \mathbb{C}_X)) \tag{4.25}$$

This is a contravariant exact functor satisfying $L^{**} \cong L$. By

$$(\mathcal{O}_X \otimes_{\mathbb{C}_X} L)^* \cong \mathcal{O}_X \otimes_{\mathbb{C}_X} L^*$$

and (4.21) we get form Proposition 4.23 the following.

Proposition 4.24. *Let X be a complex manifold. Then we have an equivalence*

$$\mathrm{Conn}(X) \cong \mathrm{Loc}(X)^{\mathrm{op}} \qquad (M \rightsquigarrow \mathrm{Sol}(M)). \tag{4.26}$$

Moreover, we have

$$\mathrm{Sol}(M) \cong DR(M)^*. \tag{4.27}$$

For a group Γ we denote by $\mathrm{Rep}(\Gamma)$ the category of $\mathbb{C}[\Gamma]$-modules, where $\mathbb{C}[\Gamma]$ denotes the group algebra of Γ over $\mathbb{C}$. We also note the following well-known fact.

Proposition 4.25. *Let X be a connected complex manifold, and let $\pi_1(X, x_0)$ be the fundamental group of X with origin $x_0 \in X$. Then we have an equivalence*

$$\mathrm{Loc}(X) \cong \mathrm{Rep}(\pi_1(X, x_0)) \tag{4.28}$$

of categories.

The correspondence is given as follows. Let L be a local system. The stalk L_{x_0} turns out to be a $\mathbb{C}[\pi_1(X, x_0)]$-module by the following action of $\pi_1(X, x_0)$. Let $[\gamma]$ be an element of $\pi_1(X, x_0)$ represented by a path $\gamma : [0,1] \to X$. By $\gamma^{-1}L \cong \mathbb{C}_X \otimes_{\mathbb{C}} L_{x_0}$ we have a canonical isomorphism

$$L_{x_0} \cong (\gamma^{-1}L)_0 \cong (\gamma^{-1}L)_1 \cong L_{x_0}.$$

The action of $[\gamma]$ is given by this isomorphism.

Example 4.26. Let

$$P = \sum_{i=0}^{r} a_i(x) \left(\frac{d}{dx}\right)^i$$

be a holomorphic differential operator defined on a open subset U of $\mathbb{C}$. We assume that the top coefficient a_r is a not identically zero. Set $Y = \{y \in U \mid a_r(y) = 0\}$, $X = U \setminus Y$, and consider the differential equation: $Pu = 0$ on X for an unknown function u. The corresponding D_X-module $M = D_X/D_X P$ belongs to $\mathrm{Conn}(X)$ since

$$M \cong (\mathcal{O}_X)^r \qquad \left(\overline{\sum_{i=0}^{r-1} f_i(x)(\tfrac{d}{dx})^i} \longleftrightarrow (f_0, \dots, f_{r-1})\right).$$

4.6 Functors for coherent D-modules

We state some functorial properties of coherent D-modules. We first note that the functors (3.24) and (3.25) do not preserve D^b_c in general.

Exercise 4.27. (1) Compute $Lf^* D_{\mathbb{C}}$ for $f : \mathrm{pt} \to \mathbb{C}$ to construct an example such that Lf^* does not preserve D^b_c.
(2) Compute $\int_f D_{\mathbb{C}}$ for $f : \mathbb{C} \to \mathrm{pt}$ to construct an example such that $\int_f$ does not preserve D^b_c.

The following is almost obvious.

Lemma 4.28. *Let $f : X \to Y$ be a smooth morphism of complex manifolds or smooth algebraic varieties over $\mathbb{C}$. Then (3.24) induces*

$$Lf^* : D^b_c(D_Y) \to D^b_c(D_X). \tag{4.29}$$

Using the notion of characteristic varieties we have a deeper result for a general morphisms f as follows. For a morphism $f : X \to Y$ of complex manifolds or smooth algebraic varieties over $\mathbb{C}$ consider the canonical morphisms

$$T^*X \xleftarrow{\rho_f} X \times_Y T^*Y \xrightarrow{\varpi_f} T^*Y.$$

Let $M \in \mathrm{Mod}_c(D_Y)$. We say that f is non-characteristic with respect to M if we have

$$\varpi_f^{-1}(\mathrm{Ch}(M)) \cap \rho_f^{-1}(T^*_X X) \subset X \times_Y T^*_Y Y.$$

Theorem 4.29. *Let $f : X \to Y$ be a morphism of complex manifolds or smooth algebraic varieties over $\mathbb{C}$. Assume that f is non-characteristic with respect to a coherent D_Y-module M.*

(1) $H^k(Lf^*(M)) = 0$ *for* $k \neq 0$.
(2) $H^0(Lf^*(M)) \in \mathrm{Mod}_c(D_X)$.
(3) $\mathrm{Ch}(H^0(Lf^*(M))) = \rho_f \varpi_f^{-1}(\mathrm{Ch}(M))$.
(4) $f^{-1} \mathrm{Sol}(M) \cong \mathrm{Sol}(H^0(Lf^*(M)))$.

This is a vast sophisticated generalization of the classical Cauchy-Kovalevski type theorem given in Kashiwara's master thesis [12], [20]. For $M^\bullet \in D^b_c(D_X)$ we set $\mathrm{Ch}(M^\bullet) = \bigcup_k \mathrm{Ch}(H^k(M^\bullet))$.

Theorem 4.30. *Let $f : X \to Y$ be a proper morphism of complex manifolds or smooth algebraic varieties over $\mathbb{C}$. Then* (3.25) *induces*

$$\int_f : D^b_c(D_X) \to D^b_c(D_Y). \tag{4.30}$$

Moreover, we have $\mathrm{Ch}(\int_f M^\bullet) \subset \varpi_f \rho_f^{-1}(\mathrm{Ch}(M^\bullet))$.

Let us define the duality functor for D-modules. A natural candidate for the "dual" of a D_X-module M is given by $\mathcal{H}om_{D_X}(M, D_X)$. Here, we regard the target D_X as a left D_X-module by the left multiplication of D_X on D_X. Then $\mathcal{H}om_{D_X}(M, D_X)$ is naturally endowed with a structure of a right D_X-module by the right multiplication of D_X;

$$(\varphi \cdot P)(m) := \varphi(m)P \qquad (\varphi \in \mathcal{H}om_{D_X}(M, D_X), P \in D_X, m \in M).$$

In order to define the duality inside the world of left D-modules we consider the correspondence given in Theorem 2.10. Namely, we consider the left D_X-module

$$\mathcal{H}om_{D_X}(M, D_X) \otimes_{\mathcal{O}_X} \Omega_X^{\otimes -1}.$$

In fact we actually consider the corresponding functor

$$\begin{aligned} \mathbb{D} : D^b(D_X) &\to D^b(D_X)^{\mathrm{op}} \\ &(M \rightsquigarrow R\mathcal{H}om_{D_X}(M, D_X) \otimes_{\mathcal{O}_X} \Omega_X^{\otimes -1}[\dim X]) \end{aligned} \tag{4.31}$$

between derived categories. The reason for the shift $[\dim X]$ will become clear later (see Theorem 5.7 below).

Proposition 4.31. *$\mathbb{D}$ sends $D^b_c(D_X)$ to $D^b_c(D_X)^{\mathrm{op}}$. Moreover, we have $\mathbb{D} \circ \mathbb{D} = \mathrm{Id}$ on $D^b_c(D_X)$.*

Lemma 4.32. *For $M \in \mathrm{Conn}(X)$ we have $\mathbb{D}M = M^*(= \mathcal{H}om_{\mathcal{O}_X}(M, \mathcal{O}_X))$.*

4.7 Involutivity of characteristic varieties

Now we state a very important theorem on the dimensions of the characteristic varieties of coherent D_X-modules. Note that the cotangent bundle T^*X is a $2n$-dimensional variety, where $n = \dim X$.

Theorem 4.33. *Let M be a coherent D_X-module. Then any connected component Λ of $\mathrm{Ch}(M)$ satisfies $\dim \Lambda \geq \dim X$.*

This is a consequence of the following result.

Theorem 4.34. *Let M be a coherent D_X-module. Then $\mathrm{Ch}(M)$ is an involutive (or a coisotropic) subvariety of the symplectic manifold T^*X.*

Let us explain terminologies from symplectic geometry employed in Theorem 4.34.

Definition 4.35. Let M be a complex manifold or a smooth algebraic variety over $\mathbb{C}$. Assume that we are given a 2-form $\omega \in \wedge^2\Omega^1_M$. We say that M is a symplectic manifold if the following conditions are satisfied:

(1) $d\omega = 0$.
(2) For any $m \in M$ the anti-symmetric bilinear form on the tangent space $(TM)_m$ defined from ω is non-degenerate.

By (2) M is automatically even-dimensional (if there exists a non-degenerate anti-symmetric bilinear form on a vector space V, then V is even-dimensional).

Example 4.36. The cotangent bundle T^*X of a manifold X is a symplectic manifold. The corresponding 2-form ω is given as follows. Note that the tangent space of T^*X at (x,ξ) is identified with $(TX)_x \oplus (T^*X)_x$ ($x \in X$, $\xi \in (T^*X)_x$). Under this identification we have

$$\omega_{(x,\xi)}((a,\alpha),(b,\beta)) = \langle a,\beta\rangle - \langle b,\alpha\rangle.$$

In terms of a local coordinate $(U, \{x_i\}_{i=1}^n)$ of X we can describe ω as follows. Consider the natural local coordinate $(T^*U, \{x_i\}_{i=1}^n \cup \{\xi_i\}_{i=1}^n)$ of T^*X. Then we have

$$\omega = \sum_{i=1}^n dx_i \wedge d\xi_i.$$

Definition 4.37. Let (M,ω) be a symplectic manifold, and let N be a subvariety of M. Denote by N_{reg} the smooth locus of N.

- We say that N is isotropic if for any $p \in N_{\mathrm{reg}}$ we have $(TN)_p \subset (TN)_p^\perp$.
- We say that N is involutive (or coisotropic) if for any $p \in N_{\mathrm{reg}}$ we have $(TN)_p \supset (TN)_p^\perp$.
- We say that N is Lagrangean if for any $p \in N_{\mathrm{reg}}$ we have $(TN)_p = (TN)_p^\perp$.

Here, we set $(TN)_p^\perp = \{v \in (TM)_p \mid \omega_p(v,(TN)_p) = \{0\}\}$.

By definition we have the following.

Lemma 4.38. *Let (M,ω) be a symplectic manifold, and let N be a subvariety of M.*

(1) *If N is isotropic, we have* $\dim N \leq \frac{1}{2}\dim M$.
(2) *If N is involutive, we have* $\dim N \geq \frac{1}{2}\dim M$.
(3) *If N is Lagrangean, we have* $\dim N = \frac{1}{2}\dim M$.

Hence Theorem 4.34 implies Theorem 4.33. There are several proofs of Theorem 4.34, all of which are very involved;

- analytic proof by Sato-Kawai-Kashiwara [29] using microlocal analysis,
- algebraic proof by Gabber [9],
- geometric proof by Kashiwara-Schapira [23].

5 Regular holonomic D-modules

5.1 Holonomic D-modules

Definition 5.1. A coherent D_X-module M is called holonomic if $\mathrm{Ch}(M)$ is a Lagrangean subvariety of T^*X, or if $M = 0$ (namely $\mathrm{Ch}(M) = \emptyset$). We denote by $\mathrm{Mod}_h(D_X)$ the category of holonomic D_X-modules. We also set $D^b_h(D_X) = D^b_{\mathrm{Mod}_h(D_X)}(\mathrm{Mod}(D_X))$.

Let us consider the D_X-module

$$M = D_X / \sum_{i=1}^{r} D_X P_i \tag{5.1}$$

corresponding to the differential equation

$$P_1 u = P_2 u = \cdots = P_r u = 0. \tag{5.2}$$

We have seen that the characteristic variety $\mathrm{Ch}(M)$ of M roughly coincides with the common zero set of the principal symbols of $P_1, \dots, P_r$. Hence $\mathrm{Ch}(M)$ becomes smaller and smaller when r becomes larger and larger. But Theorem 4.33 asserts that $\dim \mathrm{Ch}(M)$ cannot become smaller than $\dim X = \frac{1}{2} \dim T^*X$ unless $M = 0$. Therefore, we can say that holonomic D_X-modules are coherent D_X-modules of the smallest possible size. In other words holonomic D_X-modules are coherent D_X-modules corresponding to differential equations with strongest possible constraints on the unknown function.

Example 5.2. (1) Integrable connections are holonomic. In particular, $\mathcal{O}_X$ is a holonomic D_X-module.
(2) For a submanifold Y of X the D_X-module $\mathcal{B}_{Y|X}$ is holonomic.
(3) Assume $\dim X = 1$. Then a coherent D_X-module M is holonomic if and only if for any $m \in M$ there exists some $P \in D_X$, $P \neq 0$ such that $Pm = 0$. In particular, any D_X-module of the form $D_X/D_X P$ with $P \neq 0$ is holonomic.

We have the following stability theorems.

Theorem 5.3. *Let $f : X \to Y$ be a morphism of complex manifolds or smooth algebraic varieties over $\mathbb{C}$. Then* (3.24) *induces*

$$Lf^* : D^b_h(D_Y) \to D^b_h(D_X). \tag{5.3}$$

Theorem 5.4. *Let $f : X \to Y$ be a morphism of complex manifolds or smooth algebraic varieties over $\mathbb{C}$. We assume that f is proper in the analytic setting. Then* (3.25) *induces*

$$\int_f : D^b_h(D_X) \to D^b_h(D_Y). \tag{5.4}$$

Remark 5.5. Theorem 5.3, and Theorem 5.4 in the algebraic setting for non-proper f are quite surprising considering the fact that those functors do not preserve D^b_c in general.

Remark 5.6. When f is a non-proper morphism in the analytic setting, it is not a good idea to consider the functor $\int_f$ even on D^b_h. For example let us consider $f : X = \mathbb{C} \setminus \{0\} \hookrightarrow Y = \mathbb{C}$. In the algebraic setting we have $H^0(\int_f \mathcal{O}_X) = f_*\mathcal{O}_X = \mathcal{O}_Y[x^{-1}]$, which is holonomic. However, in the analytic setting we have $H^0(\int_f \mathcal{O}_X) = f_*\mathcal{O}_X$, which is very large and not even coherent since it contains all functions holomorphic outside of the origin, like $e^{1/x}$.

Theorem 5.7. *Let X be a complex manifold or a smooth algebraic variety over $\mathbb{C}$. Then the duality functor* (4.31) *preserves* $\mathrm{Mod}_h(D_X)$. *Namely, for $M \in \mathrm{Mod}_h(D_X)$ we have $H^i(\mathbb{D}M) = 0$ for $i \neq 0$, and $H^0(\mathbb{D}M) \in \mathrm{Mod}_h(D_X)$. Hence we have functors*

$$\mathbb{D} : \mathrm{Mod}_h(D_X) \to \mathrm{Mod}_h(D_X), \qquad \mathbb{D} : D^b_h(D_X) \to D^b_h(D_X). \tag{5.5}$$

5.2 Solutions

As we saw in 5.1 holonomic D-modules are coherent D-modules corresponding to differential equations with strongest possible constraints on the unknown function. Hence the solution space should be smallest possible for holonomic D-modules. So it is natural to relate holonomicity with the finite-dimensionality of the solution space. In fact we have the following deep result (Kashiwara [14]).

Theorem 5.8. *Let X be a complex manifold. Let $M^\bullet \in D^b_h(D_X)$. Then for any $i \in \mathbb{Z}$ the $\mathbb{C}_X$-module $H^i(\mathrm{Sol}(M^\bullet))$ is constructible.*

Definition 5.9. Let X be a complex manifold.

(1) A decomposition $X = \bigsqcup_{\alpha \in A} X_\alpha$ of X into locally closed submanifolds is called a stratification of X if
   (a) for any $x \in X$ there exists an open neighborhood U of x such that $\sharp\{\alpha \in A \mid X_\alpha \cap U \neq \emptyset\} < \infty$,
   (b) for any $\alpha \in A$ there exists some $B \subset A$ such that $\overline{X_\alpha} = \bigcup_{\beta \in B} X_\beta$.

(2) A $\mathbb{C}_X$-module K is called constructible if there exists a stratification $X = \bigsqcup_{\alpha \in A} X_\alpha$ such that $K|_{X_\alpha}$ is a local system for any $\alpha \in A$. We denote by $\mathrm{Mod}_{con}(\mathbb{C}_X)$ the category of constructible $\mathbb{C}_X$-modules. We also set $D^b_{con}(\mathbb{C}_X) = D^b_{\mathrm{Mod}_{con}(\mathbb{C}_X)}(\mathrm{Mod}(\mathbb{C}_X))$.

We have the duality functor

$$\mathbf{D} : D^b_{con}(\mathbb{C}_X) \to D^b_{con}(\mathbb{C}_X)^{\mathrm{op}} \qquad (\mathbf{D}K^\bullet = R\mathcal{H}om_{\mathbb{C}_X}(K^\bullet, \mathbb{C}_X)) \tag{5.6}$$

satisfying $\mathbf{D} \circ \mathbf{D} = \mathrm{Id}$.

Theorem 5.10. *Let X be a complex manifold. For $M^\bullet \in D^b_h(D_X)$ we have* $\mathbf{D}(\mathrm{Sol}(M^\bullet)) \cong DR(M^\bullet)$.

By Theorem 5.8 and Theorem 5.10 we have two functors

$$\mathrm{Sol} : D^b_h(D_X) \to D^b_{con}(\mathbb{C}_X)^{\mathrm{op}}, \qquad DR : D^b_h(D_X) \to D^b_{con}(\mathbb{C}_X) \tag{5.7}$$

for a complex manifold X.

Next we describe what happens in the algebraic setting.

Definition 5.11. Let X be a smooth algebraic variety over $\mathbb{C}$.

(1) A decomposition $X = \bigsqcup_{\alpha \in A} X_\alpha$ of X into locally closed smooth subvarieties is called a stratification of X if the corresponding decomposition $X_{\mathrm{an}} = \bigsqcup_{\alpha \in A} (X_\alpha)_{\mathrm{an}}$ is a stratification of X_{an}.

(2) A $\mathbb{C}_{X_{\mathrm{an}}}$-module K is called algebraically constructible if there exists a stratification $X = \bigsqcup_{\alpha \in A} X_\alpha$ such that $K|_{(X_\alpha)_{\mathrm{an}}}$ is a local system for any $\alpha \in A$. We denote by $\mathrm{Mod}_{con}(\mathbb{C}_X)$ the category of algebraically constructible $\mathbb{C}_{X_{\mathrm{an}}}$-modules. We also set $D^b_{con}(\mathbb{C}_X) = D^b_{\mathrm{Mod}_{con}(\mathbb{C}_X)}(\mathrm{Mod}(\mathbb{C}_{X_{\mathrm{an}}}))$.

For a smooth algebraic variety X over $\mathbb{C}$ the duality functor $\mathbf{D} : D^b_{con}(\mathbb{C}_{X_{\mathrm{an}}}) \to D^b_{con}(\mathbb{C}_{X_{\mathrm{an}}})^{\mathrm{op}}$ preserves $D^b_{con}(\mathbb{C}_X)$. Hence we have the duality functor

$$\mathbf{D} : D^b_{con}(\mathbb{C}_X) \to D^b_{con}(\mathbb{C}_X)^{\mathrm{op}} \qquad (\mathbf{D}K^\bullet = R\mathcal{H}om_{\mathbb{C}_{X_{\mathrm{an}}}}(K^\bullet, \mathbb{C}_{X_{\mathrm{an}}})) \tag{5.8}$$

for a smooth algebraic variety X over $\mathbb{C}$. It satisfies $\mathbf{D} \circ \mathbf{D} = \mathrm{Id}$.

Theorem 5.12. *Let X be a smooth algebraic variety over $\mathbb{C}$. Let $M^\bullet \in D^b_h(D_X)$. Then for any $i \in \mathbb{Z}$ the $\mathbb{C}_{X_{\mathrm{an}}}$-module $H^i(\mathrm{Sol}(M^\bullet))$ is algebraically constructible.*

By Theorem 5.12 and Theorem 5.10 we have two functors

$$\mathrm{Sol} : D^b_h(D_X) \to D^b_{con}(\mathbb{C}_X)^{\mathrm{op}}, \qquad DR : D^b_h(D_X) \to D^b_{con}(\mathbb{C}_X) \tag{5.9}$$

for a smooth algebraic variety X over $\mathbb{C}$.

5.3 Regular holonomic D-modules

Let Y be a finite subset of $\mathbb{P}^1$, and set $X = \mathbb{P}^1 \setminus Y$. Assume that we are given a local system L on X. Then the classical Riemann-Hilbert problem (Hilbert's 21st problem) asks whether there exists an ordinary linear differential equation on X with regular singularities at points of Y such that the solution sheaf coincides with L. We will discuss higher dimensional analogue of this problem in the following. After Deligne's contribution [8] to this problem in the algebraic case Kashiwara formulated the following conjecture around 1975.

Conjecture 5.13. *Let X be a complex manifold. Then there exists a good subcategory $\mathrm{Mod}_{rh}(D_X)$ of $\mathrm{Mod}_h(D_X)$ such that (5.7) induces equivalences*

$$\mathrm{Sol} : D^b_{rh}(D_X) \simeq D^b_{con}(\mathbb{C}_X)^{\mathrm{op}}, \qquad DR : D^b_{rh}(D_X) \simeq D^b_{con}(\mathbb{C}_X) \tag{5.10}$$

of categories, where $D^b_{rh}(D_X) = D^b_{\mathrm{Mod}_{rh}(D_X)}(\mathrm{Mod}(D_X))$.

This conjecture was subsequently solved by Kashiwara himself [17], [19] (Theorem 5.15 below).

Definition 5.14. Let X be a complex manifold and let $M \in \mathrm{Mod}_h(D_X)$. We say that M is regular holonomic if there exists locally a good filtration F of M such that

$$P \in F_pD_X,\ \sigma_p(P)|_{\mathrm{Ch}(M)} = 0 \Longrightarrow PF^kM \subset F^{k+p-1}M \ (\forall k).$$

We denote by $\mathrm{Mod}_{rh}(D_X)$ the category of regular holonomic D_X-modules. We also set $D^b_{rh}(D_X) = D^b_{\mathrm{Mod}_{rh}(D_X)}(\mathrm{Mod}(D_X))$.

In the one-dimensional case the ordinary linear differential equation corresponding to objects of $\mathrm{Mod}_{rh}(D_X)$ are those which have regular singularities at boundaries.

Theorem 5.15. *Let X be a complex manifold. Then* (5.7) *induces equivalences*

$$\mathrm{Sol}: D^b_{rh}(D_X) \simeq D^b_{con}(\mathbb{C}_X)^{\mathrm{op}}, \qquad DR: D^b_{rh}(D_X) \simeq D^b_{con}(\mathbb{C}_X). \tag{5.11}$$

The equivalence (5.11) is called the Riemann-Hilbert correspondence. We note that another proof of this correspondence was also obtained by Mebkhout [27]. For the more detailed historical comments, compare the foreword by Schapira in [20].

Let us also describe the corresponding results in the algebraic setting. Let X be a smooth algebraic variety over $\mathbb{C}$. We say that $j: X \hookrightarrow \overline{X}$ is a smooth completion of X if $\overline{X}$ is a smooth complete algebraic variety over $\mathbb{C}$ and j is an open dense embedding.

Proposition 5.16. *Let X be a smooth algebraic variety over $\mathbb{C}$, and let $M \in \mathrm{Mod}_h(D_X)$. Then the following conditions are equivalent.*

(1) *For some smooth completion $j: X \hookrightarrow \overline{X}$ of X we have $(\int_j M)^{\mathrm{an}} \in D^b_{rh}(D_{X_{\mathrm{an}}})$.*
(2) *For any smooth completion $j: X \hookrightarrow \overline{X}$ of X we have $(\int_j M)^{\mathrm{an}} \in D^b_{rh}(D_{X_{\mathrm{an}}})$.*

Definition 5.17. Let X be a smooth algebraic variety over $\mathbb{C}$, and let $M \in \mathrm{Mod}_h(D_X)$. We say that M is regular holonomic if the equivalent conditions in Proposition 5.16 are satisfied. We denote by $\mathrm{Mod}_{rh}(D_X)$ the category of regular holonomic D_X-modules. We also set $D^b_{rh}(D_X) = D^b_{\mathrm{Mod}_{rh}(D_X)}(\mathrm{Mod}(D_X))$.

Theorem 5.18. *Let X be a smooth algebraic variety over $\mathbb{C}$. Then* (5.7) *induces equivalences*

$$\mathrm{Sol}: D^b_{rh}(D_X) \simeq D^b_{con}(\mathbb{C}_X)^{\mathrm{op}}, \qquad DR: D^b_{rh}(D_X) \simeq D^b_{con}(\mathbb{C}_X). \tag{5.12}$$

This result was deduced from the original analytic version by Brylinski [6]. Later, Beilinson-Bernstein developed the corresponding theory for algebraic D-modules based on much simpler arguments (see [3]).

We have the following stability theorems for regular holonomic D-modules.

Theorem 5.19. *Let $f: X \to Y$ be a morphism of complex manifolds or smooth algebraic varieties over $\mathbb{C}$. Then* (5.3) *induces*

$$Lf^*: D^b_{rh}(D_Y) \to D^b_{rh}(D_X). \tag{5.13}$$

Theorem 5.20. *Let $f: X \to Y$ be a morphism of complex manifolds or smooth algebraic varieties over $\mathbb{C}$. We assume that f is proper in the analytic setting. Then* (5.4) *induces*

$$\int_f : D^b_{rh}(D_X) \to D^b_{rh}(D_Y). \tag{5.14}$$

By Theorem 5.20 we have

$$\mathcal{B}_{Z|X} \in \mathrm{Mod}_{rh}(D_X) \tag{5.15}$$

for a closed submanifold Z of X.

Theorem 5.21. *Let X be a complex manifold or a smooth algebraic variety $\mathbb{C}$. Then the duality functor* (5.5) *induces*

$$\mathbb{D} : \mathrm{Mod}_{rh}(D_x) \to \mathrm{Mod}_{rh}(D_X), \qquad \mathbb{D} : D^b_{rh}(D_x) \to D^b_{rh}(D_X). \tag{5.16}$$

Let us describe functors for constructible sheaves corresponding, via the Riemann-Hilbert correspondence, to functors for regular holonomic D-modules.

Let $f : X \to Y$ be a morphism of complex manifolds (resp. smooth algebraic varieties over $\mathbb{C}$). Then the exact functor

$$f^{-1} : \mathrm{Mod}(\mathbb{C}_Y) \to \mathrm{Mod}(\mathbb{C}_X) \quad (\text{resp. } (f_{\mathrm{an}})^{-1} : \mathrm{Mod}(\mathbb{C}_{Y_{\mathrm{an}}}) \to \mathrm{Mod}(\mathbb{C}_{X_{\mathrm{an}}}))$$

induces a functor

$$f^{-1} : D^b(\mathbb{C}_Y) \to D^b(\mathbb{C}_X) \quad (\text{resp. } (f_{\mathrm{an}})^{-1} : D^b(\mathbb{C}_{Y_{\mathrm{an}}}) \to D^b(\mathbb{C}_{X_{\mathrm{an}}}))$$

between derived categories. It is known that it preserves D^b_{con}, and hence we have a functor

$$f^{-1} : D^b_{con}(\mathbb{C}_Y) \to D^b_{con}(\mathbb{C}_X) \tag{5.17}$$

between constructible derived categories.

Theorem 5.22. *Let $f : X \to Y$ be a morphism of complex manifolds or smooth algebraic varieties over $\mathbb{C}$. Then we have the following commutative diagram.*

$$\begin{array}{ccc} D^b_{rh}(D_Y) & \xrightarrow{\mathrm{Sol}} & D^b_{con}(\mathbb{C}_Y) \\ {\scriptstyle Lf^*}\downarrow & & \downarrow{\scriptstyle f^{-1}} \\ D^b_{rh}(D_X) & \xrightarrow{\mathrm{Sol}} & D^b_{con}(\mathbb{C}_X). \end{array}$$

Let $f : X \to Y$ be a *proper* morphism of complex manifolds (resp. a (not necessarily proper) morphism of smooth algebraic varieties over $\mathbb{C}$). Then the left exact functor

$$f_* : \mathrm{Mod}(\mathbb{C}_X) \to \mathrm{Mod}(\mathbb{C}_Y) \quad (\text{resp. } (f_{\mathrm{an}})_* : \mathrm{Mod}(\mathbb{C}_{X_{\mathrm{an}}}) \to \mathrm{Mod}(\mathbb{C}_{Y_{\mathrm{an}}}))$$

induces a functor

$$Rf_* : D^b(\mathbb{C}_X) \to D^b(\mathbb{C}_Y) \quad (\text{resp. } R(f_{\mathrm{an}})_* : D^b(\mathbb{C}_{X_{\mathrm{an}}}) \to D^b(\mathbb{C}_{Y_{\mathrm{an}}}))$$

between derived categories. It is known that it preserves D^b_{con}, and hence we have a functor

$$Rf_* : D^b_{con}(\mathbb{C}_X) \to D^b_{con}(\mathbb{C}_Y) \tag{5.18}$$

between constructible derived categories.

Theorem 5.23. *Let $f : X \to Y$ be a proper morphism of complex manifolds or a (not necessarily proper) morphism of smooth algebraic varieties over $\mathbb{C}$. Then we have the following commutative diagram.*

$$\begin{array}{ccc} D^b_{rh}(D_X) & \xrightarrow{DR} & D^b_{con}(\mathbb{C}_X) \\ {\scriptstyle \int_f}\downarrow & & \downarrow{\scriptstyle Rf_*[\dim X - \dim Y]} \\ D^b_{rh}(D_Y) & \xrightarrow{DR} & D^b_{con}(\mathbb{C}_Y). \end{array}$$

Theorem 5.24. *Let X be a complex manifold or a smooth algebraic variety over $\mathbb{C}$. Then we have the following commutative diagram.*

$$\begin{array}{ccc} D^b_{rh}(D_X) & \xrightarrow{DR} & D^b_{con}(\mathbb{C}_X) \\ {\scriptstyle \mathbb{D}}\downarrow & & \downarrow{\scriptstyle \mathbf{D}} \\ D^b_{rh}(D_X) & \xrightarrow{DR} & D^b_{con}(\mathbb{C}_Y). \end{array}$$

5.4 Perverse sheaves

Note that $\mathrm{Mod}_{rh}(D_X)$ is an abelian subcategory of the derived category $D^b_{rh}(D_X)$. Hence in view of the Riemann-Hilbert correspondence

$$D^b_{rh}(D_X) \cong D^b_{con}(\mathbb{C}_X)$$

there should exists an abelian subcategory of $D^b_{con}(\mathbb{C}_X)$ corresponding to $\mathrm{Mod}_{rh}(D_X)$. This is one of the motivations for the theory of perverse sheaves (Beilinson-Bernstein-Deligne [2]), to be explained below.

Definition 5.25. Let X be a complex manifold or a smooth algebraic variety over $\mathbb{C}$. Let $K^\bullet \in D^b_c(\mathbb{C}_X)$. We say that $K^\bullet$ is a perverse sheaf if it satisfies the following conditions.

(1) $H^i(K^\bullet) = 0$ for $i < 0$, and $\operatorname{codim} \operatorname{Supp}(H^i(K^\bullet)) \geq i$ for $i \geq 0$.
(2) $H^i(\mathbf{D}K^\bullet) = 0$ for $i < 0$, and $\operatorname{codim} \operatorname{Supp}(H^i(\mathbf{D}K^\bullet)) \geq i$ for $i \geq 0$.

We denote by $\mathrm{Perv}(\mathbb{C}_X)$ the category of perverse sheaves.

Example 5.26. Let X be a complex manifold, and let $L \in \mathrm{Loc}(X)$. Then we have $\mathbf{D}L = \mathcal{H}om_{\mathbb{C}_X}(L, \mathbb{C}_X) = L^* \in \mathrm{Loc}(X)$. Hence we have $L \in \mathrm{Perv}(\mathbb{C}_X)$.

Example 5.27. Let X be a complex manifold, and let Y be a closed submanifold of X. We denote by $i : Y \hookrightarrow X$ the embedding. Take $L \in \mathrm{Loc}(Y)$. Then we have $\mathbf{D}(i_*L) = i_*L^*[-2\mathrm{codim}Y]$. Hence we have $i_*L[-\mathrm{codim}Y] \in \mathrm{Perv}(\mathbb{C}_X)$.

Theorem 5.28. *Let X be a complex manifold or a smooth algebraic variety over $\mathbb{C}$. Then* (5.11), (5.12) *induce equivalences*

$$\mathrm{Sol} : \mathrm{Mod}_{rh}(D_X) \cong \mathrm{Perv}(\mathbb{C}_X)^{\mathrm{op}}, \quad DR : \mathrm{Mod}_{rh}(D_X) \cong \mathrm{Perv}(\mathbb{C}_X) \qquad (5.19)$$

of abelian categories.

Example 5.29. Let X be a complex manifold, and let Y be a closed submanifold of X. We denote by $i : Y \hookrightarrow X$ the embedding. Then we have

$$\mathrm{Sol}(\mathcal{B}_{Y|X}) \cong DR(\mathcal{B}_{Y|X}) \cong i_*\mathbb{C}_Y[-\mathrm{codim} Y].$$

5.5 Intersection cohomologies

Let X be a complex manifold or a smooth algebraic variety over $\mathbb{C}$, and let Z be a closed subvariety of X. If Z is smooth, then we have a regular holonomic D_X-module $\mathcal{B}_{Z|X}$. We want to construct a similar regular holonomic D_X-module even in the case Z is not smooth. Denote by Z_{sing} the singular locus of Z. Since $Z \setminus Z_{\mathrm{sing}}$ is a closed smooth subvariety of $X \setminus Z_{\mathrm{sing}}$, we have a regular holonomic $D_{X\setminus Z_{\mathrm{sing}}}$-module $\mathcal{B}_{Z\setminus Z_{\mathrm{sing}}|X\setminus Z_{\mathrm{sing}}}$. We have a canonical extension of $\mathcal{B}_{Z\setminus Z_{\mathrm{sing}}|X\setminus Z_{\mathrm{sing}}}$ to whole X by the following.

Theorem 5.30. *Let X and Z be as above. Then there exists a unique extension ${}^{\pi}\mathcal{B}_{Z|X} \in \mathrm{Mod}_{rh}(D_X)$ of $\mathcal{B}_{Z\setminus Z_{\mathrm{sing}}|X\setminus Z_{\mathrm{sing}}}$ which has no coherent subquotients supported in Z_{sing}.*

We call ${}^{\pi}\mathcal{B}_{Z|X} \in \mathrm{Mod}_{rh}(D_X)$ the minimal extension of $\mathcal{B}_{Z\setminus Z_{\mathrm{sing}}|X\setminus Z_{\mathrm{sing}}}$. Define ${}^{\pi}\mathbb{C}_Z \in D^b_{con}(\mathbb{C}_X)$ by

$$DR({}^{\pi}\mathcal{B}_{Z|X}) = {}^{\pi}\mathbb{C}_Z[-\mathrm{codim} Z]. \tag{5.20}$$

Then the intersection homology group $IH^i(Z)$ of Z (Goresky-MacPherson [10]) is defined by

$$IH^i(Z) = \begin{cases} H^i(R\Gamma(X, {}^{\pi}\mathbb{C}_Z)) & (X \text{ is a complex manifold}), \\ H^i(R\Gamma(X_{\mathrm{an}}, {}^{\pi}\mathbb{C}_Z)) & (X \text{ is a smooth algebraic variety}). \end{cases} \tag{5.21}$$

Proposition 5.31. ${}^{\pi}\mathbb{C}_Z[-\mathrm{codim} Z] \in \mathrm{Perv}(\mathbb{C}_X)$ *is characterized by the following properties.*

(1) ${}^{\pi}\mathbb{C}_Z[-\mathrm{codim} Z]|_{X\setminus Z_{\mathrm{sing}}} \cong \mathbb{C}_{Z\setminus Z_{\mathrm{sing}}}[-\mathrm{codim} Z]$.
(2) $H^i({}^{\pi}\mathbb{C}_Z[-\mathrm{codim} Z]) = 0$ *for* $i < \mathrm{codim} Z$, *and* $\mathrm{codim}\,\mathrm{Supp}(H^i({}^{\pi}\mathbb{C}_Z[-\mathrm{codim} Z])) > i$ *for* $i > \mathrm{codim} Z$.
(3) $H^i(\mathbf{D}({}^{\pi}\mathbb{C}_Z[-\mathrm{codim} Z])) = 0$ *for* $i < \mathrm{codim} Z$, *and* $\mathrm{codim}\,\mathrm{Supp}(H^i(\mathbf{D}({}^{\pi}\mathbb{C}_Z[-\mathrm{codim} Z]))) > i$ *for* $i > \mathrm{codim} Z$.

6 Application to representation theory

6.1 *D*-modules on the flag manifold

Let G be a connected semisimple algebraic group over $\mathbb{C}$ and let B be a Borel subgroup of G. The homogeneous space $X = G/B$ is a smooth projective algebraic variety, called the flag manifold for G. We will consider D-modules on X in the following. Let $\mathfrak{g}$ be the Lie algebra of G and let $U(\mathfrak{g})$ be its enveloping algebras.

We first explain how D_X-modules are related to $U(\mathfrak{g})$-modules. By differentiating the action of G on X we obtain a Lie algebra homomorphism

$$\mathfrak{g} \to \Gamma(X, \Theta_X) \qquad (a \mapsto \partial_a). \tag{6.1}$$

Regarding G and X as a complex Lie group and a complex manifold the homomorphism (6.1) is described as

$$(\partial_a(f))(x) = \frac{d}{dt} f(\exp(-ta)x)|_{t=0} \qquad (a \in \mathfrak{g}, f \in \mathcal{O}_X, x \in X).$$

However, in the algebraic setting, which we are going to use in this section, (6.1) can be described as

$$\pi^*(\partial_a(f)) = a(\pi^* f) \qquad (a \in \mathfrak{g}, f \in \mathcal{O}_X),$$

where $\pi : G \to X$ is given by $g \mapsto gB$, and $\mathfrak{g}$ is regarded as the Lie algebra consisting of *right* invariant vector fields on G. By the universal property of the enveloping algebra $U(\mathfrak{g})$ we have a unique algebra homomorphism

$$U(\mathfrak{g}) \to \Gamma(X, D_X) \tag{6.2}$$

extending (6.1). Denote by $\mathfrak{z}(\mathfrak{g})$ the center of $U(\mathfrak{g})$. By restricting the canonical algebra homomorphism

$$\varepsilon : U(\mathfrak{g}) \to \mathbb{C} \qquad (\mathfrak{g} \ni a \mapsto 0, 1 \mapsto 1)$$

to $\mathfrak{z}(\mathfrak{g})$ we obtain an algebra homomorphism

$$\chi_0 : \mathfrak{z}(\mathfrak{g}) \to \mathbb{C}. \tag{6.3}$$

Theorem 6.1. *The algebra homomorphism* (6.3) *is surjective, and its kernel is generated as an ideal by* $\operatorname{Ker} \chi_0$. *Hence we have*

$$\Gamma(X, D_X) \cong U(\mathfrak{g})/U(\mathfrak{g}) \operatorname{Ker} \chi_0.$$

Denote by $\operatorname{Mod}(\mathfrak{g}, \chi_0)$ the category of $U(\mathfrak{g})$-modules M such that $z \in \mathfrak{z}(\mathfrak{g})$ acts on M by $\chi_0(z)\,\mathrm{id}$, namely $U(\mathfrak{g})$-modules with central character χ_0. By Theorem 6.1 we obtain a left exact functor

$$\Gamma(X, \bullet) : \operatorname{Mod}(D_X) \to \operatorname{Mod}(\mathfrak{g}, \chi_0) \qquad (M \rightsquigarrow \Gamma(X, M)), \tag{6.4}$$

and a right exact functor

$$D_X \otimes_{U(\mathfrak{g})} (\bullet) : \operatorname{Mod}(\mathfrak{g}, \chi_0) \to \operatorname{Mod}(D_X) \qquad (V \rightsquigarrow D_X \otimes_{U(\mathfrak{g})} V). \tag{6.5}$$

We have the following result due to Beilinson-Bernstein [1].

Theorem 6.2. *The functors* (6.4) *and* (6.5) *induce an equivalence*

$$\operatorname{Mod}_c(D_X) \cong \operatorname{Mod}_f(\mathfrak{g}, \chi_0) \tag{6.6}$$

of categories, where $\operatorname{Mod}_f(\mathfrak{g}, \chi_0)$ *denotes the category of finitely generated* $U(\mathfrak{g})$*-modules with central character* χ_0.

The equivalence (6.6) is called the Beilinson-Bernstein correspondence.

6.2 Character formula of highest weight modules

There exists a series of irreducible $U(\mathfrak{g})$-modules, called irreducible highest weight modules. Any finite-dimensional irreducible $U(\mathfrak{g})$-module is a highest weight module, and its character is classically known by Weyl's character formula; however, we have also infinite-dimensional irreducible highest weight modules, and the determination of their characters was a big problem in representation theory around 1970's. A breakthrough to this problem came around 1980. Kazhdan-Lusztig formulated a conjectural character formula in [25]. Subsequently the conjecture was independently solved by Brylinski-Kashiwara [7] and Beilinson-Bernstein [1] using D-modules on the flag manifold. We give an account of this result in the following.

Irreducible highest weight modules belonging to $\mathrm{Mod}_f(\mathfrak{g}, \chi_0)$ are parametrized by the Weyl group W. We denote the irreducible highest weight module corresponding to $w \in W$ by $L_w \in \mathrm{Mod}_f(\mathfrak{g}, \chi_0)$. It is a unique irreducible quotient of the standard module $V_w \in \mathrm{Mod}_f(\mathfrak{g}, \chi_0)$, called the Verma module. Denote by $\mathrm{Mod}_f^B(\mathfrak{g}, \chi_0)$ the subcategory of $\mathrm{Mod}_f(\mathfrak{g}, \chi_0)$ consisting of B-equivariant objects. Then the Grothendieck group $K(\mathrm{Mod}_f^B(\mathfrak{g}, \chi_0))$ has two bases $\{[L_w]\}_{w\in W}$ and $\{[V_w]\}_{w\in W}$. The following result was conjectured by Kazhdan-Lusztig [25], and proved by Beilinson-Bernstein [1] and Brylinski-Kashiwara [7].

Theorem 6.3. *There exists an isomorphism*

$$K(\mathrm{Mod}_f^B(\mathfrak{g}, \chi_0)) \cong \mathbb{Z}[W]$$

of $\mathbb{Z}$-modules, such that

$$[V_w] \leftrightarrow (-1)^{\ell(w)} w, \qquad [L_w] \leftrightarrow (-1)^{\ell(w)} \sum_{y \in W} P_{y,w}(1) y,$$

where $\ell : W \to \mathbb{Z}_{\geq 0}$ is the length function, and $P_{y,w} \in \mathbb{Z}[q]$ for $y, w \in W$ is the Kazhdan-Lusztig polynomial introduced in [25].

Since the characters of Verma modules are easily determined, the characters of irreducible highest weight modules can be obtained from the above Theorem. In the rest of this note we give an outline of the proof of Theorem 6.3.

Kazhdan-Lusztig polynomials are combinatorially defined in terms of the Hecke algebra; however, there is also a geometric interpretation of them as follows. Recall that X denotes the flag manifold G/B. It is well-known that the B-orbits on X are also parametrized by the Weyl group W. Namely, we have a stratification

$$X = \bigsqcup_{w \in W} X_w \qquad (X_w = BwB/B)$$

of X by B-orbits on X. It is known that X_w is a locally closed subset of X isomorphic to $\mathbb{C}^{\ell(w)}$. Its closure $\overline{X}_w$ is a possibly singular variety, called a Schubert variety. We have $\overline{X}_w = \bigsqcup_{y \in W'} X_y$ for some $W' \subset W$. Recall that $\mathrm{Perv}(\mathbb{C}_X)$ denotes the category of perverse sheaves on X. We denote by $\mathrm{Perv}^B(\mathbb{C}_X)$ the subcategory of $\mathrm{Perv}(\mathbb{C}_X)$ consisting of B-equivariant objects. Then the Grothendieck group $K(\mathrm{Perv}^B(\mathbb{C}_X))$ has two bases $\{[\mathbb{C}_{X_w}[-(n-\ell(w))]\}_{w\in W}$ and $\{[{}^{\pi}\mathbb{C}_{\overline{X}_w}[-(n-\ell(w))]\}_{w\in W}$. Here $n = \dim X$.

Theorem 6.4. *For $w, y \in W$ we have*

$$\sum_{i \in \mathbb{Z}} \dim H^i({}^\pi\mathbb{C}_{\overline{X}_w})_{yB} q^{i/2} = P_{y,w}(q).$$

This implies the following

Theorem 6.5. *There exists an isomorphism*

$$K(\mathrm{Perv}^B(\mathbb{C}_X)) \cong \mathbb{Z}[W]$$

of $\mathbb{Z}$-modules, such that

$$[\mathbb{C}_{X_w}[-(n-\ell(w))] \leftrightarrow (-1)^{\ell(w)} w, \quad [{}^\pi\mathbb{C}_{\overline{X}_w}[-(n-\ell(w))] \leftrightarrow (-1)^{\ell(w)} \sum_{y \in W} P_{y,w}(1) y.$$

In view of Theorem 6.5 it is natural in proving Theorem 6.3 to expect some correspondence between highest weight modules and intersection cohomology sheaves of the Schubert varieties. In fact, this was realized by Brylinski-Kashiwara [7] and Beilinson-Bernstein [1] by employing *D*-modules on the flag manifold as the missing link.

highest weight modules	*D*-modules on the flag manifold	intersection cohomology groups

More precisely, they obtained the following results on *D*-modules on flag manifolds. Recall that $\mathrm{Mod}_{rh}(D_X)$ denotes the category of regular holonomic D_X-modules. We denote by $\mathrm{Mod}^B_{rh}(D_X)$ the subcategory of $\mathrm{Mod}_{rh}(D_X)$ consisting of B-equivariant objects. For $w \in W$ we have an irreducible object ${}^\pi\mathcal{B}_{\overline{X}_w|X} \in \mathrm{Mod}^B_{rh}(D_X)$. We also set

$$M_w = \mathbb{D}(H^0(\int_{i_w} \mathcal{O}_{X_w})) \in \mathrm{Mod}^B_{rh}(D_X),$$

where $i_w \colon X_w \hookrightarrow X$ is the embedding. Then the Grothendieck group $K(\mathrm{Mod}_{rh}(D_X))$ has two bases $\{[M_w]\}_{w \in W}$ and $\{[{}^\pi\mathcal{B}_{\overline{X}_w|X}]\}_{w \in W}$. By the Riemann-Hilbert correspondence we have

$$DR({}^\pi\mathcal{B}_{\overline{X}_w|X}) = {}^\pi\mathbb{C}_{\overline{X}_w}[-(n-\ell(w))], \qquad DR(M_w) = \mathbb{C}_{X_w}[-(n-\ell(w))], \quad (6.7)$$

and hence we obtain an isomorphism

$$[DR] : K(\mathrm{Mod}^B_{rh}(D_X)) \to K(\mathrm{Perv}^B(\mathbb{C}_X)) \quad (6.8)$$

such that $[{}^\pi\mathcal{B}_{\overline{X}_w|X}] \leftrightarrow [{}^\pi\mathbb{C}_{\overline{X}_w}[-(n-\ell(w))]$, $[M_w] \leftrightarrow [\mathbb{C}_{X_w}[-(n-\ell(w))]$.

By the above argument Theorem 6.3 is a consequence of the following result of [1] and [7].

Theorem 6.6. *There exists an equivalence*

$$\mathrm{Mod}_f^B(\mathfrak{g}, \chi_0) \cong \mathrm{Mod}_{rh}^B(D_X)$$

of categories such that

$$V_w \leftrightsquigarrow M_w, \qquad L_w \leftrightsquigarrow {}^{\pi}\mathcal{B}_{\overline{X}_w|X} \qquad (w \in W).$$

Beilinson-Bernstein [1] established a more general correspondence (the Beilinson-Bernstein correspondence (6.6)) and deduced Theorem 6.6 from it, while Brylinski-Kashiwara [7] proved this directly.

Remark 6.7. Some generalizations of Theorem 6.3 to Kac-Moody algebras were obtained by Kashiwara and the author subsequently (see [24] and its references).

References

[1] A. A. Beilinson and J. Bernstein, Localisation de $\mathfrak{g}$-modules, C. R. Acad. Sci. Paris Sér. I Math. **292** (1981), no. 1, 15–18.

[2] A. A. Beilinson, J. Bernstein, and P. Deligne, Faisceaux pervers, Analysis and topology on singular spaces, I (Luminy, 1981), Astérisque, **100** (1982), 5–171.

[3] J. Bernstein, Algebraic theory of D-modules, unpublished notes.

[4] J.-E. Björk, Rings of Differential Operators, North-Holland Mathematical Library, vol. 21, 1979.

[5] J.-E. Björk, Analytic D-modules and applications, Mathematics and its Applications, vol. 247, Kluwer Academic Publishers Group, Dordrecht, 1993.

[6] J.-L. Brylinski, Transformations canoniques, dualité projective, théorie de Lefschetz, transformations de Fourier et sommes trigonométriques, Géométrie et analyse microlocales, Astérisque **140**, **141** (1986), 3–134.

[7] J.-L. Brylinski and M. Kashiwara, Kazhdan-Lusztig conjecture and holonomic systems, Invent. Math. **64** (1981), no. 3, 387–410.

[8] P. Deligne, Équations différentielles à points singuliers réguliers, Lecture Notes in Math., vol. 163, Springer-Verlag, Berlin, 1970.

[9] O. Gabber, The integrability of the characteristic variety, Amer. J. Math. **103** (1981), no. 3, 445–468.

[10] M. Goresky and R. MacPherson, Intersection homology. II, Invent. Math. **72** (1983), no. 1, 77–129.

[11] R. Hotta, K. Takeuchi and T. Tanisaki, D-modules, perverse sheaves, and representation theory, Progr. Math., vol. 236, Birkhäuser, Boston, Basel, Berlin 2008.

[12] M. Kashiwara, Algebraic study of partial differential equations, Master's thesis, University of Tokyo, 1970, (Japanese).

[13] M. Kashiwara, Index theorem for a maximally overdetermined system of linear differential equations, Proc. Japan Acad. **49** (1973), 803–804.

[14] M. Kashiwara, On the maximally overdetermined system of linear differential equations. I, Publ. Res. Inst. Math. Sci. **10** (1974/75), 563–579.

[15] M. Kashiwara, B-functions and holonomic systems. Rationality of roots of B-functions, Invent. Math. **38** (1976/77), no. 1, 33–53.

[16] M. Kashiwara, On the holonomic systems of linear differential equations. II, Invent. Math. **49** (1978), no. 2, 121–135.

[17] M. Kashiwara, Faisceaux constructibles et systèmes holonômes d'équations aux dérivées partielles linéaires à points singuliers réguliers, Séminaire Goulaouic-Schwartz Exp. No. 19, 1979–1980 (French), École Polytech., Palaiseau, 1980.

[18] M. Kashiwara, Systems of microdifferential equations, Progr. Math., vol. 34, Birkhäuser Boston Inc., 1983, Based on lecture notes by Teresa Monteiro Fernandes translated from the French.

[19] M. Kashiwara, The Riemann-Hilbert problem for holonomic systems, Publ. Res. Inst. Math. Sci. **20** (1984), no. 2, 319–365.

[20] M. Kashiwara, Algebraic study of systems of partial differential equations (Master thesis, University of Tokyo, December 1970), Mém. Soc. Math. France (N.S.) **63**, 1995, translated by A. D'Agnolo and J.-P. Schneiders.

[21] M. Kashiwara, D-modules and microlocal calculus, Translations of Mathematical Monographs, vol. 217, American Mathematical Society, 2003, Translated from the 2000 Japanese original by Mutsumi Saito, Iwanami Series in Modern Mathematics.

[22] M. Kashiwara and T. Kawai, On holonomic systems of microdifferential equations. III. Systems with regular singularities, Publ. Res. Inst. Math. Sci. **17** (1981), no. 3, 813–979.

[23] M. Kashiwara, P. Schapira, Sheaves on manifolds, Grundlehren der Mathematischen Wissenschaften, vol. 292, Springer-Verlag, Berlin, 1990.

[24] M. Kashiwara and T. Tanisaki, Characters of irreducible modules with non-critical highest weights over affine Lie algebras, Representations and quantizations (Shanghai, 1998), High. Educ. Press, Beijing, 2000, 275–296.

[25] D. Kazhdan and G. Lusztig, Representations of Coxeter groups and Hecke algebras, Invent. Math. **53** (1979), no. 2, 165–184.

[26] D. Kazhdan and G. Lusztig, Schubert varieties and Poincaré duality, Geometry of the Laplace operator (Proc. Sympos. Pure Math., Univ. Hawaii, Honolulu, Hawaii, 1979), Proc. Sympos. Pure Math., XXXVI, Amer. Math. Soc., 1980, 185–203.

[27] Z. Mebkhout, Une équivalence de catégories; Une autre équivalence de catégories, Compositio Math. **51** (1984), no. 1, 51–62; ibid. **51** (1984), no. 1, 63–88.

[28] F. Pham, Singularités des systèmes différentiels de Gauss-Manin, Progr. Math., vol. 2, Birkhäuser, Boston, Mass., 1979.

[29] M. Sato, T. Kawai, and M. Kashiwara, Microfunctions and pseudo-differential equations, Hyperfunctions and pseudo-differential equations (Proc. Conf., Katata, 1971; dedicated to the memory of André Martineau), Lecture Notes in Math., vol. 287, Springer, Berlin, 1973, 265–529.

Made in United States
Orlando, FL
04 November 2026

90200277R00127